INTERPRETATION OF AIRPHOTOS AND REMOTELY SENSED IMAGERY

INTERPRETATION OF AIRPHOTOS AND REMOTELY SENSED IMAGERY

Robert H. Arnold
Salem State College

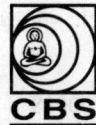

CBS Publishers & Distributors Pvt. Ltd.
New Delhi • Bengaluru • Chennai • Kochi • Mumbai • Pune
Hyderabad • Kolkata • Nagpur • Patna • Vijayawada

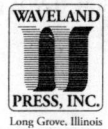

WAVELAND
PRESS, INC.
Long Grove, Illinois

Interpretation of Airphotos and Remotely Sensed Imagery

Waveland ISBN: 978-1-57766-353-9

CBS Reprint: 2015

CBS ISBN: 978-81-239-2666-7

Published by:
Satish Kumar Jain for CBS Publishers & Distributors Pvt. Ltd.,
4819/XI Prahlad Street, 24 Ansari Road, Daryaganj, New Delhi - 110002
delhi@cbspd.com, cbspubs@airtelmail.in • www.cbspd.com
Ph.: 23289259, 23266861, 23266867 • Fax: 011-23243014

Corporate Office: 204 FIE, Industrial Area, Patparganj, Delhi - 110 092
Ph: 49344934 • Fax: 011-49344935
E-mail: publishing@cbspd.com • publicity@cbspd.com

Branches:
• *Bengaluru:* 2975, 17th Cross, K.R. Road, Bansankari 2nd Stage, Bengaluru - 70 Ph: +91-80-26771678/79 • Fax: +91-80-26771680
  E-mail: cbsbng@gmail.com, bangalore@cbspd.com
• *Chennai:* No. 7, Subbaraya Street, Shenoy Nagar, Chennai - 600030
  Ph: +91-44-26681266, 26680620 • Fax: +91-44-42032115
  E-mail: chennai@cbspd.com
• *Kochi:* 36/14, Kalluvilakam, Lissie Hospital Road, Kochi - 682018
  Ph: +91-484-4059061-65 • Fax: +91-484-4059065
  E-mail: cochin@cbspd.com
• *Mumbai:* 83-C, Dr. E. Moses Road, Worli, Mumbai - 400018
  Ph: +91-9833017933, 022-24902340/41 • E-mail: mumbai@cbspd.com
• *Pune:* Bhuruk Prestige, Sr. No. 52/12/2+1+3/2,
  Narhe, Haveli (Near Katraj-Dehu Road Bypass), Pune - 411041
  Ph: +91-20-64704058/59, 32342277 • E-mail: pune@cbspd.com

Representatives:

• Hyderabad: 0-9885175004   • Kolkata: 0-9831437309, 0-9051152362
• Nagpur: 0-9021734563   • Patna: 0-9334159340
• Vijayawada: 0-9000660880

Printed at: India Binding House, Noida, U.P.

Dedication

To Doris Singleton Arnold and Lynn Waters Arnold
for your loving support all these years

To Jennifer whose talent and dedication
to art promise a bright future

CONTENTS

13. Digital Image Processing of Remotely Sensed Data

14. Environmental Monitoring and Assessment

Appendixes

PREFACE 2004 ————————

This reissue represents a major improvement. All of the aerial photographs and satellite images in the book are now presented on the enclosed CD. Not only is it possible to view these images on your computer screen, but all of the images that were originally in color are presented on the CD in color. Because of this format, the reader can zoom in on a photo up to the limit of the resolution, thereby adding considerable information to aid in image interpretation. After many years of instruction using paper copies, in the last few years I utilized digital, computerized images in my teaching. This led easily into digital analysis and the study of geographic information systems.

Robert H. Arnold, PhD
Professor Emeritus of Cartography
Salem State College
Salem, Massachusetts

PREFACE

This laboratory manual represents the fruits of many years of instruction in the interpretation of aerial photographs and other remotely sensed imagery. Many of the ideas incorporated here owe their genesis to the author's use of the leading textbooks in the field over many years and I extend my gratitude to the creators of those sources. These books and other materials are listed in the Selected Bibliography.

The impact of the many hundreds of students who investigated this field under my guidance is evident in most of the exercises. Their insights and suggestions have helped shape my approach to image interpretation over the past twenty years.

The exercises deal initially with fundamental concepts that are involved in working with maps and aerial images. The intent is to allow anyone with training in a field that utilizes airphotos or other remotely sensed images to enlist this manual as an aid in learning techniques of interpretation.

Succeeding exercises gradually increase in level of difficulty and sophistication while assisting the user to learn more advanced methods of interpretation. The ultimate goal is to train junior and senior college students, graduates, and professionals to extract information from images efficiently so that image interpretation becomes a useful tool in their particular field of investigation.

The guidance and expertise of Robert McConnin and his staff is gratefully acknowledged as is the support given to me by Salem State College. Finally, my thanks and sincere appreciation are extended to Glenn Young for his patience and skill in producing the line diagrams in this publication. Many people have contributed to this effort in countless ways, and so I share any credit with them. Any errors that might have been made through commission or omission are mine alone and corrections will be welcomed.

Robert H. Arnold, PhD
Professor of Cartography
Salem State College
Salem, Massachusetts

OVERVIEW
THE FIELD OF REMOTE SENSING

TERMINOLOGY

remote sensing
camera obscura
photographic process
negative-positive process
wet plate photography
dry plate process
Kodak No. 1

black and white photography
color photography
color infrared photography
photograph
image
airphoto interpretation
photogrammetry

HISTORICAL PERSPECTIVE

If one defines **remote sensing** as the examination, measuring, and analysis of an object without being in contact with it, then the field of remote sensing has its roots far back in the history of the Earth's inhabitants. People have been studying the surface of the Earth, the Moon, and other heavenly bodies of our universe from afar for many centuries with the unaided human eye.

A device that served as an important step in the development of the modern camera is the **"camera obscura"** which can be traced back to the time of Aristotle—about 350 BC. This allowed for the projection of an outdoor scene through a pinhole into a darkened room. Much later a lens was added to collect more light and this was embodied in a portable box used by architects and artists to trace landscapes and buildings. The creation of lenses also made it possible to magnify views through use of telescopes.

The invention of the **photographic process** was shared by many individuals who conducted experiments with a wide variety of materials and chemical substances through the 1700s and well into the 19th century. This eventually led to the announcement in 1839 of the **negative-positive process** by William Henry Fox Talbot before the Royal Institution of Great Britain. This is the same basic process in use today.

Following this was a period of experimentation and the use of **wet plate photography** in which glass plates were coated with an emulsion that was light-sensitive only while still moist. This is the process employed by Matthew Brady and others who immortalized scenes of the American Civil War, and by William Henry Jackson when he photographed the American West. It was an arduous undertaking because of the weight of the plates and the short drying time factor.

George Eastman of Rochester, NY developed the **dry plate process** in the 1870s that used gelatin and silver bromide to coat glass plates. This led to his mass production of roll film and by 1888 he introduced the **"Kodak No. 1,"** a hand-held camera that opened the field of photography to the layman.

AERIAL PHOTOGRAPHS

In the 1860s photographers used balloons to allow them to photograph cities (e.g., Paris, France and Boston, MA), landscapes, and battlefields (e.g., the Battle of Richmond, VA during the Civil War). Kites and even trained pigeons were used to carry cameras above the earth to record aerial views. What was probably the first camera on a rocket was patented in 1903 in Germany and was designed to take photographs from 2000 feet or higher and return to Earth by parachute.

World War I provided a training ground for aerial reconnaissance from small planes and dirigibles and was the first meaningful military use of aerial photography. During World War II this use expanded greatly and became an integral part of military activities. After the war the military/national defense use of airphotos continued without interruption to the present in the monitoring of one nations' activities by another.

As is discussed in Chapter 5 the 1920s witnessed the rapid growth of aerial photography as an aid in land management, agriculture, forestry, and land use/land cover mapping. In the United States and elsewhere national agencies used airphotos as tools in the normal conduct of planning and development in many fields. In the earlier years **black and white photography** was used almost exclusively due to lower cost than **color photography.** More recently the emphasis is shifting toward **color infrared photography** in which information about growing plants is added to the recording of the visual data in a scene. CIR is discussed in Chapter 7.

SATELLITE IMAGERY (CIR)

In 1960 the first weather satellites were placed in orbit to usher in the era of **satellite imaging** of the Earth. The images of the Earth and its atmosphere as recorded from space became the most important tool available to meteorologists. See Chapter 10 for further discussion.

In July of 1972 the launch of **LANDSAT 1** by the United States began a period of continuous surveillance of the Earth's environment that is ongoing. Chapter 9 provides examples of this type of information and other chapters deal with fields in which images from space are utilized.

EXPLANATION OF TERMINOLOGY

There is need for defining some commonly used terms to avoid confusion. In this publication the term **photograph** is used when the recording device utilizes a lens to collect reflected energy in order to **directly expose** a **light-sensitive film.**

An **image** is the recording of a representation of an object that is produced by optical, electro-optical, optical mechanical, or electronic means. The radiation reflected or emitted from such an object is usually **not directly recorded** on the film.

These distinctions are becoming somewhat more difficult to maintain due to the emergence of **digital photography** in which the representation of an object is directly recorded on light-sensitive sensors instead of film. The product of the process is a digital file stored on a floppy disk that can be transferred directly to a computer and viewed on the screen. Thus far the inclusion of the word, "digital," alerts one to the procedure involved.

The **field of remote sensing** is the designated terminology to represent all of the areas of investigation within this publication. It includes the measurement and analysis of the objects without physical contact between the sensing device and the target. Such a definition encompasses aerial photography, recording reflected and emitted electromagnetic radiation, radar imaging, and thermography, all of which are treated in this volume. Remote sensing also includes the study of gravity fields, magnetic fields, and acoustic energy which are not dealt with here.

The **interpretation of airphotos** is a subfield of remote sensing that focuses upon the data produced by recording reflected energy. The field of **photogrammetry** concentrates on obtaining reliable measurements through photography. It is bound by a set of mathematical relationships that exist between the photographic target and the camera and allows for determination of photo scale and photo object measurement. Chapter 1 begins the discussion of some of these points.

chapter 1

CONCEPTS OF SCALE

TERMINOLOGY

scale	verbal scale
linear units of measure	scale by comparison
areal units of measure	arithmetic proportion
representative fraction	resolution
large scale vs. small scale	detection
graphic or bar scale	

BACKGROUND INFORMATION

The concept of **scale** is one of the more important controlling factors that impact upon the interpretation of airphotos and satellite images. Scale may be defined as a statement of the relationship between the size of a map or image in relation to the area of Earth surface portrayed. This association may be displayed using several different techniques.

METHODS OF REPRESENTING MAP AND IMAGE SCALE

Representative Fraction (RF). This is a numerical statement of the scale relationship, e.g., **RF 1:50,000.** In this example one unit of **linear** distance on the map or image portrays 50,000 of the **same linear units** on the Earth's surface. The units may be inches, feet, yards, centimeters, or meters, etc., as long as the units on both sides of the colon are the same.

A comparison of different RF values will yield the **linear** relationship between two maps or images. An actual ground linear distance will be displayed on a RF of 1:25,000 as a line twice as long as it would appear on a RF of 1:50,000. Note, however, that the **areal** associations do not have the same relative values for these same two scales.

In the example shown below, the maps are all the same physical dimensions but the scales vary. The linear relationship between point A and point B on each map is double from Map 1 to

Linear and Areal Relationships at Different Scales

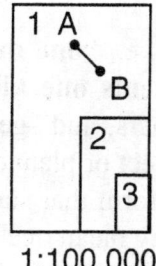 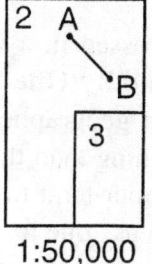 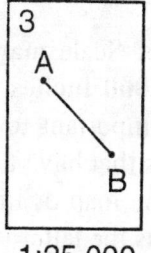

1:100,000   1:50,000   1:25,000

Figure 1–1. Representative fractions.

1

Map 2 and four times from Map 1 to Map 3. However, the **areal** association is very different. Map 1 covers four times the area of Map 2 and sixteen times the area of Map 3. These are important points to remember in comparing maps with images or with the Earth's surface.

Large Scale vs. Small Scale. These terms are important to the use and understanding of maps and images. Unfortunately, they are too frequently misunderstood. The terms apparently originated during the period when **RF** values were written as fractions. The examples in Figure 1–1 would have been expressed on maps as shown below. Note that the **value** of each fraction is an indication of its scale. **Large scale** maps and images portray greater amounts of detail for localized areas. **Smaller scale** maps and images show less detail for regional or even whole world areas. The airphoto in Figure 1–3 and the map in Figure 1–4 are large scale examples. The Landsat TM images exhibited later in this manual are small scale. (Unfortunately, the users of scale concepts don't always agree on the precise cutoff between large, medium, and small scale. In addition, such values suggested for maps vary from those suggested for images.)

$$\frac{1}{100,000} \quad \text{smallest scale}$$

$$\frac{1}{25,000} \quad \text{largest scale}$$

Graphic or Bar Scale. A common method of indicating scale is the use of a linear scale marked off in units of distance measure. This technique allows one to perform straight line calculation of distance between two points by using a ruler to note the length from point to point and then through comparison with a printed bar or graphic scale to convert the measured length directly to feet, yards, miles, meters, or kilometers. The example below is similar to those printed on large scale topographic maps such as shown in Figure 1–4.

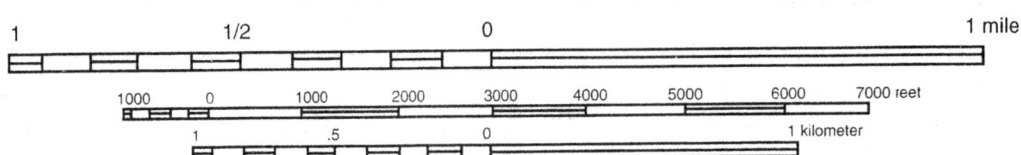

Figure 1–2. Scale 1:24,000.

In utilizing the concept of scale, one should take note that enlarging or reducing a map photographically does not affect the accuracy of a graphic scale, but it does render the RF invalid. That is, the RF is a statement of the relationship between the map and the ground area it portrays. If you modify the size of one of the components, you have changed the nature of the scale association. If the map and the accompanying graphic scale are enlarged or reduced by the same amount, the scale representation is maintained.

Verbal Scale. Scale may also be expressed in a verbal form, i.e., **"one inch on the map represents 100,000 inches on the ground." "One meter represents one kilometer on the ground."** It is important to be aware that geographers, cartographers, and geologists tend to use verbal scales that have a different meaning than that of an architect or planner when they are viewing the same map or image. The former tend to use a verbal form that states the RF relationship, whereas the latter use forms such as "one to 100" when they mean one inch on the map portrays 100 feet on the ground. This would yield quite a different result in making measurements on maps and images.

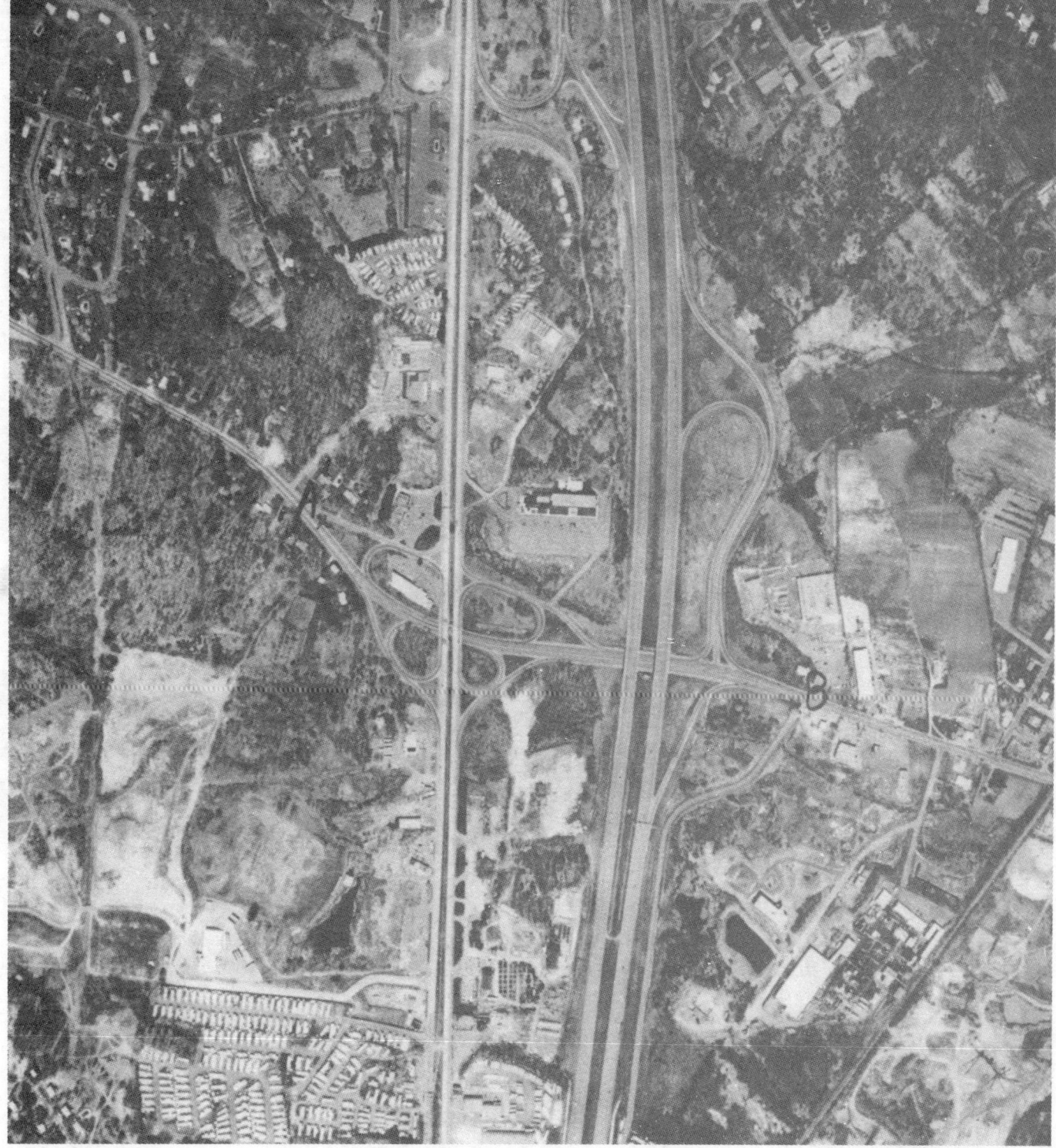

Figure 1–3. Junction of U.S. 1, I-95, and Route 114, 5/3/80.
Photo courtesy J. W. Sewall Co., Old Town, ME, and Town of Danvers, MA.

Determining Scale by Comparison

Occasionally one is confronted with a map or image for which the scale is unknown or is in question. There is a simple technique that may be employed to determine that scale relationship. The example shown in Figures 1–3 and 1–4 is in the town of Danvers, MA which is located 20 miles north of Boston. The map and aerial photograph display the intersections of U.S. 1, I-95, and state Route 114, an area that will be examined in more detail in a later chapter. Note the location of points **A** and **B** on each figure.

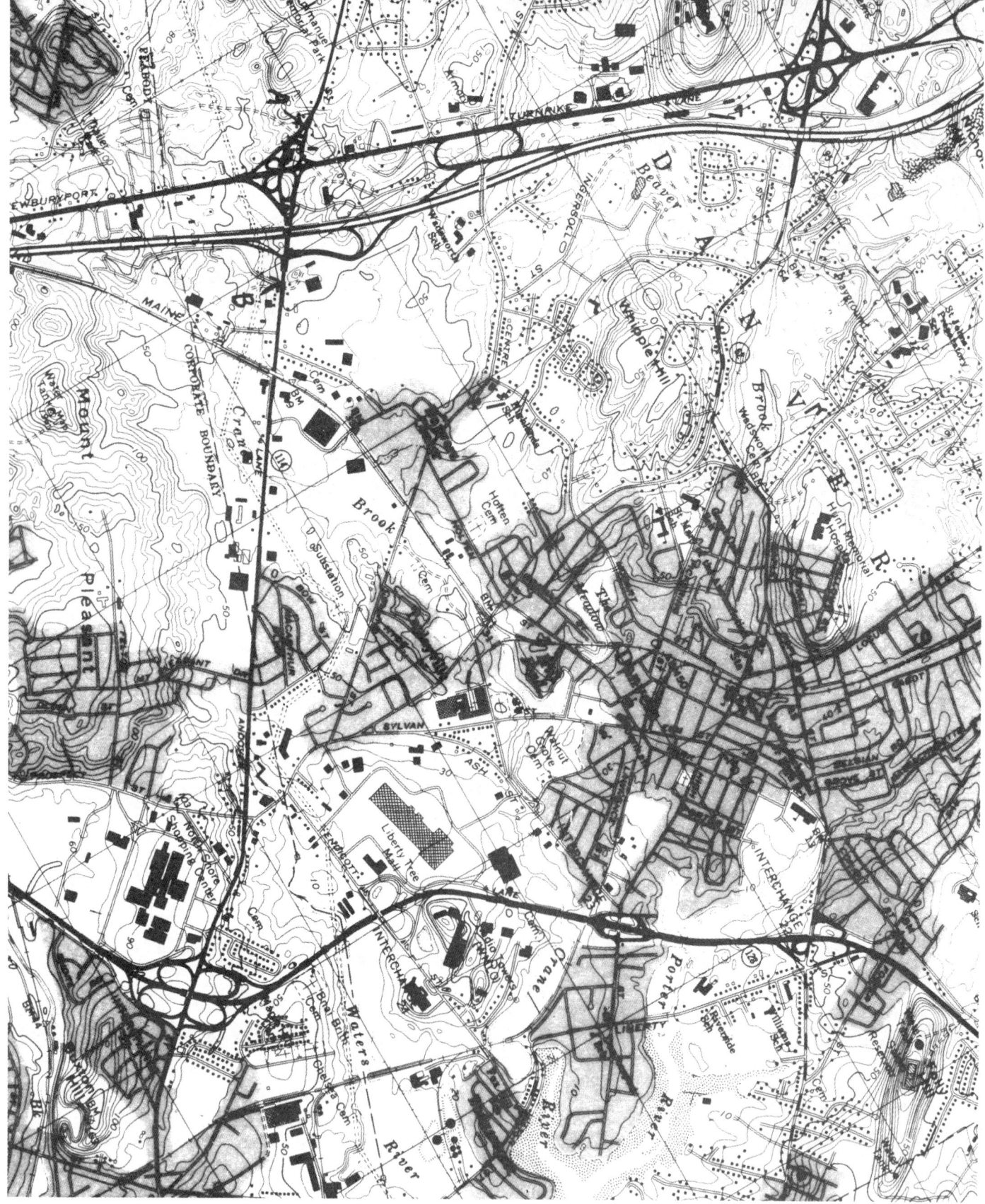

Figure 1–4. U.S. Geological Survey topographic quadrangle, Salem, MA (part) photorevised 1979. RF 1:25,000.

1. Locate the same two points on each map or image of a particular Earth surface area. They should be easy to pinpoint on both representations. Examples here are road junctions.

2. If you are using a vertical aerial photo as part of your scale determination, try to select points so that the line between them will pass through or close to the center (principal point) of the photo. This will produce a more accurate result as it will lessen the effects of normal image displacement on the photo. Principal point is shown here with a "plus" sign.

3. Measure the **straight-line distances** as carefully as possible with an engineer's scale or good quality ruler. Then insert your measured values into the equation below. Note that **line AB** = distance from point **A** to point **B** on map of known scale, and **line A′B′** = distance between the same two ground locations on the image of unknown scale.

4. The resulting arithmetic ratio can then be solved by cross-multiplication as shown. The value determined for **X** in the equation will be the denominator for the unknown RF.

Arithmetic Proportion.

$$\frac{AB}{\frac{1}{25,000}} = \frac{A'\ B'}{\frac{1}{X}}$$

$$\frac{1.12\ \text{in}}{\frac{1}{25,000}} = \frac{3.59\ \text{in}}{\frac{1}{X}} \qquad\qquad \frac{3.59\ \text{in}}{25,000} = \frac{1.12\ \text{in}}{X}$$

$$3.59\ X = 1.12\ (25,000) \qquad\qquad\qquad X = 7800$$

Therefore, unknown RF of airphoto is 1:7800.

RESOLUTION AND DETECTION FACTORS

The quality of an image will always play a major limiting role in the amount of information that can be acquired. There are two important parameters that help to determine the level of quality.

Resolution. This term may be defined as the ability to separate or resolve two adjacent objects on an image. Usually two lines or linear features that are parallel are used as an example. As the quality of resolution increases, the linear features can be seen as separate objects. As resolution deteriorates, the two objects would more likely be observed as one feature.

The quality of resolution in aerial photography is a function of the size and optical characteristics of the lens, the grain size in the film emulsion (smaller grains mean better resolution), and target properties. This relationship may be expressed by the following formula:

$$Rg = \frac{(Rs)(f)}{H}$$

Rg = ground resolution, line pairs/m
Rs = system resolution, line pairs/mm
f = camera focal length, mm
H = height of camera above ground, m

Detection. The smallest object or feature that can be seen on an image is an indication of the quality of detection. Recognition or identification is not required. Usually it is possible to detect very small photographic representations of Earth surface phenomena. These areas on the image are so small they are difficult to map from a practical point of view. That is, the diameter of the object may be less than the width of an inked line on an overlay to denote its presence.

Both **resolution** and **detection** are impacted by other image characteristics such as **contrast, tone,** and **color.** If the object on the image is the same gray tone or hue as the background, it becomes much more difficult to resolve or detect.

Exercise 1   Scale and Image Quality Factors

1. Measure the distances in Figure 1–3 on page 3 and perform the calculations to determine the RF scale of the photo. If your results agree with those stated above, try the same procedure with maps and/or photos of your local area. You might start by using two maps of different but known scales to be sure your work is correct. Note that very small differences in measurement can produce large variations in RF values. Try to be as precise as possible.

2. Examine Figure 1–3. In regard to the characteristics of resolution and detection, how might your choice of points be affected? Are some locations easier to pinpoint because of contrast between object and background? Note that any photo printed in a book tends to have a smaller range of shades of gray than a photographic print because of the need to print photos as halftone images. Try this examination of resolution and detection characteristics on any photos that haven't been produced on an offset press to observe this important phenomenon.

3. Review the formula presented on page 5. If you were working with an aerial camera that had a 152 mm focal length, acquired photographs from an altitude of 3000 meters above ground datum, and the system resolution was set at 30 line pairs/mm, what would the ground resolution be?

chapter 2
DIRECT MEASURE OF LINEAR DISTANCE AND AREA

TERMINOLOGY

linear distance vs. area	sampling procedures
Civil Engineer's Scale	transects
paper straight edge	dot grids
map measurer	compensating polar planimeter

MEASUREMENT ON MAPS AND IMAGES

A common first step in the interpretation and analysis of an image is the determination of the **linear distance** between photographic features or obtaining the **area** of all or part of the image itself. Being able to find the circumference around a lake as well as its total area is often an important type of procedure in image analysis.

In this chapter the focus will be on the measurement of straight line and irregular linear distances on photos/maps along with determination of map or image areas. These results may then be converted to "real world" line or areal values by conversion tables.

MEASURING LINEAR DISTANCE

As was demonstrated in Chapter 1, it is a simple matter to use a ruler or similar device to measure the straight line distance between two points on a map or photo. This value could then be converted to actual Earth surface distance in meters, feet, etc., utilizing the graphic or bar scale on a map or the RF value for either a map or image.

One instrument commonly used for making such measurements is referred to as a **Civil Engineer's Scale,** a ruler that is presented in six scales of 10, 20, 30, 40, 50, and 60 parts to an inch. The tenths and fiftieths scales are especially useful with images as the values measured may be converted easily to the decimal system.

It is a more difficult matter to measure accurately the length of an irregular course such as the shoreline of a lake or a trail through a wooded area. There are several techniques that may be employed to carry out direct measurement.

Use of a Paper Straight Edge. It is possible to obtain rough measurements of an irregular, linear course by employing a plain piece of bond paper or similar material. Fold the sheet in half lengthwise to give the edge added strength where it is double thickness. Place the folded edge against the linear course that is to be measured and mark the beginning point at one end of the paper using a sharp pencil.

Align the fold against the course to be followed and mark straight segments successively along the paper's edge, turning the paper to follow the path. Mark the end of the course on the paper when you have completed the final straight section. Now place the folded edge of the paper against a bar or graphic scale to read the distance directly in meters, feet, miles, etc. If you are working with an image rather than a map, the measured distance on the paper may be converted to meters or feet using the RF value following the procedure explained in Chapter 1.

Use of a Map Measurer. There are a number of companies listed in map-related catalogs that manufacture a device commonly called a **map measurer.** (Also referred to as an opisometer or map wheel.) This is a relatively simple instrument that easily fits in your hand and is used to trace an irregular course on a map or photograph. It consists of a small tracing wheel on the end of the instrument which is connected by gears to large dials that may be marked in centimeters, inches, feet, or other units. See Figure 2–1.

Care should be exercised if you use or purchase a map measurer—since they are utilized by architects, planners, foresters, geographers, geologists, and others, the dials are marked in many different units of measure, some of which will not be appropriate for your needs. An inexpensive version can be obtained for less than $10. More precise, sturdier units are available at higher prices.

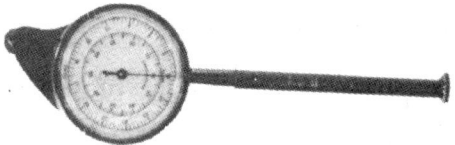

Figure 2–1. Map measurer. Courtesy Avery/Berlin.

Representative fraction (scale)	Meters per centimeter	Centimeters per kilometer	Hectares per square centimeter	Feet per inch	Inches per mile	Acres per square inch
1:1,000	10	100.00	0.01	83.33	63.66	0.16
1:2,000	20	50.00	0.04	166.67	31.68	0.64
1:3,000	30	33.33	0.09	250.00	21.12	1.43
1:4,000	40	25.00	0.16	333.33	15.84	2.55
1:5,000	50	20.00	0.25	416.67	12.67	3.99
1:10,000	100	10.00	1.00	833.33	6.34	15.94
1:15,000	150	6.67	2.25	1,250.00	4.22	35.87
1:20,000	200	5.00	4.00	1,666.67	3.17	63.77
1:25:000	250	4.00	6.25	2,083.33	2.53	99.64
1:50,000	500	2.00	25.00	4,166.67	1.27	398.56
1:75,000	750	1.33	56.25	6,250.00	0.84	896.75
1:100,000	1,000	1.00	100.00	8,333.33	0.63	1,594.22
Method of calculation*	$\dfrac{RFD}{100}$	$\dfrac{100,000}{RFD}$	$\dfrac{(m/cm)^2}{10,000}$	$\dfrac{RFD}{12}$	$\dfrac{63,360}{RFD}$	$\dfrac{(ft/in)^2}{43,560}$

* RFD refers to the representative fraction denominator. After Avery & Berlin, 4th ed.

Figure 2–2. Scale conversions for maps and vertical photographs.

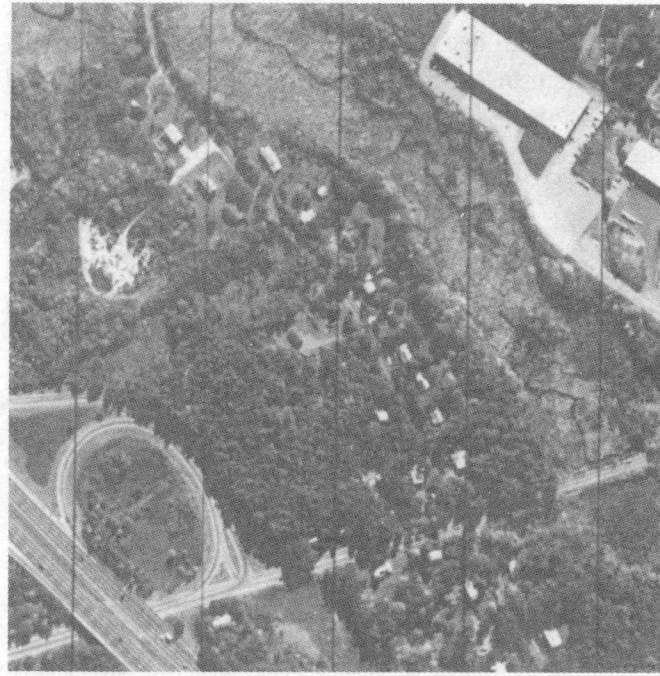

Land Use Types	Transect Inches	%
Forest (F)	6.1	36
Wetland (W)	3.5	20
Residential (R)	3.3	19
Transportation (T)	2.4	14
Industrial/Other(I)	1.8	10

Figure 2–3. Area sampling with transect lines.

The map measurer is very simple to use but results will vary based upon the care taken in its operation. Try the device on a short stretch of the map or image to be measured to be sure that the dial is recording in the proper direction, i.e., increasing in values. Set the dial at the zero mark and carefully trace the course you wish to measure. Record your result to the greatest accuracy allowed by the dial and then repeat the procedure two more times. Calculate the average of your three readings. If the dial reads in centimeters or inches and you desire a result in kilometers or miles, consult the table in Figure 2–2 for the conversion factor.

MEASURING AREAS ON MAPS AND IMAGES

There are several techniques that may be used to yield the area of all or a portion of a map or image. Which approach is employed is dependent upon the level of accuracy required and the time available to carry out the procedure. These are **sampling procedures.** That is, a part of the universe or area is measured under the assumption that it is representative of the whole.

Use of Transect Technique. Perhaps the least time consuming approach that promises some predictable level of accuracy involves the construction of **transects** across the area of the map/image to be measured. How accurate this procedure is depends upon the number of transects employed. That is, more transects will yield more precise results but at a cost of greater time expended. Fewer such lines is accomplished quickly but at the expense of accuracy.

This procedure is carried out by drawing a series of evenly spaced and parallel lines across a map or image, or perhaps on a transparent overlay that may be placed on the map/image. See Figure 2–3 above. The lines may be vertical, horizontal, or diagonal as long as they are equidistant from each other.

The area represented on the map/image is separated into a few delineated zones or subareas on the basis of land use or some such criterion. The example used here is an airphoto that has been divided into several land use categories.

This is a sampling technique that requires measuring the length of each transect line that crosses the area of a particular map class, e.g., industrial use. The total centimeters or inches of industrial use is divided by the total length of all the transect lines together to yield the estimated percentage of the image devoted to industrial land use. The assumption is that the areas under the lines that were sampled represent all of the rest of the image, and obviously this is not always absolutely true.

The same procedure is followed until all land uses or categories have been measured and their areal percentages calculated. While this technique is fast and easily accomplished, a drawback to this approach is that the nature of the arrangement of the lines may introduce a bias in the results.

That is, the lines might align with a street pattern or agricultural field pattern that would yield a higher percentage for that type of use than a random type of sampling procedure. Consider the arrangement of streets in Manhattan or many other city areas and the grid pattern of midwest farmlands.

Area Sampling with Dot Grid Overlays. Another technique for determining areas on maps/images involves the use of **dot grids.** Figure 2–4 shows a dot grid similar to those used by the USDA, Soil Conservation Service to determine areas of land use on airphotos. The grid is "dropped" or laid on the photo so that the lines of the overlay are not purposely aligned with any land use pattern of lines in the photo. The intent is to lessen or remove bias in measurement.

Based upon the number of dots per square centimeter or per square inch, each dot represents a number of hectares or acres on the photo. When the scale of the photo is known the conversion values in Figure 2–2 may be used to determine these amounts. Each dot in Figure 2–4 represents 0.16 hectares or 0.4 acres.

Appendix B contains a printed version of a dot grid that is suitable for use in producing a transparent overlay on a photocopy machine. That grid is an example of random dot alignment but other grids with a regular pattern are often used. That grid will be utilized in exercises at the end of this chapter.

Compensating Polar Planimeter. This instrument assists in the measurement of areas on maps, images, etc. to the nearest one-hundredth of a square centimeter or square inch. The procedure calls for tracing the circumference of an area so that a complete circuit is made ending at the starting point. As is the case in most types of measuring procedures, the area should be circumscribed at least three times and the average value obtained to acquire the most accurate result. The tracing MUST proceed clockwise for this particular instrument to produce the correct result.

Note that the planimeter may be set to read in metric or English units. The largest dial (at center) reads in tens of units, the dial in the left window reads in single digits and tenths on the right side, and the vernier scale on the left side of that window yields values in hundredths. The results obtained may be converted to hectares or acres using Figure 2–2.

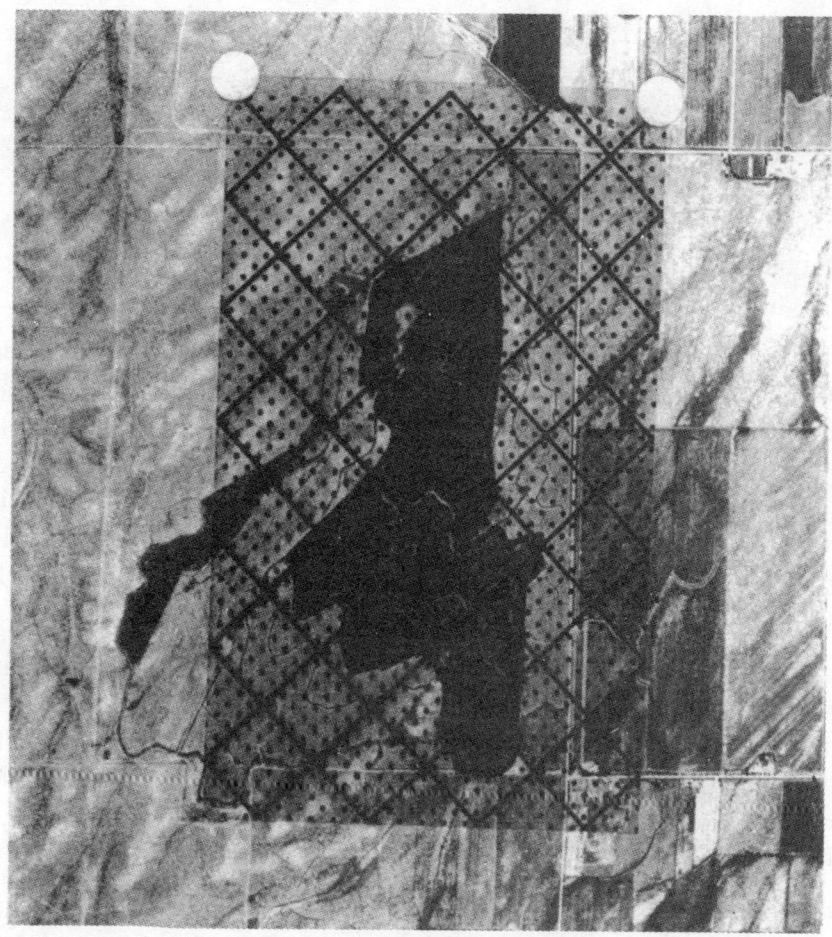

Figure 2–4. Dot grid overlay on vertical airphoto (photo scale = 1:20,000). Each grid square is 1 cm on a side and contains 25 dots.
Courtesy Avery/Berlin.

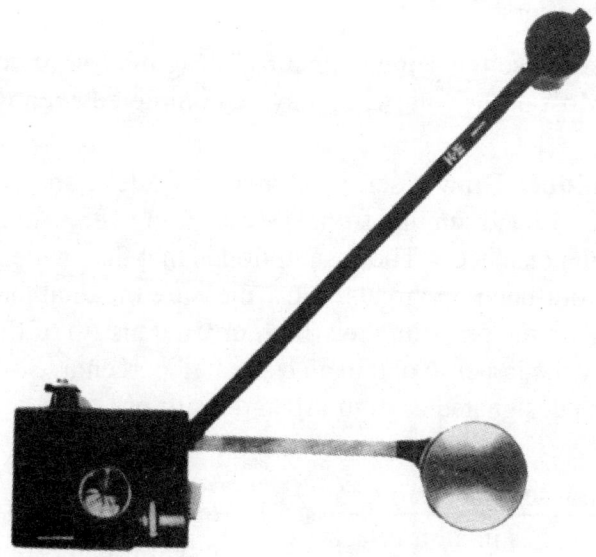

Figure 2–5. Compensating polar planimeter.
Courtesy Avery/Berlin.

Exercise 2 Direct Measurement on Maps and Images

1 **Paper Straight Edge.** Using the technique described on pages 3–5 measure the straight line distance between points A′ and B′ on the airphoto in Figure 1–3. This value should be approximately 3.6″ which equals 2340 feet at the RF 1:7800. Now use the same method to measure the distance along the center of the major road adjacent to these two points. This value should be close to 3.8″ or 2470 feet at this scale. Note that a 0.2″ difference in measurement yields an actual ground distance of 130 feet.

Obviously short distance measurements are more susceptible to error with approximation techniques such as this one. Also, it should be pointed out that whatever personal bias you normally exhibit in the direct measurement process will be more important as a factor the shorter the distance measured. Use this technique on some other maps or images where you can verify the straight line distance with a ruler before you attempt to measure an irregular route. The distance between two towns on a road map is a good test as the actual miles or kilometers of road travel are often given on the map between specific points.

2. **Map Measurer.** You might check your results in #1 above utilizing this instrument. Again it should be noted that the percentage of error will likely be high for such small measurements. A better test of the map measurer would be to obtain a large scale topographic map of your local area and then trace an irregular course around a lake or wooded area shown in green on the map. If you complete a circuit by ending at your starting point, the same course may be utilized for the polar planimeter.

Figure 1–4 on page 4 is a black and white copy of a portion of the Salem, MA USGS topographic map utilized previously. Trace the course of the roads marked US 1, 62, 128, and 114, turning at the center of each intersection between the roads listed rather than following interchange roads. Measure the circumference of this course by paper straight edge method and then with a map measurer. How do your results compare? Which result is most likely closer to the true distance? If one inch on this map equals 2.53 miles, how many miles long is the route? Can you convert the distance to kilometers?

3. **Area Measurement.** Use the same course from Figure 1–4 to compare techniques of area measure. Most of the following may be completed even if you lack some of the tools.

   a. **Transect Technique.** Draw a series of seven parallel transects at 1/2″ intervals parallel to Route 114 and running from US 1 to Route 128. Add up the total length of the transect line distances. The assumption is that this value represents the total area within the four bounding roads. Now measure the total length of all the transect lines that cross the gray tint area (high urban density) of the map. Divide the smaller figure by the larger to obtain an estimated percentage of the area inside the four roads that is designated as high urban density use.

$$\frac{\text{Sum transect inches in gray}}{\text{Total inches in all transects}} = \begin{array}{l}\text{\% of area within bounding} \\ \text{roads classed as densely} \\ \text{settled urban (gray)}\end{array}$$

b. **Dot Grid Overlay Use.** Produce a black on a clear transparency of the dot grid in Appendix B using a photocopy machine and transparency masters. Use the grid two times in the same manner: "drop" the grid on Figure 1–4 so that it covers the area within the four major roads; count the total number of dots inside the bounding roads; count the total number of dots falling on the gray tint area (count every other dot lying on the edge) and divide the smaller number of dots by the larger to obtain the percentage of the area within the roads that is devoted to high urban density.

$$\frac{\text{Sum of dots within gray area}}{\text{Total dots within four roads}} = \begin{array}{l}\text{\% of area within 4 roads}\\ \text{classed as densely}\\ \text{settled urban (gray)}\end{array}$$

Is the result different when the grid alignment on the map varies? If so, what might account for the variation?

How do these results compare with the transect figure?

Which of these techniques seems to be the most reliable?

Would the density of dots in a grid be a factor?

c. **Compensating Polar Planimeter.** Photocopy page 4 or remove it from the manual to facilitate use of the planimeter if you have access to one. Trace the circumference of the roads in Figure 1–4 moving the tracing arm in a clockwise direction. Read the scales to determine the area within the roads to the nearest one-hundredth of a square inch. Now trace the gray tint area in the same fashion. Convert both totals to square miles using the values in Figure 2–2. Divide the smaller value by the larger to obtain the percentage of high urban density area within the roads.

$$\frac{\text{Area within gray}}{\text{Area within 4 roads}} = \begin{array}{l}\text{\% of area within 4 roads}\\ \text{classed as densely}\\ \text{settled urban (gray)}\end{array}$$

How does your result compare with the earlier methods?

What factors seem to affect the operation and accuracy of the planimeter?

chapter 3
PARAMETERS OF IMAGE SIGNATURES

TERMINOLOGY

signature identification
electromagnetic radiation
electromagnetic spectrum
wavelength
frequency
aerosols
attenuation
absorption
atmospheric transmission
atmospheric windows
bands
photographic spectrum
visible spectrum
reflected infrared
thermal infrared
albedo

culture
signature characteristics
  size
  shape
  pattern
  texture
  contrast
  shadow
  volume
  orientation
  tone/gray level
  hue and chroma
topographic inversion
natural color
false color
positive identification

THE SIGNATURE CONCEPT IN IMAGE INTERPRETATION

Perhaps the most important, recurring theme of this manual is the theory that phenomena on or near the Earth's surface display a particular combination of characteristics that may be utilized in their identification. These features may be recorded by remote sensing devices, such as aerial cameras, and then will appear on the resulting images. It is believed that by establishing what the specific mix of characteristics is for an object or phenomenon, one can consistently and reliably establish the correct identification.

The material which follows is a summary of many of the most important types of information that may be used in **signature identification.** Some of these parameters are subjective in nature, but many of them are suitable for quantification and so may be easily measured or digitized for computer-assisted analysis.

ELECTROMAGNETIC RADIATION

Figure 3–1, **Characteristics of the Electromagnetic Spectrum** (EMS) on the following page provides considerable information related to energy that is generated in space, primarily by the Sun, and then intercepted by the Earth and its atmosphere. The nature of this energy and its

Characteristics of the Electromagnetic Spectrum

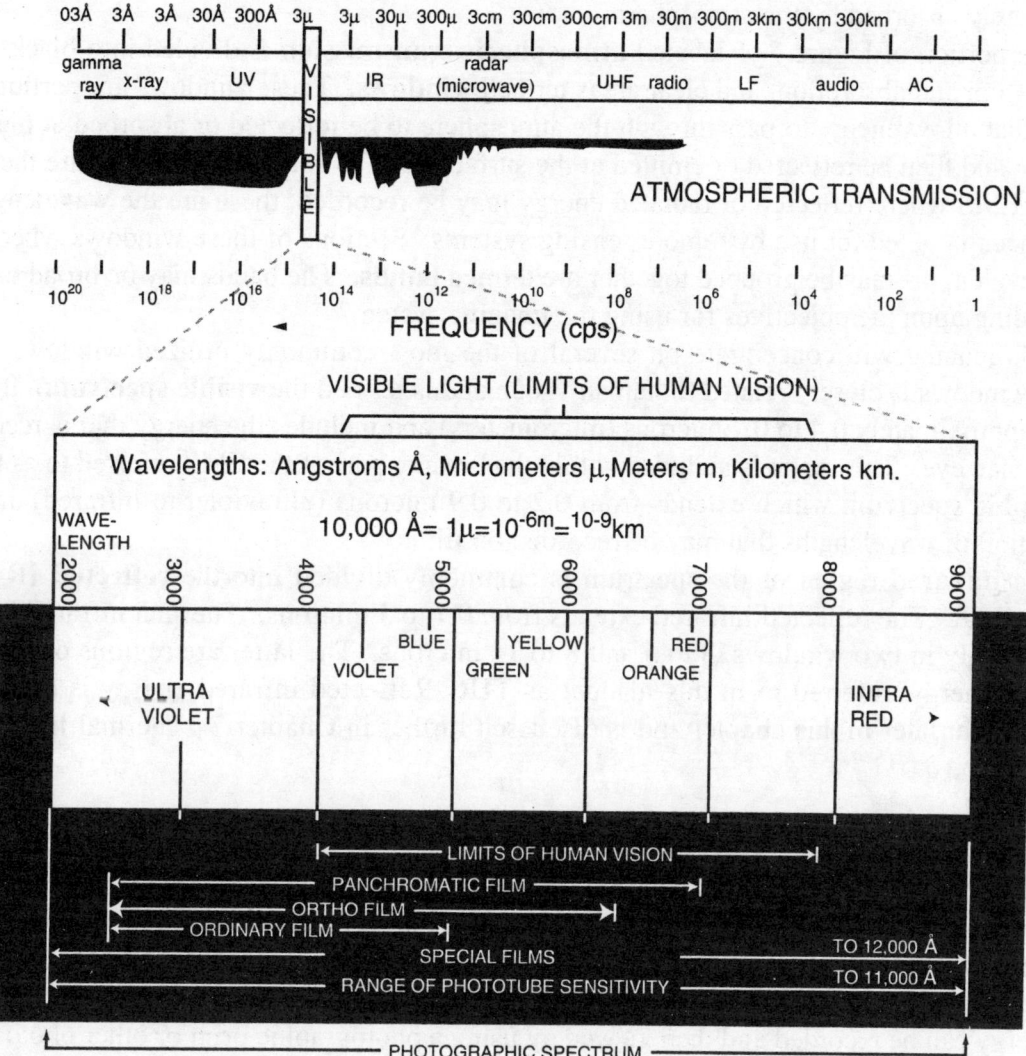

Figure 3–1. Characteristics of the electromagnetic spectrum.

characteristics are critical to the understanding of remote sensing systems and the images they produce. The discussion here is brief by necessity, but further amplification is readily available in a number of sources (see Appendix E).

The upper portion of Figure 3–1 illustrates the arrangement of electromagnetic energy according to **wavelength.** The waves are measured from crest to crest or trough to trough and are in the form of sine curves similar to waves at the ocean's surface. This energy can pass through the ether of space and the atmosphere of the Earth in this form. The shortest wavelengths are at the left and the longest wavelengths are shown to the right. Relative sizes of wavelengths are compared in the center of the diagram.

Note that the **frequency** of wave movement, the number of wave cycles passing a point per second, is in inverse relation to wavelength. These characteristics of length and frequency of waves have great bearing upon the specific types of energy which can pass through the Earth's atmosphere.

Our atmosphere contains molecules of a number of gases along with minute solid and liquid particles termed **aerosols.** The relationship between the diameters of these molecules and particles compared to the wavelengths of electromagnetic energy helps to explain the **attenuation** and

absorption of energy by the Earth's atmosphere. That is, which wavelengths of energy may pass through the Earth's atmosphere to reach the Earth's surface vs. the waves that are partially or completely absorbed by the atmosphere.

The portion of Figure 3–1 labeled **atmospheric transmission** is divided into black areas of attenuation and absorption, and clear areas termed **windows.** These windows are portions of the EMS that allow energy to pass through the atmosphere to be reflected or absorbed at the Earth's surface and then be reflected or emitted at the surface back to space. Since these are the regions of the EMS where reflected or radiated energy may be recorded, these are the wavelengths that have been targeted for use by remote sensing systems. Sections of these windows where different wavelengths may be grouped together are termed **bands.** The bands may be broad or narrow depending upon the objectives for using the imaging system.

This manual will concentrate on several of the most commonly utilized windows. One of these windows is closely related to human vision and is termed the **visible spectrum.** It extends from approximately 0.4 to 0.7 microns (micrometers) and includes the energy that is received by the human eye. This range is included within the broader part of the EMS referred to as the **photographic spectrum** which extends from 0.2 to 0.9 microns (ultraviolet to infrared) and is the collection of wavelengths that may be recorded on film.

The **infrared** region of the spectrum is commonly divided into the **reflected IR** and the **thermal IR.** The reflected infrared extends from 0.7 to 3 microns. Thermal infrared is recorded primarily in two windows, 3 to 5 and 8 to 14 microns. The latter are regions of radiated or emitted energy, referred to in this manual as **TIR.** Reflected infrared energy is compared to visible light later in this chapter and is discussed further in Chapter 7. Thermal IR is pursued in Chapter 10.

SPECTRAL SIGNATURES

All objects and phenomena at the Earth's surface reflect or absorb radiant energy from the Sun. Humans can perceive this reflected radiation within the visible part of the spectrum referred to above. Photographic film may be used to extend the range of the "visible" EMS wherein reflected energy can be recorded and then viewed by using a photographic print or other photo product.

Each Earth surface feature has its own array of reflectance characteristics which are controlled by the nature of the feature. The total amount of sun energy striking an object is the sum of the energy that is reflected, absorbed, and transmitted. The percentage of the energy that is reflected is termed the **albedo** and is a useful characteristic in identification. Objects with a high albedo are bright, e.g. snow, and those with a low albedo are dark as is the case with most wetland and floodplain soils.

In later chapters the spectral signatures of Earth surface features will be shown to vary depending upon the bands of the EMS used to record the features. That is, the spectral signature of an object in one band may be quite different in another band since reflection and absorption characteristics vary with wavelength. For example, growing vegetation reflects strongly within the near IR region and considerably weaker within the visible bands of EMS.

This variation in spectral reflectance is a powerful tool in the use of multispectral satellite imagery. Comparison of spectral signatures by band allows for the identification of a feature or its elimination from consideration once signatures are known.

Figure 3–2 on the following page displays a number of aerial views of small parts of the Earth's surface that have been impacted by human activities. Such artifacts are referred to as examples of **culture** to distinguish them from totally natural landscapes. Note that there is a decided variation in albedo within the photos and from one photo to another. That information will be used as an aid in exercises at the end of this chapter.

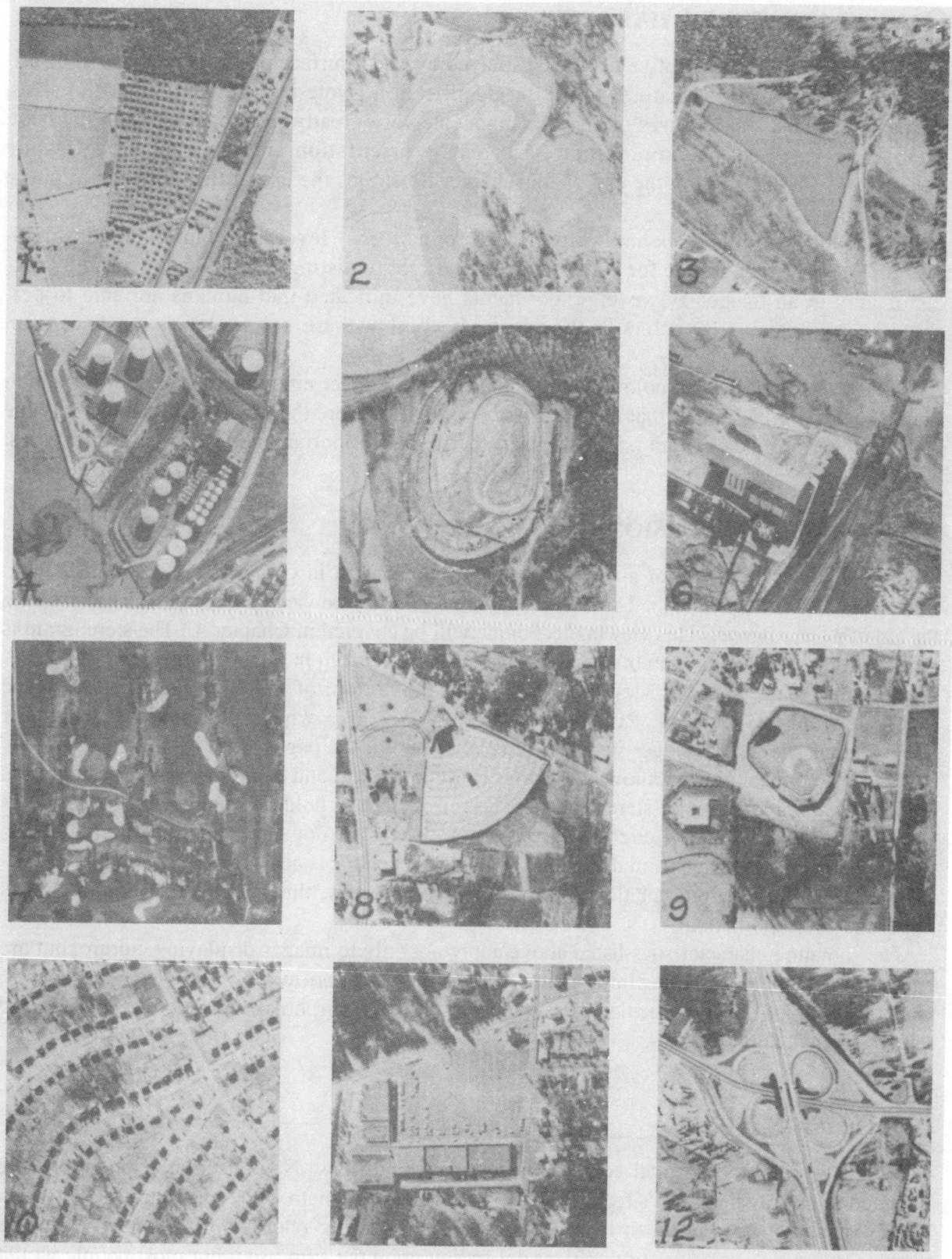

Figure 3–2. Cut-out views from vertical aerial photographs. Avery, 1968.
Courtesy of Abrams Aerial Survey Corp.

OTHER SIGNATURE CHARACTERISTICS

There are many other descriptive features displayed by Earth surface phenomena that can be utilized to create specific signatures to aid in identification. Note in Figure 3–2 that the **sizes, shapes, patterns,** and **textures** of photo components vary greatly from one image to the next. In addition, **contrast, shadow, volume,** and feature **orientation** are all important in feature delineation. The exercise at the end of this chapter illustrates the use of these criteria in photo identification.

When working with black and white images **tone** or **gray level** is an important consideration. It is generally possible for normal human eyesight to distinguish between 16 and 32 levels of gray in an image. However, experiments have indicated that humans are able to perceive approximately 100 colors of the spectrum. Within these colors, levels or shades can also be discerned.

These facts indicate that color images contain much greater amounts of visible information than is the case with images that are black and white. Therefore the **hue** (wavelength, e.g., blue 0.4–0.5 microns) and **chroma** (shade or strength) are also important **signature characteristics.**

TOPOGRAPHIC INVERSION

Figure 3–3 is a stereogram of a section of the Grand Canyon in Grand Canyon National Park, Arizona from a collection at the University of Illinois. (It may be viewed three dimensionally with the aid of a stereoscope. This viewing technique will be covered in Chapter 4.) The stereogram is displayed with north at the top of the page as is often the custom in publications.

Note that if the image is viewed with north at the bottom of the page (upside down), the topography appears changed. Rivers appear to flow in valleys whereas the same rivers appear to flow along the tops of ridges when north is at the top of the page.

This is an optical illusion known as **topographic inversion** and is an important consideration whether examining single or stereo images. It is important to hold an image with the **shadows facing the viewer** whenever circumstances permit. If you are not sure which orientation is correct, try viewing the image from different directions. When the shadows face you, you will experience the appropriate presentation of topography. Obviously, this characteristic of images is important to keep in mind.

The signature characteristics listed above apply not only to images displaying culture, but are equally important in extracting information from images that portray natural vistas or landscapes. Chapters 8, 9, and 12 in particular will include a number of airphotos and satellite images used to investigate natural features.

COLOR SIGNATURES

Color images may be **natural color** or **false color** renditions. Natural color photographs are intended to portray the Earth's features as humans would see them. One may purchase different brands of photographic film to record a scene. The resulting photographic print may appear warmer (more red) or colder (more blue) depending upon the film used, but most people would perceive the photos as realistic in their presentation.

Figure 3–4 (see Plate I) is a natural color aerial photograph of the northwest portion of San Francisco immediately southwest of the Golden Gate Bridge. This photo was acquired at a nominal scale of 1 inch = 1200 feet (RF 1:14,400) on April 1, 1994. That date is near the end of the rainy period of the year, so that vegetative growth is active.

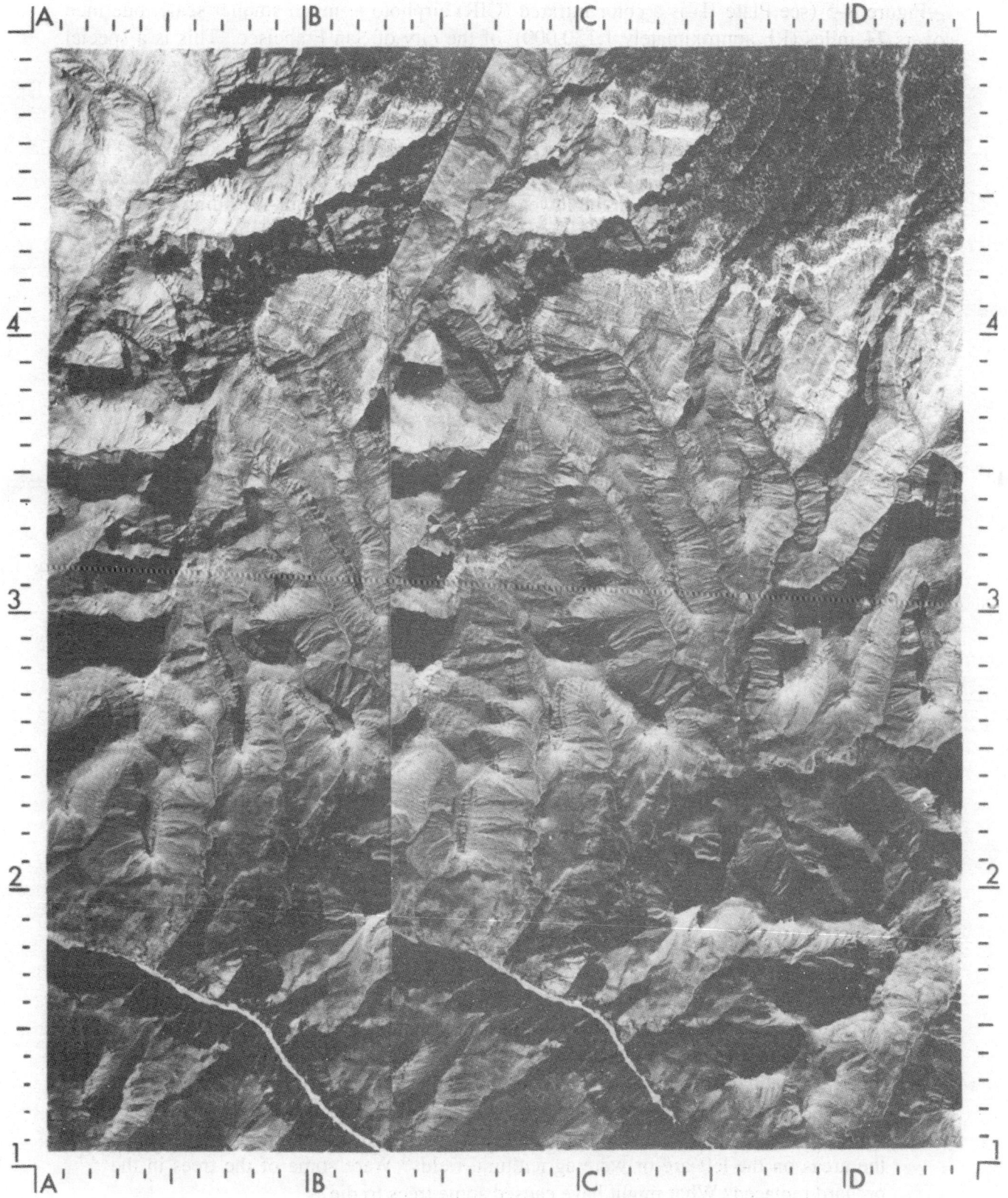

GRAND CANYON
Grand Canyon National Park, Arizona
October 11, 1955
RF = 1:54,400   H = 30,000 feet

Stereogram No. 528
Prepared from USGS photography
by the University of Illinois
Committee on Aerial Photography

Figure 3–3. Courtesy University of Illinois, GIS Laboratory.

Figure 3–5 (see Plate II) is a color infrared (CIR) airphoto at much smaller scale, one inch covers 2+ miles (RF approximately 1:130,000), of the city of San Francisco. This is a special form of false color photo which has considerable scientific value since growing vegetation will appear in various shades of red in the photographic reproduction of the scene.

Note that the central business district (CBD), to the right of the photo center near the west end of the bridge to Oakland, has very little vegetation and is mostly light blue in color. The small, black spots are shadows of buildings in the financial district. Compare those colors with the appearance of the residential areas in the lower left corner near the ocean. Winding streets lined with homes have some grass and trees showing blue and red intermixed.

These color differences hold great significance in image interpretation. The fact that CIR aerials usually contain more extractable information than do color airphotos explains the rapid increase in use of CIR aerial film. This topic is further examined in Chapter 7, "Color Infrared Photography

Exercise 3   Establishing Signature Characteristics

General Procedure

It should be noted here that all of the interpreter's knowledge relating to locations and patterns of distribution on the Earth's surface can play a role in breaking down the detail of an image. Therefore, people with different travel experiences and learned information about place will likely attain varying results in image interpretation, other factors being equal.

Regardless of the level of expertise of the image user, it is recommended that the following approach be pursued in order to extract the greatest amount of information. Image examination should begin with **establishing positive identification** of the most general or obvious features, e.g., separating the area shown into land vs. water, or natural landscape vs. that impacted by people, etc. This allows the interpreter to proceed from the known with some feeling of confidence.

Read any information that accompanies the image, noting especially the date of acquisition, and location shown. Are there any features that give an indication of scale? If a map of the area is accessible, keep it handy. In general, one should make available every type of information that will make positive identification most likely. Approach the image interpretation as though you were being paid for results and your job depended upon success.

Black and White Airphotos

1. **Pattern, Shapes, Tone.** Examine Figure 3–2 on page 17. The #1 view shows an orchard, a wooded area, several agricultural fields and what appears to be a divided highway. Are all the trees in the orchard the same age? How can you tell? What might be distinguishing characteristics of older trees as opposed to younger trees on such an airphoto? Explain. Why are some of the rows of trees parallel to the road and some are not? Which feature was present first, orchard or road? What indicates that the areas on the left are or were agricultural fields? Were some of the trees in the orchard replaced? What might have caused some trees to die?

   The information needed to answer these questions with a high probability of success is present in the photo view. You don't need to answer all of these questions correctly, but thinking about achieving solutions will help to develop a good approach to image identification.

2. **Shape, Texture, Tone.** The #3 view shows a water body in a setting that appears rural. Is the water body natural or is it a reservoir? What evidence supports your answer?

3. **Shape, Texture, Shadow, Tone.** The #4 picture contains some tanks. Some of the tanks hold flammable liquids and some do not. How can you differentiate? What types of transportation serve this location? What is the evidence? Is there any part of the photo that indicates land that is not used? What is the evidence? What photo information indicates that the area on the left side is water?

4. **Orientation, Pattern, Texture, Tone, Shape.** Photo #7 shows part of a golf course. Can you identify putting greens, sand traps, fairways, and rough areas? What specific information did you use to arrive at your answers? Once you establish the signature characteristics for a golf course, it will be readily apparent on most airphotos. Most golf courses display similar characteristics even though the particular layouts are often different.

5. **Shape, Orientation, Pattern, Texture, Shadow.** In photo #8 the dominant feature is a drive-in movie theatre. This type of use represents a feature that was common for several decades in the United States. What photo characteristics identify the use? Which use was likely present first, the drive-in or the nearby houses? Explain. If you were to examine airphotos acquired years ago, how might the age of the photos impact your success in feature identification? Explain. Keep in mind that successful interpretation of features appearing in current or recent landscapes or scenes often requires examination and interpretation of airphotos acquired many years previously.

6. **Size, Shape, Pattern, Orientation, Texture, Tone, Shadow.** The view in #9 contains a baseball field that is enclosed by a fence or wall. What information is present in the photo that could lead you to determine whether the playing field or the surrounding uses were present first? What information is present to tell you whether this field is used by players at major league, minor league, high school, or Little League levels?

7. **Pattern, Texture, Orientation.** Photo #10 shows an area of single-family, residential housing. Most such housing built prior to World War II is aligned along streets that are laid out in a grid and that meet at right angles. Housing that was constructed since 1946 often displays irregular street patterns, similar house and lot sizes, and cul-de-sac dead ends. Which category do you think this photo matches? How old were these houses when the picture was taken? What information in the picture indicates fairly recent construction relative to the date of the photography?

This is a unique signature in the United States that may be identified as single-family, residential housing by texture and pattern as well as orientation to a city center. This is often the case even when the individual houses cannot be resolved. Examine some of the smaller scale airphotos in this manual, such as Figure 3–5, to verify this fact.

8. Photo view #11 portrays a small shopping mall with several different stores. What evidence is present to indicate this? How can one tell that this is not a factory? What can you say about the age of the surrounding housing? Evidence?

9. The most prominent feature in view #12 is a highway interchange or cloverleaf. This is another unique signature of developed societies that may be identified even on very

small scale satellite imagery. Test that proposition with images in this manual. What are the characteristics that make this an easily discernible phenomenon?

10. Examine Figure 3–3 on page 19. Now that you know this is an example of "topographic inversion" consider the possible impact upon image interpretation. How might your ability to examine the drainage pattern, define slopes, or interpret the physical landscape be impacted by this illusion? Explain.

Color Airphotos

11. Now compare the photographs shown in Figures 3–4 and 3–5. While the scale differences impact the resolution and detection capabilities of an interpreter, the color variations are of more interest here. Compare the areas of vegetation versus the water surfaces. Which photo, the natural color or the CIR displays greater contrast between the water and vegetation? Which photo exhibits greater contrast between vegetation and dense urban areas?

The natural color photo may be interpreted more readily due to familiarity with colors seen around you every day, but the level of contrast between photographic features is often weak especially on a hazy day. Figure 3–4 is a high quality aerial product with excellent detail and yet the contrast is greater in the CIR photo, Figure 3–5. Becoming familiar with typical color signatures shown on color infrared photography obviously will reap great benefit for the image interpreter.

chapter 4

PHOTOGRAMMETRIC CONSIDERATIONS

TERMINOLOGY

photogrammetry	low oblique
vertical airphoto	high oblique
oblique airphoto	pocket stereoscope
fiducial marks	mirror stereoscope
principal point (PP)	corresponding principal
nadir	point (CPP)
parallax	average photo base length (P)
sidelap	differential parallax (dp)
forward overlap	stereogram
stereovision	contact prints
stereopair	A,B,C location method

PHOTOGRAMMETRY DEFINED

The art and science of obtaining reliable measurements from photographs is termed **photogrammetry.** In carrying out the interpretation of airphotos and other remotely sensed images, it is often of value to know the scale of the image, distances between points shown thereon, and the altitude at which the image was acquired. All of these values are interrelated in a series of photogrammetric relationships that may be expressed as mathematical formulae or in graphic diagrams. Some of the more important aspects that impact image interpretation are discussed below.

VERTICAL VS. OBLIQUE AIRPHOTOS

An aerial camera can be positioned to photograph the ground from any angle above the horizon. In Figures 4–1 and 4–2 some of the characteristics of a **vertical airphoto** are illustrated. Such photos are acquired with the camera pointing vertically down to the Earth's surface on a line which, if extended, would reach the center of the Earth. This is a view that is utilized in airphoto analysis most frequently as it allows for the greatest data extraction based upon photogrammetric principles. For the average person an **oblique photo,** in which the view of the ground is other than vertical, is more familiar and more easily comprehended. However, in oblique photography photo scale is constantly changing from the foreground to the distant background. **Low oblique photos** are those where the camera position is closer to a vertical view, whereas **high oblique photos** usually include the horizon and the camera angle is approaching a horizontal line of sight. The emphasis in this manual is on the use of vertical photography due to the greater amount of information that may be extracted through measurement as well as interpretive techniques.

COMPONENTS OF THE VERTICAL AIRPHOTO

Figure 4–1 is a diagram that displays some of the important characteristics of a vertical, aerial photograph. Built into the frame of a typical aerial camera are **fiducial marks** that are photographed each time an image is recorded. These marks are of varying shape or form and may appear in the corners or the middle of each side of the photo or in all eight locations. The purpose of these marks is to allow the user to locate the precise center of the photograph, the **principal point (PP).**

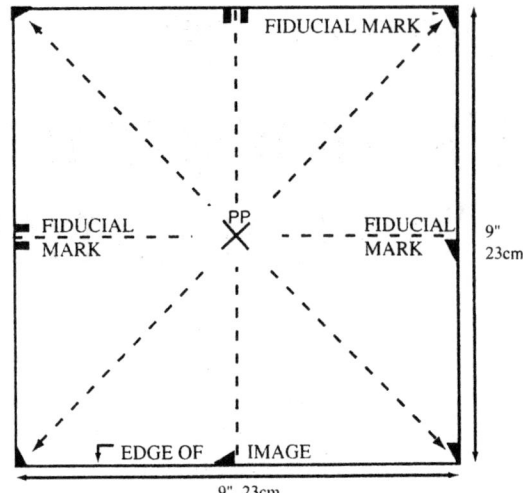

Figure 4–1. Features of a vertical aerial photograph.

CAMERA–GROUND RELATIONSHIPS

The principal point is assumed to be the point on the photo that shows the location of the **nadir** on the ground. The nadir is the plumb point, the ground location that lies beneath the aircraft on a vertical line between the camera and the center of the Earth. This location for each airphoto is important in determining height measurement of objects in the photo. It is much easier to assume the PP and the nadir lie at the same point on the photo than to try to find the nadir on the ground.

Figure 4–2 is a diagram that illustrates the relation of the points mentioned above as well as graphically portraying the concept of **parallax,** or image displacement. Parallax occurs in any

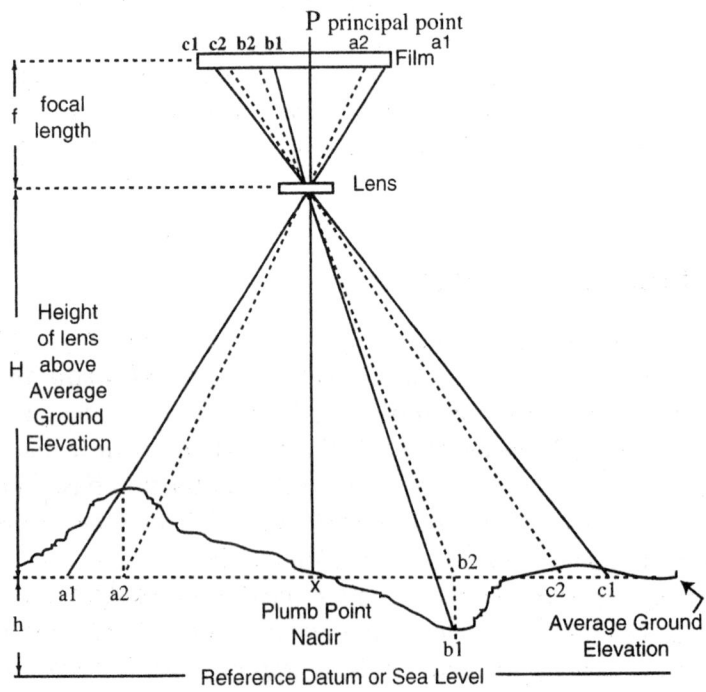

Figure 4–2. Image displacement on a vertical airphoto.

vertical airphoto of ground features that lie above or below the average ground elevation. As can be seen in the diagram, features that extend above the average ground elevation are displaced on the airphoto away from the PP, while those that drop below that datum are displaced toward the PP and center of the photo.

The amount of displacement of objects on the airphoto is a function of the height of the objects and the scale of the photo. This relationship may be expressed in a mathematical formula and allows for the determination of the height of photo features when several parameters are known. This will be pursued later in this chapter.

TYPICAL OVERFLIGHT CHARACTERISTICS

Most aerial photography is flown to produce a mosaic of airphotos that cover an area on the ground in a series of parallel flight lines. The photos overlap each other within and between flight lines. The amount of overlap that is produced relates to the intended purpose of the coverage. In Figure 4–3 the **sidelap** between successive flight lines is illustrated. The percentage of overlap is usually planned to be from 25 to 40% to be certain no areas are left unphotographed.

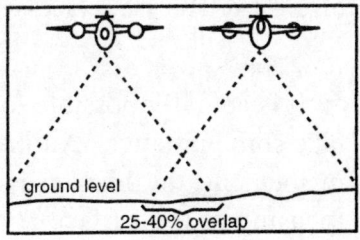

Figure 4–3. Adjacent flight lines yield sidelap.

Forward overlap within each flight line is planned to be from 60 to 70% as shown in Figure 4–4. This is accomplished by varying the speed of the film past the shutter of the camera.

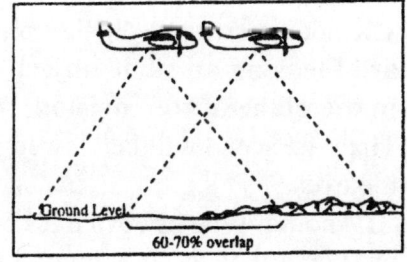

Figure 4–4. Shutter speed produces forward overlap.

The flight plan to produce the photo mosaic is displayed in Figure 4–5. The most common and efficient pattern has the plane flying across the area in one of the cardinal directions of the compass, reversing its path 180 degrees and crossing again and so on. The airphotos are numbered consecutively from the first exposure to the last.

AIRPHOTOS AND STEREOVISION

What seems to be excessive overlapping of aerials within a flight line is carried out so that photos taken in succession may be viewed in three dimension, i.e. 3D. This allows an interpreter to set up a **stereopair** of airphotos to recreate a view similar to that which a person with normal eyesight would have seen from the plane at the time of acquisition.

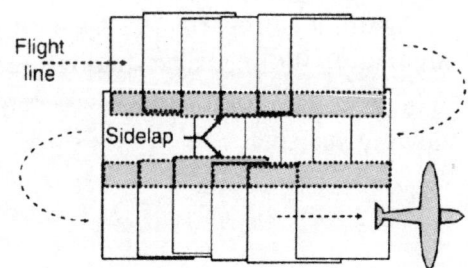

Figure 4–5. Mosaic of photos over mission area.

The human brain receives two images of a scene viewed from two different vantage points. The brain assimilates these views and creates one image along with depth perception. The same operation can be accomplished with two stereo photos and a **pocket stereoscope.**

Figure 4–6 illustrates the correct procedure for using a pocket stereoscope to view consecutive airphotos in a given flight line. The area of photo overlap may be viewed with photo 1 on top and then another area of overlap can be viewed with photo 2 on top. **Note that the shadows should face the interpreter, if possible.** Lenses of these devices are normally only two or three power.

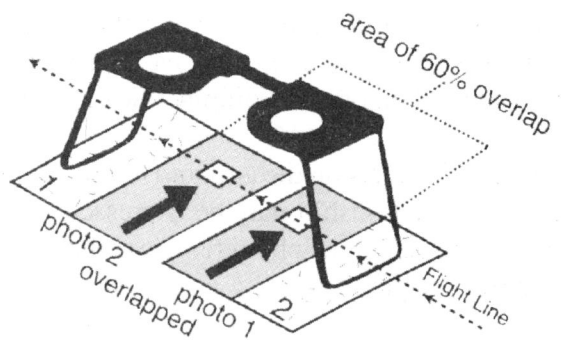

Figure 4–6. Use of a pocket stereoscope.

It is actually possible to see photos in 3D without using the pocket stereoscope, but it does take some practice. An important consideration in stereoviewing relates to the depth of field in focusing the human eye. Normal viewing of a photo on a table or desk would call for focusing at the surface level of the photo. **In stereoviewing the eyes focus at infinity and the line of sight of one eye is parallel to the other.**

Note that this is not a normal technique of vision and it can produce severe strain and headache. First time stereovision should be limited to fifteen or twenty minutes. To acquire the type of distant focus needed an exercise similar to this might be followed: Look through a window that has venetian blinds or a wood framework around the glass panes. **When you are focusing on some object well beyond the window, the blinds or the wood frame are in the plane of stereovision.** This requires the interpreter to "look right through" the photos. Then the scene will be viewed in three dimension and buildings and trees will pop up towards the eyes.

Another type of device for 3D viewing is the **mirror stereoscope.** Figure 4–7 is a diagram of such a device. It is much more expensive than a pocket version, but it does allow for the photos to be viewed separately, for more of the photos to be seen in 3D, and for greater amounts of light to reach the surface of the photos. One can also add a binocular attachment that will magnify the view 10× or more.

Again, it is important that the photos be two separate views taken of the same area in succession in a flight line. The shadows in the photos must face the viewer if at all possible. Directing light onto the photo surface will remove the problem of the interpreter's shadow falling on the viewing surface.

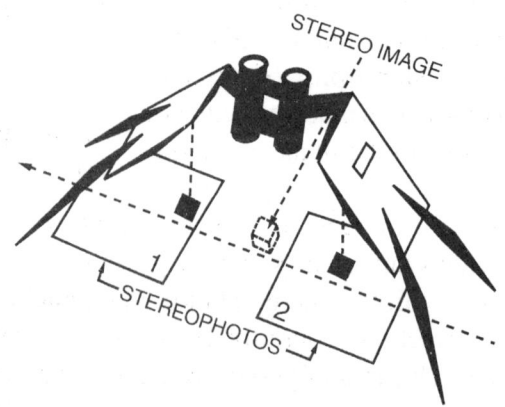

Figure 4–7. Use of a mirror stereoscope.

CAMERA–ALTITUDE–SCALE RELATIONSHIPS

There is a precise relationship between the focal length of an aerial camera, altitude of the plane above the ground datum, and the resulting airphoto scale. This may be represented as a mathematical formula or in graphic form.

Figure 4–8 is a graph that portrays these relationships for several selected focal lengths. The most common focal lengths in aerial cameras are 6″ (15.2 cm), 8.25″ (21 cm), and 12″ (30.5 cm). Another focal length often used is 2.8″ (7 cm) in the 70 mm camera.

If the flight altitude above the ground and the focal length of the camera are known, then the scale (RF) may be determined from this graph. It is possible to find the third variable when two of the three parameters are known values. This information may be used in mission planning or problem solving after the airphotos have been acquired.

DIRECT DETERMINATION OF OBJECT HEIGHTS

Because of the precise relationships between photogrammetric parameters such as aircraft height, focal length, photo scale, and object displacement, it is possible to measure the height of objects shown on airphotos. The following section deals with such mensuration techniques.

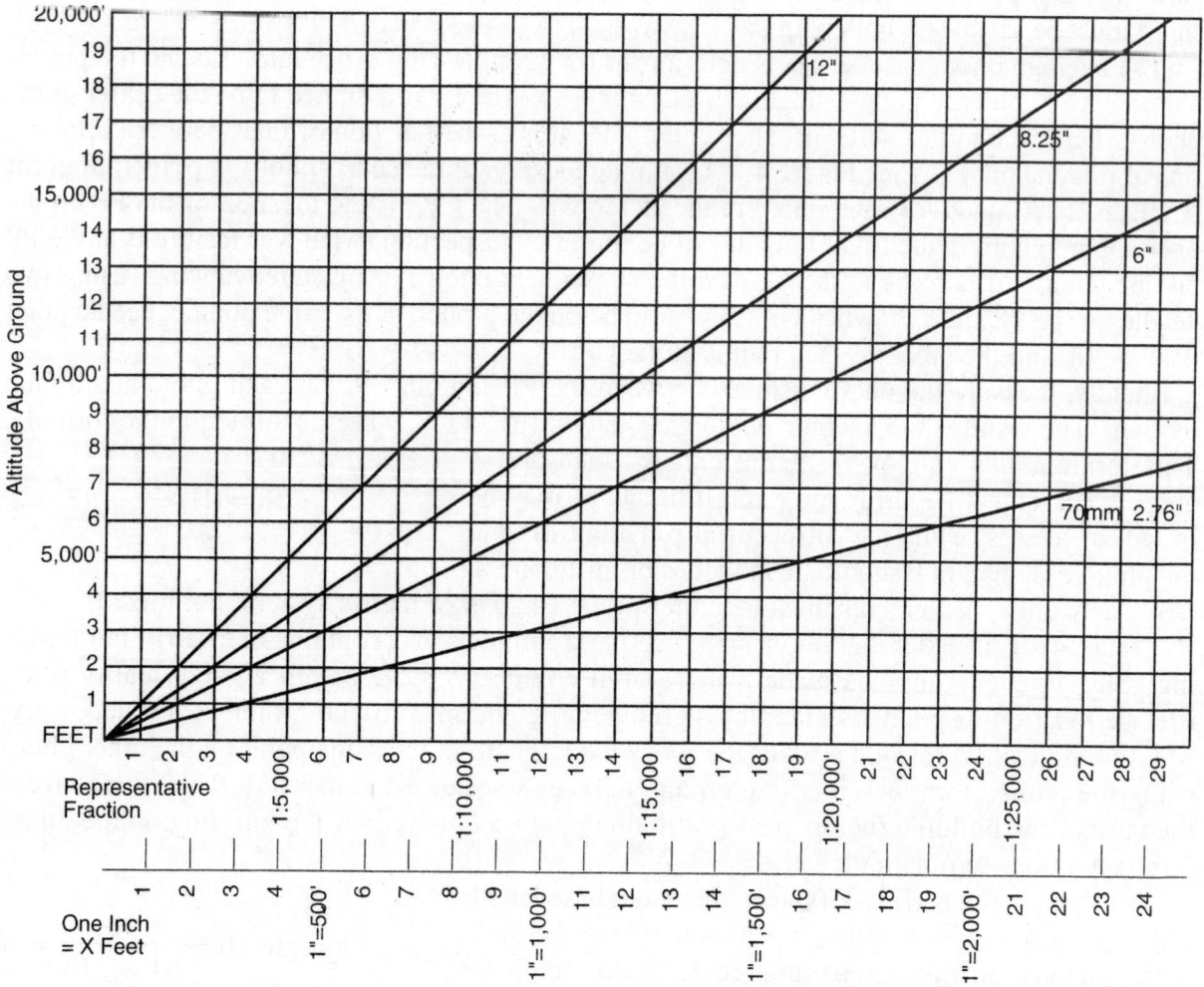

Figure 4–8. Altitude, focal length, scale relationships.

Single photo method. Under certain conditions it is possible to measure the height of some objects using only one airphoto. If the object to be measured is not at the center of the photo and the top and bottom of the object are clearly visible, then the object height can be measured using the photo scale (RF) and flying height.

In the formula at right, **H** = the flying height above the ground, **d** = the length of the displaced object being measured, **r** = the radial distance from the PP to the top of the object, and **h** = the height of the object. Note that d and r must be in the same units and that different units are employed in this formula. That is, if d and r are in inches, H and h will be in feet. (If d and r are in mm, then H and h will be in meters).

$$h = \frac{d}{r} \ (H)$$

Figure 4–9 is an airphoto that meets the requirements listed above. The height of the tank that is left of center in the photo may be established by substituting these values in the formula. Height of the aircraft above the ground is 3000 feet, the length of 0.09 the displaced object (tank) is 0.09 inches, and the radial distance from the PP to the top of the tank is 2.20 inches. Therefore the height of the tank is 122.7 feet.

$$h = \frac{0.09}{2.20} = 3000 \ \text{ft.}$$

Stereopair parallax method. This method for determining the height of objects requires two overlapping airphotos in the same flight line, the known height of the aircraft above ground datum, and **the average photo base length** (which is substituted for the absolute stereoscopic parallax when photos are assumed to be vertical). The top and bottom of the object to be measured must be visible on both photos.

The average photo base length is determined by the following procedure. Locate the principal point (PP) on each photo using the fiducial marks. The PP is assumed to be the center of the photo when the photo is vertical. The center is located at the junction of lines between opposite sets of fiducial marks. (See Figure 4–1.) Next the location of the **corresponding principal point (CPP)** is determined for each photo in the stereopair. The CPP is the location of the PP on the overlapping photo in the flight line. It can be found by inspection (whatever feature is at the PP on one photo will also be at the CPP on the subsequent photo) or by stereoviewing (using two needles at the PP and CPP when both appear to be on the ground at the same point, a needle point is at the PP and the other needle point is at the CPP).

Finally, measure the photo distance between the PP and the CPP on each photo and divide by two. The result is the average photo base length (P) and may be substituted in the formula. This formula also employs different units of measure as above. The object height h and aircraft height H may be in feet or meters, while the **differential parallax dP** and the photo base length P should be in inches or millimeters.

$$h = (H) \frac{dp}{P} + dP$$

Obviously all measurements must be either in the English or metric system, not mixed.

Figure 4–10 is a stereogram of an office park complex that contains one fairly tall building. Measurements may be made using a civil engineer's scale or other good quality ruler. Precise location of points is aided by stereoviewing and magnification, but is not necessary. Measure dP from right side peak of the roof on one photo to the same point on the other photo using the 50ths of an inch marking on an engineer's scale. Now measure the distance from the base of the building (below peak) from one photo to the base of the building on the other. Subtract values to obtain dP.

The flying height (H) is 3800 feet, the photo base length P is 3.64 inches, and the differential parallax may be measured directly on the stereogram. Your result should be approximately 0.16 inches. Substituting in the formula above yields a building height of 160 feet.

$$h = (3800) \frac{0.16}{3.64 + 0.16} = 160$$

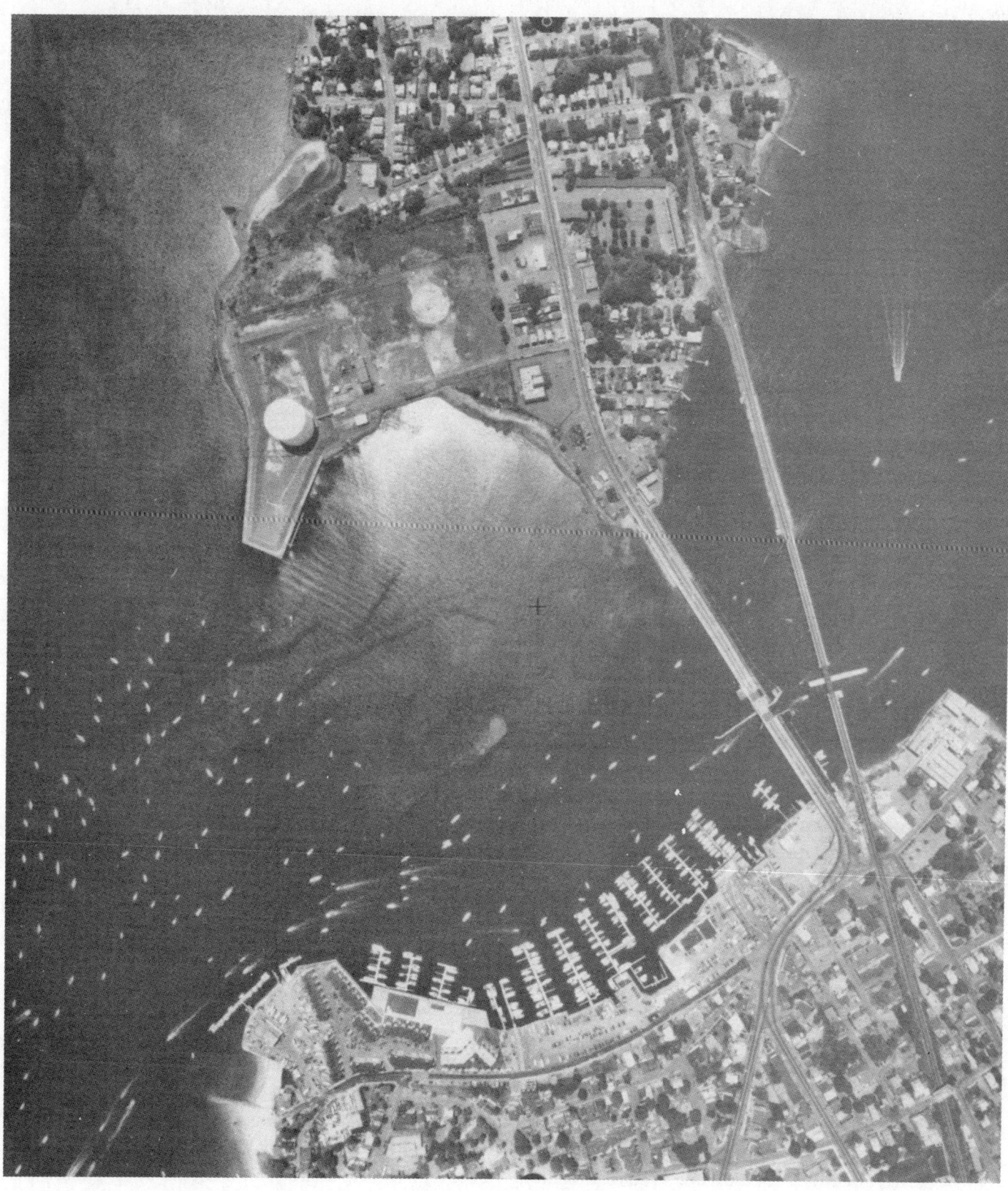

Figure 4–9. Salem-Beverly, MA Harbor 6/12/88.
Courtesy Flight Survey & Mapping Inc.

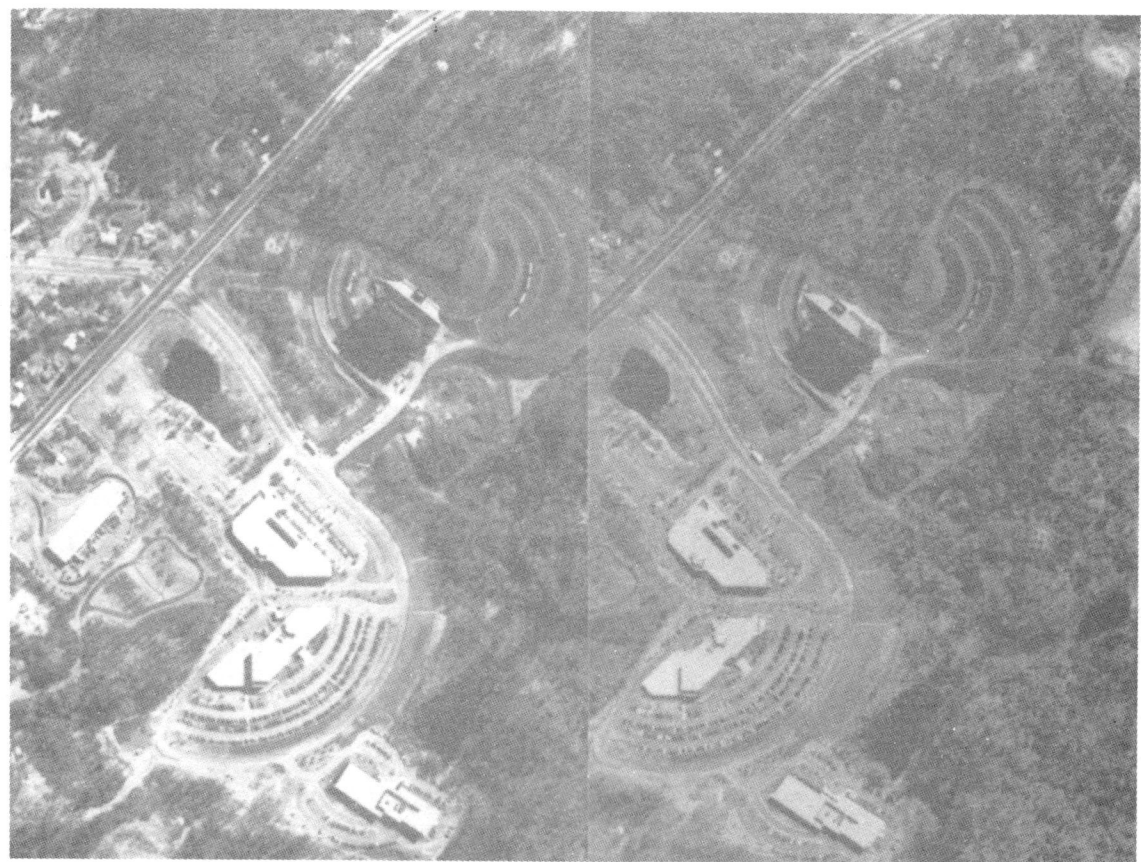

Figure 4-10. Northwoods Business Park, Danvers, MA 3/22/90.
Courtesy James W. Sewall Co. and Town of Danvers, MA

CONSTRUCTION AND USE OF A STEREOGRAM FILE

It is easy to appreciate how valuable an organized collection of stereo images would be in the training of an image interpreter. Airphotos and other images arranged by subject, feature, landform, etc., provide a resource for reference and comparison. If the airphotos are in the form of a stereopair, then they have the added value of providing a three dimensional view.

A **stereogram** is a form of learning tool that utilizes all or parts of two or more airphotos and presents them so that they are ready for stereo viewing with a pocket stereoscope or a mirror stereoscope. That is, the photos are fixed in place in the relationship required for stereoviewing.

The stereogram may consist of two complete 9″ × 9″ **contact prints** (same size as aerial negative) arranged for use with a mirror stereoscope and fixed to a piece of drafting board or similar material (see Figure 4–7). Or it can be in the form of parts of photos extracted for stereoviewing of specific features. Figure 4–10 is an example of this latter form.

An extensive file of stereograms mounted on specially constructed forms or card stock is time-consuming to produce, but a major asset in image analysis and interpretation. Construction of such a file as shown in Figure 4–11 is described in the exercise section at the end of this chapter.

Note that the edges contain numbers that may be assigned to particular categories of land use, land cover, vegetation type, or economic activity, etc. Certain holes may be punched so that a series of cards with the same type features may be extracted from a file with a needle all at once.

There are other, more formalized versions, such as that displayed in Figure 4–12. This is a stereogram card created at the University of Illinois and part of an extensive collection. Note that there are two separate images of Rock Spring Branch in St. Clair County, Illinois. There is a grid

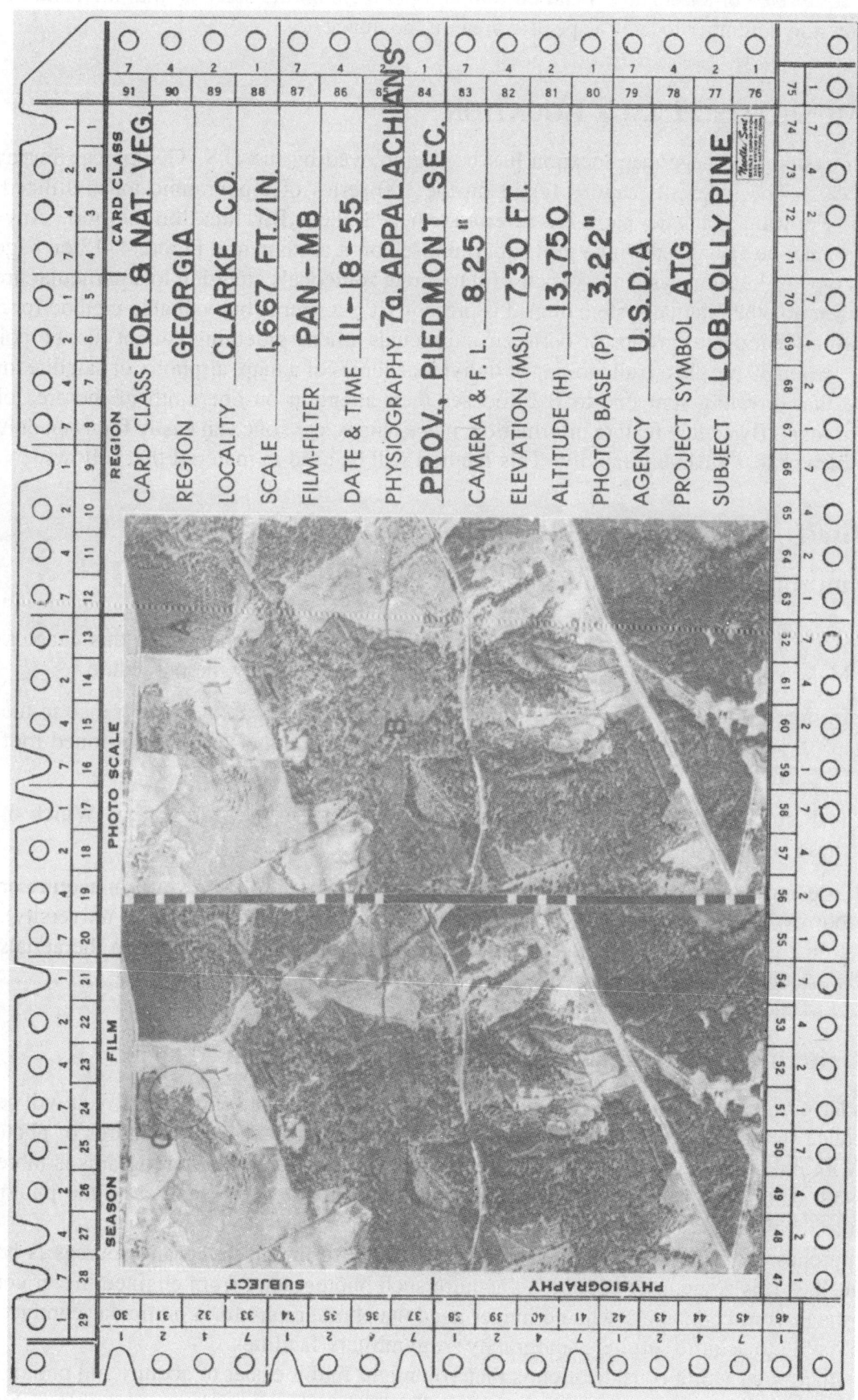

Figure 4-11. Stereogram needle-sort file card. Courtesy T.E. Avery, 1968.

system comprised of letters at top and bottom and numbers on the sides so that the combination of a letter and a number locates a specific area on the image.

COMMUNICATING IMAGE LOCATION

Another system of image/map location has been employed by the U.S. Geological Survey for many years. The index of features found on the 100 series of topographic maps utilized this method. For want of a better name it is referred to here as the **A,B,C location method.** Any map or image may be visualized with a grid superimposed on it as shown in Figure 4–13 on page 34.

This type of location system is very useful to direct someone's attention to a particular area or point when no other vicinal system or grid is present. It is a simple but valuable method for communicating locations to groups or between individuals when something cannot just be pointed out, e.g., when 30 people are all looking at individual copies of a map, airphoto, or satellite image.

Note that directing someone to B-2 focuses their attention on one/ninth of the area of the image or map. By adding feature information, place names, etc., one can easily focus on only one square inch or less area very rapidly. This method will be used in material that follows.

Exercise 4   Object Heights/Stereogram Construction

Determining Object Heights

1. Review the techniques for determining object heights on airphotos in this chapter. Make your own measurements of the examples used to verify the procedure.

2. In Figure 4–9 there are two buildings to the right of the road at the entrance to the left bridge. Can you determine the height of the building with the light toned roof (B-2 area)?

3. In Figure 4–10 there are three other office buildings in the stereogram. Which of these seems to be the tallest? How high is it?

4. You can practice these measurement techniques on any airphotos you can acquire or borrow (your town or city planning department or nearby college or university). Remember that you need to know facts about the original mission when the aerials were obtained. See parameters called for in formulae in this chapter.

Construction of Stereograms/Create Your Own Library

If you have access to aerial photography that can be cut to produce stereograms, you will be able to produce stereogram card files that will allow you to compare known signatures of photo features with unknown features to aid in identification. Such a library of stereograms is more useful if you can also obtain the original flying height, nominal photo scale, and the photo base length for each set of aerials.

If purchase of new photography from the federal government or private sources is beyond your means, it is sometimes possible to acquire such photography from engineering or consulting firms in your community who no longer need the photography for a particular contract. It is also possible to acquire surplus photography from military facilities.

Another point worth consideration is that you might find it easier to acquire old photography that is considered obsolete and of little value. Remember that it is always possible to buy the most recent aerial coverage, but older photos that are a record of previously existing conditions

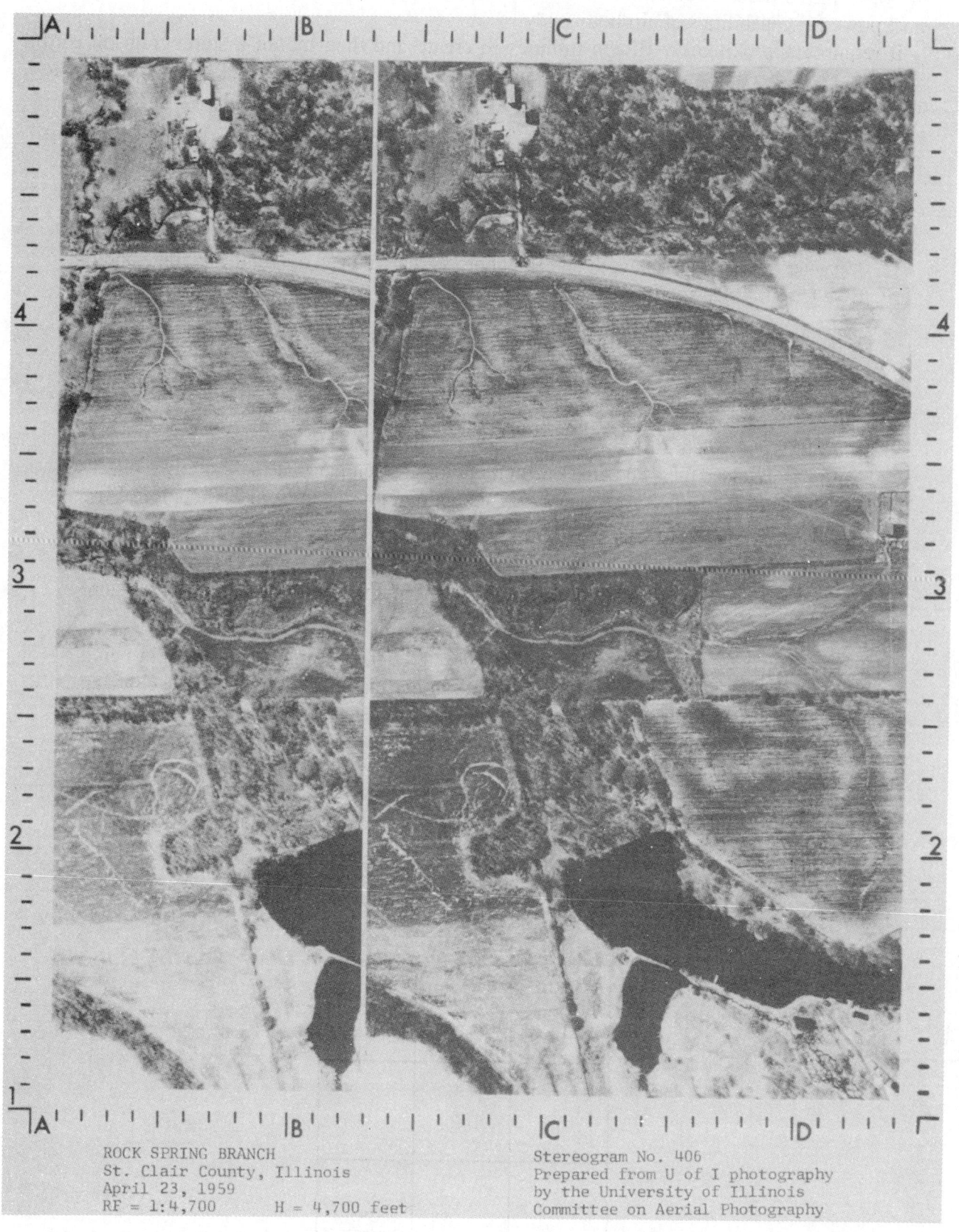

ROCK SPRING BRANCH
St. Clair County, Illinois
April 23, 1959
RF = 1:4,700   H = 4,700 feet

Stereogram No. 406
Prepared from U of I photography
by the University of Illinois
Committee on Aerial Photography

Figure 4–12. Courtesy University of Illinois, GIS Laboratory.

are sometimes difficult to obtain. Establishing changes that have occurred over time is one of the most powerful capabilities of airphoto study.

It requires approximately fifteen minutes to create and then mount a stereogram on card stock after you become familiar with the procedure. You may modify the following steps to suit your needs.

1. Using two or more overlapping 9″ × 9″ contact airphotos, establish the flight line path by locating the Principal Point (PP) and the Corresponding Principal Point (CPP) on the prints. Mark these locations just dark enough to see. A more precise method is to put a tiny needle hole at the two locations on each photo.

2. Measure the photo base length on each photo so that you can obtain the average photo base length. Measure to the nearest 1/50th inch.

$$\frac{(PP - CPP \text{ photo A}) + (PP - CPP \text{ photo B})}{2} = P$$

3. Draw two lines perpendicular to flight line 2.2″ apart that will enclose the area to be viewed. (One such strip from each photo will frequently provide for two stereograms.) Enclose the same area on each photo. Remember that these are two different views of the same surface area. (A stereogram produced from two copies of the same print does not yield any three dimensional effect.)

4. Record all information printed on the airphotos before using scissors to cut along both sets of lines. (The information may be included on the stereogram card—see Figure 4–11.)

5. Align one strip with the left margin of the stereogram card's inside margin being careful not to cover any letters or numbers in the margin. (See Appendix C for an example of this type card.) Mark the portion of the strip to be cut to a size of 2.2″ wide and 3.9″ high. Cut it and paste it in the left area of card. Shadows should face bottom of card if possible. **Note!** If photos are aligned on the stereogram card in reverse of the order in which they were photographed, it may result in topographic inversion—rivers become ridges, etc. View strips side by side before pasting to achieve the proper alignment.

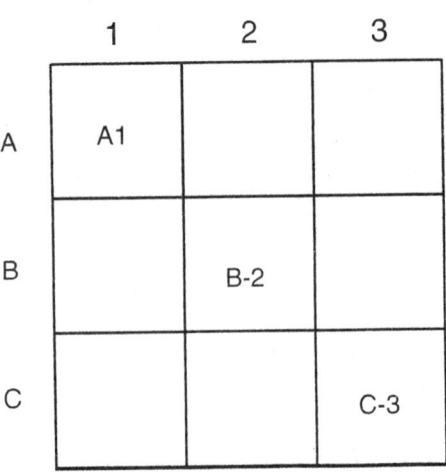

Figure 4–13. Communicating image location.

6. Align the second strip next to the first, but before cutting set up a stereoscope to be sure a three dimensional view is going to be produced. Mark the second strip on the photo and on the card for proper alignment, then cut the second strip and paste where marked on the card.

7. Fill out all available information on the data key on the right side of the card:

Project symbol	Exposure numbers
Print scale RF	Scale in feet and meters
Date and time of photography	Camera focal length
Average photo base length	Agency ordering photos
Film roll number	

8. Other information may be important for your particular needs, such as:

USGS Topographic Quadrangle name
Elevation of surface in photographs
Compass orientation
Physiographic province or type of physical features
Types of vegetation present
Designation of road or route numbers

See Appendix C for example of form and use.

Once you have completed several of these stereogram card exercises you will be able to cut the length of time needed to produce them considerably. Most likely you will find that your needs call for some variation in the manner in which you use the cards and in the information required. The setup of the airphoto pieces to produce the stereogram is fairly standard.

chapter 5

LAND USE AND LAND COVER MAPPING

TERMINOLOGY

land use
land cover
orthophoto
orthophotoscope
MacConnell's classification
USGS LU/LC classification
rectified airphotos

classification factors
mutually exclusive
image scale
seasonal variation
interpretability
homogeneous classes

LAND USE MAPPING AS A GOAL OF IMAGE INTERPRETATION

One of the most frequently produced products of image interpretation is some sort of land use or land cover map. The scientific approach to Earth surface analysis usually demands classification of the variety of phenomena distributed on the Earth. It is important to put order to the myriad detail so that additional forms of analysis may be carried out. Regional planning efforts, GIS investigations, and environmental analysis most often begin with land use/land cover mapping, and so the topic is relevant to this discussion.

ANTECEDENTS OF LAND USE/LAND COVER MAPPING

Some of the earliest examples of **Land Use** mapping in the United States can be traced to the 1920s. The Michigan Land Economic Survey was begun in 1922 and another major project was the Tennessee Valley Authority survey that followed. Both of these studies were intended as appraisals of the existing resources to aid planning efforts. The latter investigation utilized airphotos as a record keeping base for on site field work.

In the 1930s the Land Utilisation Survey of Great Britain was carried out as an inventory and classification of land uses. Then the Rural Land Classification of Puerto Rico took advantage of the methodology of the earlier studies as well as that of the British work to create an inventory used to evaluate the island's resource base.

The Soil Conservation Service of the U. S. Department of Agriculture has utilized airphotos at approximately 1:15,840 scale to carry out field mapping across the country. These products are then used to generate mesoscale maps for soil surveys.

By the 1960s many such investigations had been carried out all over the world and the methodologies of land use mapping had become well known. At about this time scientists began acquiring photographs and other imagery from space platforms. Much of this new imagery was at rather small scale and portrayed areas almost entirely devoid of human use and activity. This led to the coining of the phrase **Land Cover.** Most subsequent investigations that have employed airphotos and satellite imagery to analyze the distribution of phenomena at the Earth's surface have been termed **Land Use/Land Cover** studies. The former term is usually reserved for areas impacted by humans and the latter treats environments that appear essentially in their natural state.

SOILS MAPPING BY THE SOIL CONSERVATION SERVICE

The Soil Conservation Service is charged with the mission of providing leadership in the conservation and use of soil, water, and related resources in the United States. To that end the SCS has utilized aerial photographs as its primary tool in mapping these resources for more than fifty years.

Since 1935 the Soil Conservation Service of the USDA has been mapping the distribution of soils on a county-wide basis across the United States. The National Cooperative Soil Survey produces soil survey maps from Order 1, large scale, to Order 5, very small scale. Local county maps are typically Order 2 with a RF 1:15,840 comprised of a black and white **orthophoto** base and the soils information overlaid.

Orthophotos are **rectified, vertical airphotos** that have had most of the parallax in the original photo removed. This is carried out by means of an **orthophotoscope** that allows light passed through the original film to be redirected onto a new emulsion and thereby change the locations of image features. The orthophoto thus produced displays accurate planimetric location. This allows measurements of distances and directions to be accomplished on the photo with the same degree of accuracy as on topographic maps at similar scales.

Figure 5–1 is an example of such an orthophoto soils map for an area in Topsfield, MA that will be examined later in regard to the mapping of wetlands. Note that the orthophoto background data have been produced at approximately a 50 percent screen value so that the overlaid soils information may be seen more readily.

The initial soil mapping is carried out in the field by personnel of the Soil Conservation Service, and is done directly on large scale airphotos. The intent is to identify and delineate areas on the ground that are characterized by essentially homogeneous soil types, slope, and condition (soils developed in situ or those artificially produced by urban development).

These maps are valuable as aids in planning for the location of various forms of agriculture, to protect wetlands, and as a guide in development. Soil types indicate whether they are appropriate for leaching fields of septic systems, as water repositories for wells, or can sustain sufficient weight for certain forms of urban development.

Each county that has completed this type of survey has the airphotos available for examination at the county Agricultural Stabilization and Conservation Service (ASCS) office as well as order forms to acquire airphotos. The soils information is published in a Soil Survey which includes descriptions of the local soils and the uses for which they are appropriate.

The soil symbols on the map typically consist of three letters. The first and second letters represents the initial letters of the soil name and the third letter is an indication of slope. For example, Great Hill on the left side of Figure 5–1 is characterized by **PbC, Paxton** very sandy fine sandy loam, 8 to 15 percent slopes, by **PbD,** same soil on 15 to 25 percent slopes, and **PcE, Paxton** and **Montauk** extremely stony fine sandy loam, 25 to 45 percent slopes.

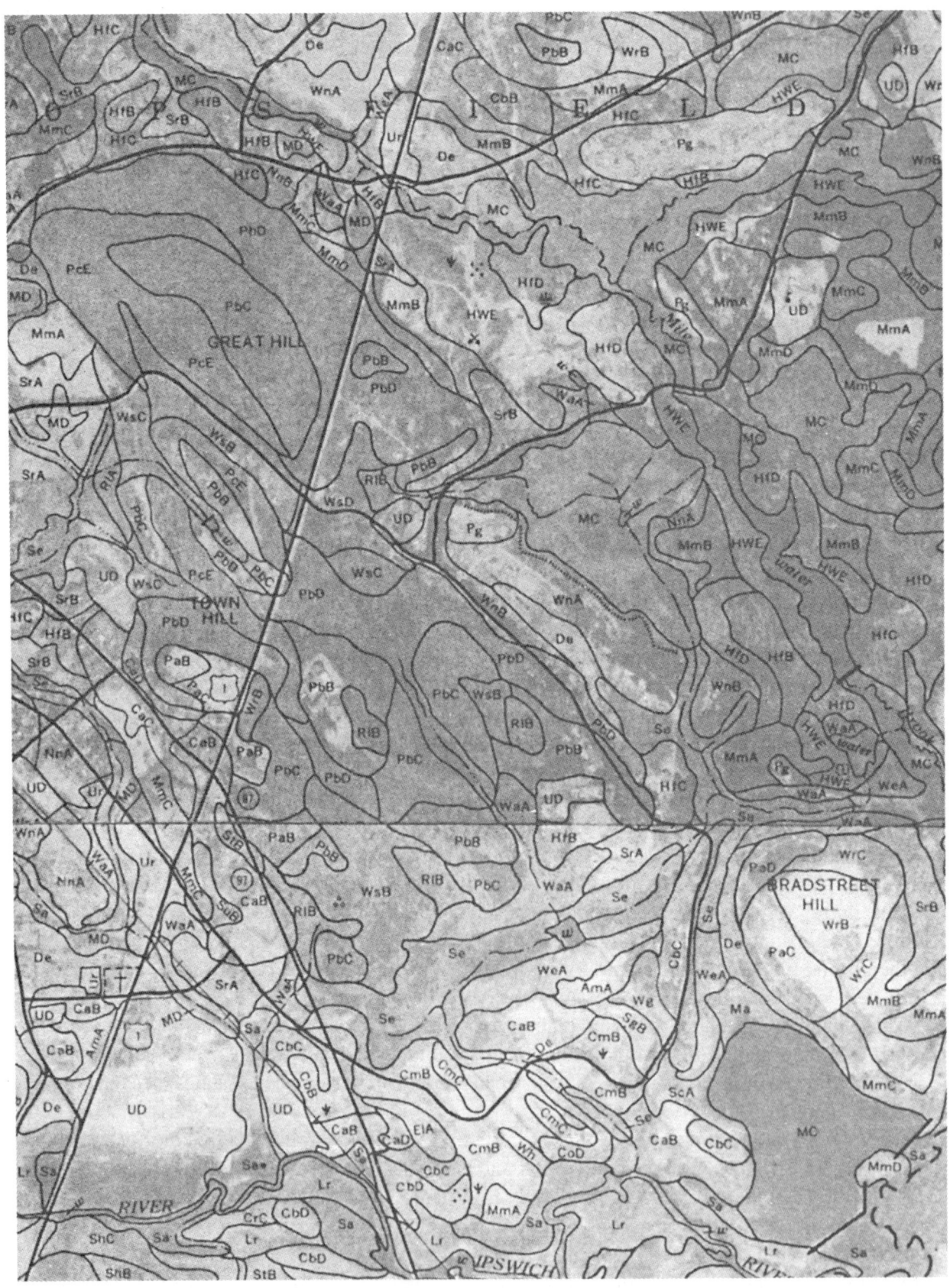

Figure 5–1. Sheet nos. 38, 42 (part). Soil Survey of Essex County, MA, northern part USDA, SCS Feb. 1981.

MacConnell's Land Use and Vegetative Cover Mapping

Professor William P. MacConnell of the Department of Forestry and Wildlife Management at the University of Massachusetts at Amherst, is lead author of a long term land use and natural resource inventory of Massachusetts. The work involves interpreting airphotos in order to classify and subdivide the landscape into homogeneous units of 3 acres or larger.

The initial objective was to create a set of maps at a scale of RF 1:24,000 that would portray the state's wildlife habitat as determined by land use and vegetation. The study was initiated in 1951, updated in 1971 and 1980, and is currently under revision. The maps produced portray the land use information superimposed over USGS topographic quadrangles in black and white.

More recently the boundaries of the areas were digitized and maps of land use and vegetative change were computer-generated. That product yields visual changes as well as quantitative summaries for areas of the various use categories for several specific years and for the changes since 1951. These data are available for each county in the state.

The photo interpreters were able to identify 26 general land use categories which were further subdivided to produce more than 100 sub-classes in early mapping. When land use change data were produced the number of general classes was reduced to 20 and a generalized land use map was created as well as a map of changes.

Land Use Category	Group Symbol	Land Use Category	Group Symbol
Urban (22 types)		Agriculture & Open Land	(11 types)
Industrial	UI	Crop Land	AC
Commercial	UC	Pasture	AP
High Density Resid.	R1	Woody Perennials	WP
Med. Density Resid.	R2	Open Land	O
Low Density Resid.	R3	Mining & Waste Disposal	(5 types)
Transportation	UT	Waste Disposal	UW
Open & Public	UO	Mining	M
Wetlands (11 types)		Outdoor Recreation Land	(15 types)
Open Water	W	Water Based	RW
Inland	FW	Participation	RP
Salt Marsh	SW	Spectator	RS
		Forestland (40 types)	F

Figure 5–2 is a composite of parts of the Georgetown and Salem, MA Topographic Quadrangles with overlaid land use types. It covers the same area as Figure 5–1 and has been photographically enlarged to the same scale to aid comparison. The interpreters were highly successful in their mapping efforts, although they did not break out forested wetlands as a separate category. Had they been able to do so this would have been most useful for the local Conservation Commissions across the state. This point will be pursued in a later chapter.

The Classification System of the USGS

In 1976 the U.S. Geological Survey published **Professional Paper 964, "A Land Use and Land Cover Classification System for Use with Remote Sensor Data."** This paper was written by James R. Anderson, Ernest E. Hardy, John T. Roach, and Richard E. Witmer. It is a revision of earlier work published as USGS Circular 671 and a product reflecting many existing widely used classification systems.

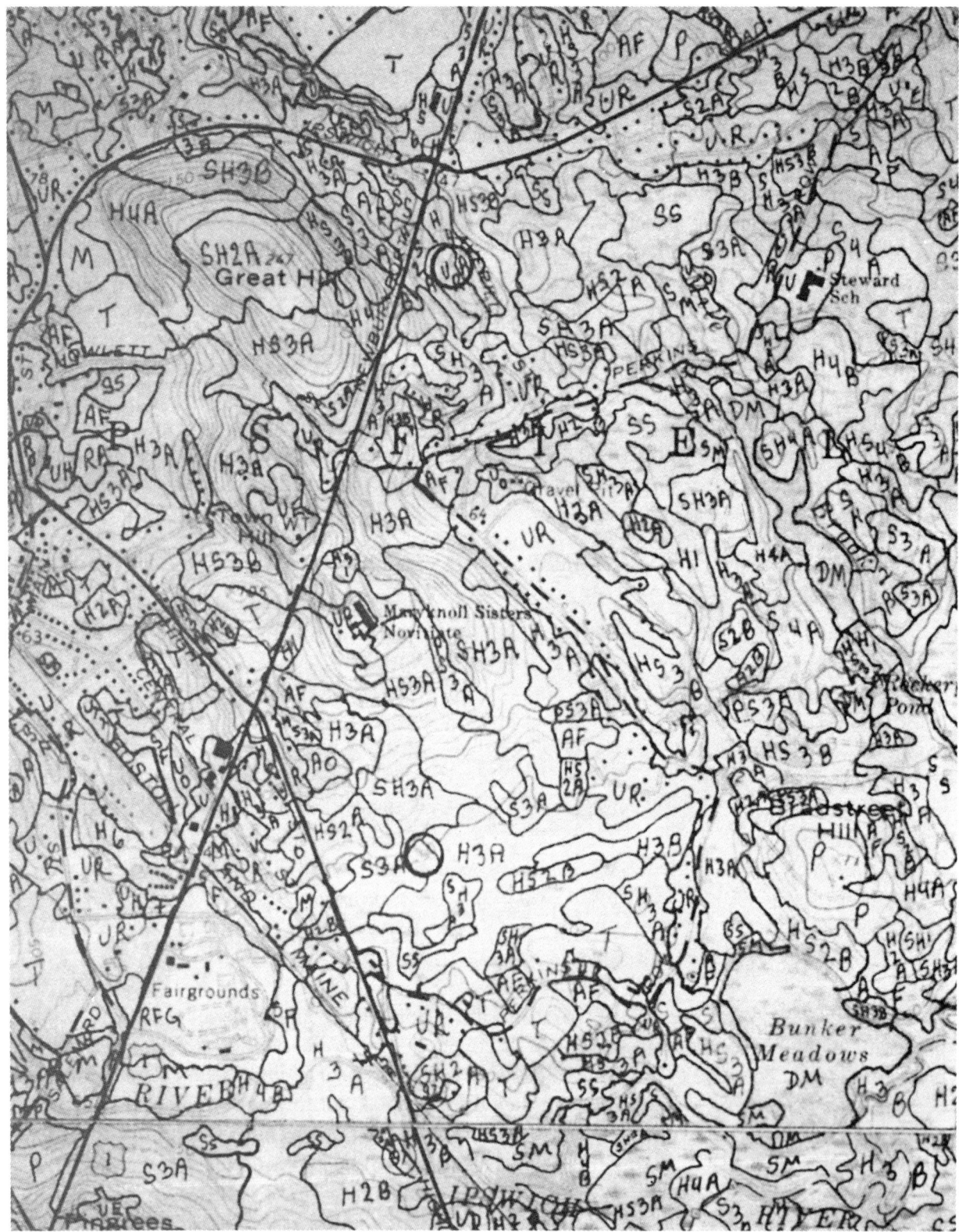

Figure 5–2. Massachusetts map down by MacConnell et. al. Salem and Georgetown, MA quadrangles (part). University of Massachusetts, School of Forestry.

The intent was to create the framework of a national classification system that could be used with data collected from satellites and aircraft as well as from more traditional sources. It was planned to be adaptable to the needs of public agencies at all levels and to be open ended so that more classes and levels could be added. This would allow the users to be compatible with a national system while serving local demands.

Several factors are involved with the implementation of a land use mapping program and this system reflects those concerns. The size of the smallest area which can be mapped is a function of image scale, resolution, and the nature of the data recorded. It is also related to the scale at which results will be presented and to the readability of the mapped information. This system consists of four levels of generalization that address these concerns.

Level I is designed to allow the classification of satellite imagery such as that from the LANDSAT or SPOT vehicles where the image scale is typically 1:500,000 or smaller. Nine categories of land use and land cover are identified as follows:

1 Urban or Built-up Land	6 Wetland
2 Agricultural Land	7 Barren Land
3 Rangeland	8 Tundra
4 Forest Land	9 Perennial Snow or Ice
5 Water	

Level II is intended for use in classifying high altitude airphotos with an approximate image scale of 1:80,000 or smaller. The 9 classes of Level I are subdivided into 37 sub-groups.

Level III is suitable for classifying image phenomena that are acquired in medium altitude photography for image scales between 1:20,000 and 1:80,000. This relates to the majority of airphotos obtained over the United States at least through 1970.

Level IV would be most useful for low level aerial photography at scales larger than 1:20,000. The specific categories could be defined by a local agency or research unit using large scale photography. This classification level is also heavily utilized to record on site field investigation results directly on aerial photographs or topographic maps.

Refer to **Appendix D** for several examples of Land Use/Land Cover Classification systems, including a more complete version of levels I, II, and III of the USGS approach. These methods of organizing image information will be utilized in interpretation exercises in this chapter and in Chapter 8.

FACTORS LIMITING LAND USE/LAND COVER MAPPING

Developing the Classification System. Any LU/LC methodology must possess several characteristics to enable its application in a wide variety of circumstances and still maintain a consistent and orderly breakdown of image phenomena. All classes at a given level must be **mutually exclusive.** That is, it cannot be possible to classify an image object into more than one class. The classification must account for all observed phenomena, suborders must be subdivisions of higher, more general levels, and all classes at a particular level should be equal in importance or weight.

Image scale. Any type of overlay mapping or classification of image features must be related to the scale of the image presentation (print, film positive, CRT display). It is virtually always possible to identify image features smaller than can be mapped appropriately. That is, the width of a pen or pencil line sometimes takes up more area than the feature mapped. How much area is covered by such a line varies with scale, but the problem of how small an area to classify and map must be dealt with.

Seasonal variation. Some land use/land cover systems may be impacted by seasonal variation in vegetation, soil moisture, or other factors. For example, mapping wetlands is usually carried out more efficiently when vegetative growth is either beginning or when the growing season is coming to an end. Therefore, the imagery should be acquired at those times and the classification should be adapted to those conditions.

Interpretability. Earth surface features vary greatly in the complexity and amount of detail from one place to another. It is possible to classify and map homogeneous stands of Douglas Fir in the Pacific northwest more easily than the mixed deciduous-coniferous forests of New England. This variation could well alter the scale of imagery utilized, the cost of the study, and the particular classification used.

Consistent results. For a classification to be utilized in a number of locations and produce comparative products, the replication of use demands flexible but well defined classes or taxons. This normally calls for testing of a classification scheme before it is employed in the final investigation.

Homogeneous classes. One of the most difficult aspects of devising and applying a classification is to determine the class limits. The goal is to create classes that display less diversity within each class than exists between classes. That is, classes should be homogeneous and different in the mapped criteria than other classes. This is easy to state and often difficult to put into practice.

Exercise 5   Creating a LU/LC Classification

1. Examine the several LU/LC classification systems that are presented in Appendix D of this manual. Consider which of these approaches might be used or modified for use in classifying some area of interest in your town, city, or area of interest. It is desirable for this exercise that you be able to visit the site.
   a. Select a specific area to study. A small area that exhibits some diversity of features is usually easier to work with initially.
   b. Devise an initial LU/LC classification system (or use one in Appendix D) that will allow you to categorize the specified area with regard to appropriate classes of use or cover.
   c. Use a topographic map, airphoto, or other representation of the area on which you can record your class boundaries.
   d. Visit the area you've selected to examine first hand the distribution of phenomena at the Earth's surface. Note the diversity of features as well as the similarity from place to place. Does your classification scheme treat all the phenomena you are interested in mapping? Are your classes comparable in weight or importance? Are the classes mutually exclusive?
   e. Modify your classification, if necessary, and divide the study area into mapped classes. Now evaluate your results. If you are satisfied with meeting all the criteria of classification mentioned above, you might try applying the final classification to another nearby location. These are steps one should pursue regardless of the nature of the features being studied.

2. Compare Figures 5–1 and 5–2 in regard to the area in the vicinity of Great Hill (labeled on both maps and left of the long, solid, almost vertical line which is U.S. Route 1). In this area many features of continental glaciation are found. Steep, well

drained, forested **drumlins** (whaleback-shaped depositional hills) alternate with low-lying, poorly drained areas characteristic of **deranged drainage** (topography rearranged by the ice sheet acting like a giant bulldozer).

If slope is indicated on the soils map (Figure 5–1) by the third letter of the class symbol, then it is possible to identify the relatively flat areas (no third letter or A) from the steeper slopes (B, C, D, and E progressively).

Soils with map symbols MC, MD, Wa, Wb, and Se are all poorly drained soils associated with low-lying areas of terraces and outwash plains. On the land use and vegetative cover map (Figure 5–2) H represents forest at least 80 percent hardwoods, S is forest at least 80 percent softwoods, HS is an area of mixed forest with the hardwoods predominating, and SH is mixed with softwoods most widespread. Some wetlands are represented by the symbols SF, SS, M, SM, DM, and B, while open water is W.

a. Using the information above place tracing paper or other overlay material on either Figure 5–1 or 5–2 with the center of your overlay approximately 3 inches to the right of the center of Great Hill and 1 inch down. Now interpret and map the area to the right and below Great Hill that is common to both figures.

b. Outline in red the steep, well drained, forested drumlin features in this area. What map information did you utilize?

c. Outline as many low-lying, wetland areas as you can find with green. Are these areas all forested?

d. Considering your mapped overlay features, where do you expect that most development would take place? Are the roads located on certain types of soil, slopes, or areas of vegetative cover?

e. Using this brief introduction can you explain what value the mapping of soils and vegetative cover might have to the economic development of an area?

chapter 6

INTERPRETATION LOG EXERCISE

TERMINOLOGY

mental library
stereogram cards
interpretation log

U.S. Public Land Survey
loessial soil
section

THE DOCUMENTATION OF SIGNATURES

Studying airphotos and other image products of remote sensing activities on a frequent or continuous basis points up the need for a record of image signatures. Having examples of known signatures at hand allows for comparison with unknown image phenomena and makes new identification easier.

There are many different ways to maintain records of image interpretation efforts. Ultimately the most important of these is the **mental library** of signatures that is developed over years of interpreting images. There is no good substitute for working with an extensive number of images of Earth surface phenomena from a particular focus of interest. This leads to instant recognition when remembered signature characteristics appear. Obviously there is a limit to the number of image features that can be committed to memory and such mental records may not be sufficient when a person is becoming familiar with a new area of interest.

Another means of maintaining records of image signatures is through the use of **stereogram cards.** (See Chapter 4 and Appendix C for information on constructing a stereogram file.) Acquiring a library of stereograms displaying signatures of interest saves considerable time in identification of features in similar, uninvestigated areas.

Stereogram cards discussed in Chapter 4 and Appendix C are designed to be used with needle sorting. This allows for many examples of specific features to be extracted from a file at once. Stereograms may be created to portray various types of vegetative cover, geomorphological features, industrial patterns, urban patterns, etc.

Another method of record keeping of image signatures was also displayed in Chapter 4 in Figure 4–12. These University of Illinois stereograms have the advantage of portraying more of the photos than is the case on a stereogram card. They are somewhat limiting in the methods of sorting possible and they don't leave as much room for printed information.

THE INTERPRETATION LOG AS A TOOL

The following procedure has been utilized for more than two decades at the university level with a number of diverse student groups. The approach is useful in that it may be adapted to any sort of image that is available.

This concept evolved because of the small amount of funding available at most colleges for the purchase of aerial photography and other imagery. The methodology involves the acquisition of images that portray a variety of Earth surface features and human activity. Large scale, small scale, black and white, color, and color infrared photos were obtained through purchase, gifts, and the surplusing activities of state and federal agencies.

Once a collection of airphotos or other images is acquired, they are sorted and placed in folders as single photos, a stereopair, or as a stereotriplet (three overlapping airphotos). The folders and the contents are each labeled with an identifying designation, usually a number. Other than that number, unless the photographic print contains some identification of place, the student must work with only the standard type of information usually printed on aerial photos.

Professional approach. Normally all of the courses at Salem State use an approach to image analysis that replicates the professional, working environment. That is, students are told to use any and all materials (except for existing LU/LC maps) to help them extract information and map the land use/land cover for the images on which they are working. This is the procedure for all image work other than the logs.

The interpretation log. This work represents the students' earliest exposure to image interpretation at SSC. It is believed that focusing upon the airphoto without ancillary information encourages the student to interpret the photo and to resist temptations to read a map of the same area and use that information in producing an interpretive overlay.

This approach is initially frustrating for most people as a normal inclination is to attack the features of greatest interest on an image. These are often the hardest to interpret and so the result is often failure. By looking for very basic and simplistic feature differentiation, e.g., land vs. water, the novice interpreter achieves instant success and encouragement. This is true even in advanced study since everyone involved in research makes the journey from beginner to expert many times.

It is advisable to proceed from the obvious to the less certain in image analysis. The more features that are positively identified, the easier it becomes to unlock the signatures of more sophisticated phenomena. Often the signature of some surface feature has several different representations, changes occasioned by differences in slope, soil, vegetation, etc. A golf course always is comprised of specific signature characteristics, but each one of the courses is modified by its setting and design.

Frequently the same feature is present in two or more locations on the same photo. It may be difficult to secure positive identification at one locus, whereas it could be obvious at some other spot. The objects and features that occur in close proximity to Earth surface phenomena are often tied to that presence in some fashion. Therefore, the more objects and features that can be identified, even those that are self-evident, the greater is the opportunity to break down an image into its component signatures.

CREATION OF A COMPREHENSIVE INTERPRETATION LOG FILE

Acquisition of airphotos and satellite images of any kind is a requisite for the following approach. The aim is to expose students to as wide a variety of image products as possible. These images are organized by some system relevant to the user, e.g., by remote sensor system or by Earth surface features, and then placed in folders. Each of these folders is referred to as an **interpretation log.**

The following procedure has been utilized, primarily with airphotos, for many years. It could be adapted for use with any type of remotely sensed imagery.

Interpretation log instructions.

1. Record the date and other identifying numbers of photo or photos. (If all such data are missing, see reverse side for log identification number.)

2. What is your overall impression of the landscape shown on the photographic image(s)? Use terms or phrases, e.g., "suburban setting of single family houses"; or "rural, coastal area with agricultural fields and tidal marsh."

3. What is your ESTIMATE of the approximate percentage of the photo (use only one, if two or more are in log) devoted to major use/cover types? For example: "water area 20%, woodland 10%, agricultural use 25%, urban 30%, industrial 12%, and transportation 3%.

4. What is the approximate scale of the photo? Do not make measurements; use your judgment and/or visual comparison with a known photo scale to arrive at the RF.

5. Is the photo vertical, high oblique, or low oblique?

6. Use the "A, B, C" method to locate features of primary importance. In most cases, A-1 to A-3 is assumed to be the top of the photo, the numbered edge. To give direction you may assume the top of the photo is "north" even though that may not be true in fact.

7. Give *specific locations of several* photo phenomena you can **positively identify** and name those features.

8. Give *specific locations of several* photo phenomena you can **tentatively identify** and name those features.

9. Give *specific location(s)* of one or more photo objects you **cannot identify** and give the description(s).

10. Do you have any questions, comments, or insights regarding this log? (You do not have to respond to this question.)

The theory behind this approach is to assist students who have little experience in interpreting airphotos in a particular field of investigation. The aim is to encourage them to make some distinctions regarding levels of difficulty in photo interpretation.

Each interpretation log is approached as though it is a container of information. The most recognizable feature on a photo may be thought of as the largest piece of data in the container and the easiest to extract. Therefore, people new to airphoto interpretation or new to such study in a particular field, should seek to identify the obvious features first and proceed to the unusual thereafter. Normally it will be found that positive identification of some features in an airphoto or satellite image will lead to the recognition of other features.

On the next page Figure 6–1 is a black and white enlargement of a portion of a CIR aerial acquired over Peoria, IL circa 1980. This photo is presented here as an **Example Log Entry** and a set of responses to the questions above are listed below. Note that the nature of some responses may vary considerably dependent upon the specific features examined by an interpreter.

The area shown in Figure 6–1 lies in the central part of the state of Illinois and is traversed by the Illinois River. Peoria was a city of about 100,000 population when the photo was taken. It lies in the heart of the "Corn Belt" and most of the land area not occupied by buildings or roads is in crops, primarily corn or soybeans.

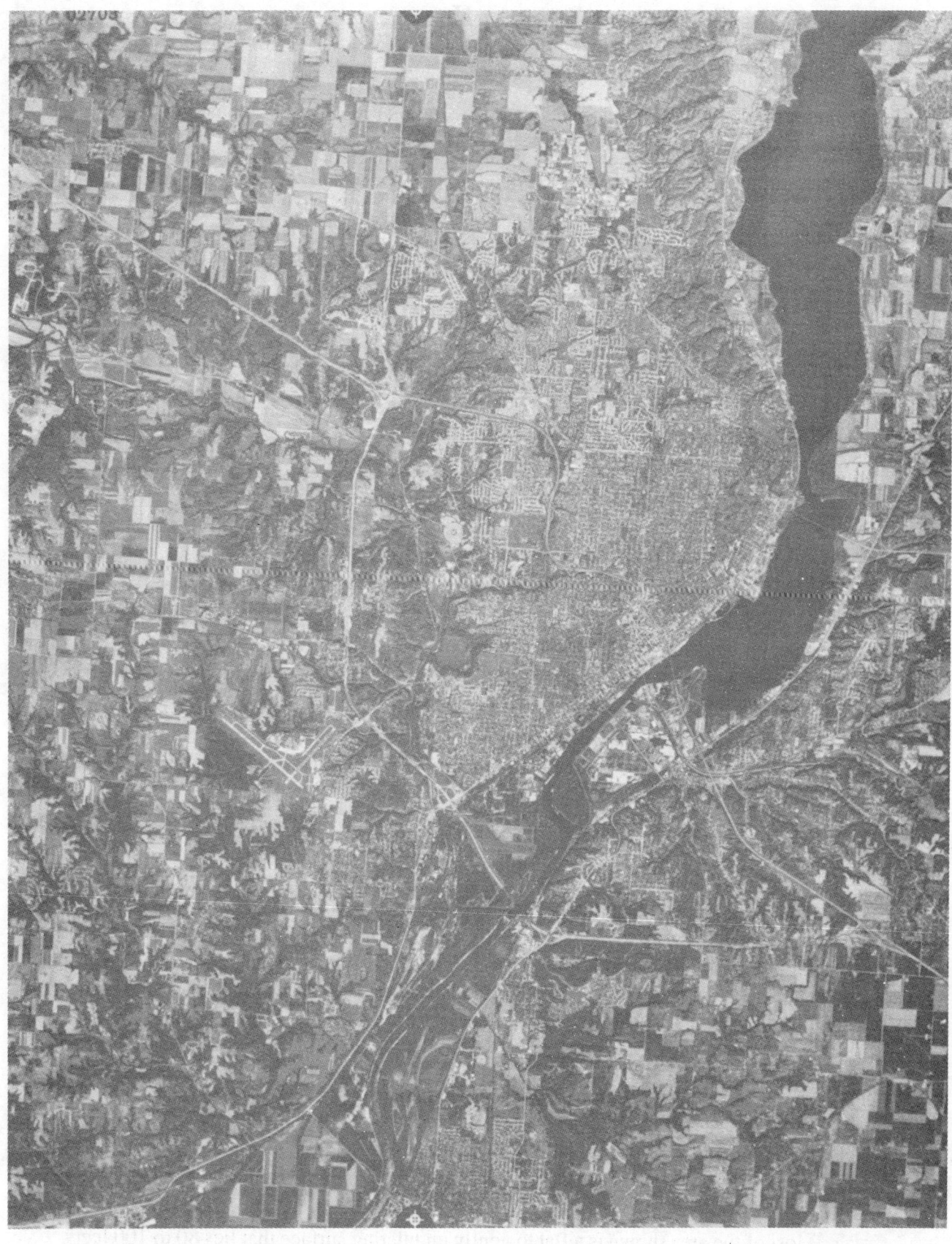

Figure 6–1. NASA high altitude CIR (shown in B&W). Peoria, IL circa 1980.

Example Log Entry for Peoria, IL

1. No identifying information. This is Log #1.

2. Area in photo is farmland dominated by medium sized city which lies on edge of a water body.

3. Urban–suburban use covers approximately   30%
   Water area is about   10%
   Forested area   10%
   Agricultural land   35%
   Land devoted to transportation   10%
   Industrial land & other   5%

4. Estimated RF 1:80,000

5. Vertical view

6. Urban–suburban area in B-2, B-3, C-2
   Water body from A-3 to B-3 to C-2
   Agriculture in A-1, A-2, B-1, C-1
   Airport in C-1

7. Airport runways and taxiways northern part of C-1
   Urban center (CBD) NW corner of C-3 next to water
   Highway interchange NE corner of B-1

8. Old residential district top of C-2 to center of B-3
   New single family housing in center of B-2
   Industrial area next to water in NE corner of C-3

9. Round object in SW corner of B-2?
   Land use in lower half of C-2 next to water?
   Collection of structures in SE corner of A-2?

10. Why is the water body wide in A-1 and narrow in C-2?
    What is the dark, branching pattern in left side of A-1?

This approach may be utilized to great advantage with images of all kinds. If good quality prints or film positives are not available initially because of cost, it is still useful to create a file of logs using airphotos and satellite images from magazine sources. There are a number of journals and magazines that print imagery as a regular feature. The main drawback to these sources is the inability to use much magnification because they are half-tone screen images.

Exercise 6   *Other Log Interpretation Questions*

1. Examine Figure 6–1. If you are given some information available from USGS Topographic Quadrangles Peoria East and West, then more facts may be gleaned from this photo.

    a. Most of the area shown is a flat to gently undulating surface that lies 80 to 100 feet above the water.

b. The water body is the Illinois River backed up into what appears as a lake by a series of locks.

c. The shorter of the two runways at the airport, the Peoria International Airport, is approximately one mile in length.

d. The darker areas west of the river in A-3 and in B-1 and C-1 are wooded. The contours show the former area to have considerable relief between the river and the agricultural land, whereas the latter areas are of less relief and below the agricultural surface.

2. Two other facts are pertinent to this area. This region is part of the United States that was covered by the **U.S. Public Land Survey** system (sometimes called **Township and Range**). Also, the surface of this region is overlain with a deep layer of **loessial soil** that enhances the soil fertility and has the capability to form almost vertical cliffs when eroded.

3. With the above information in hand, answer the following questions:

a. If the **sections** of the Township and Range system are perfect square miles that impact the layout of agricultural fields and even the street layout within Peoria, and if the shorter runway of the airport is approximately one mile in length, what is the precise scale of this photo as a Representative Fraction?

b. What is the evidence that the agricultural surface in this area is fine textured? How does the process of gullcying (headward erosion) relate to this problem?

c. Is there any evidence of a valley bluff overlooking the river? What is it?

d. Can you find any evidence of industrial land? Where is it located and why might it be found there?

chapter *7*

COLOR INFRARED PHOTOGRAPHY

TERMINOLOGY

color infrared (CIR)
false color images
Kodachrome CIR
Kodacolor Aero Reversal Film
black & white infrared
NHAP photography
NAPP photography

leaf on/leaf off
CIR window
CIR signatures
CIR film dye lot
haze penetration
CIR videography
real time CIR video

FALSE COLOR PHOTOGRAPHY

Color infrared (CIR) photography is a medium comprised of **"false color" images.** That is, the colors do not look the same as the colors most humans see when they view their surroundings. The specific colors that are present in CIR film are arbitrary and could have been represented by many different combinations of dye. The colors that were selected and are now commonly used in CIR photography represent a shift of one color band in the electromagnetic spectrum (EMS). That is, the order in the EMS is ultraviolet, blue, green, red, infrared. See Figures 7–1 and 3–1.

In normal color photography blue is displayed as blue, green as green, and red as red. However, in CIR photography objects reflecting blue in the actual scene will be black (blue light is filtered out), green reflectance will be portrayed as blue, red reflectance as green, and near infrared shows as red. (A yellow, "minus blue" filter must be used or all dye layers would be partially exposed by blue light.) Thus, there is a shift for each of the emulsion layers in CIR film to represent a slightly longer wavelength of the EMS. See Figure 7–2.

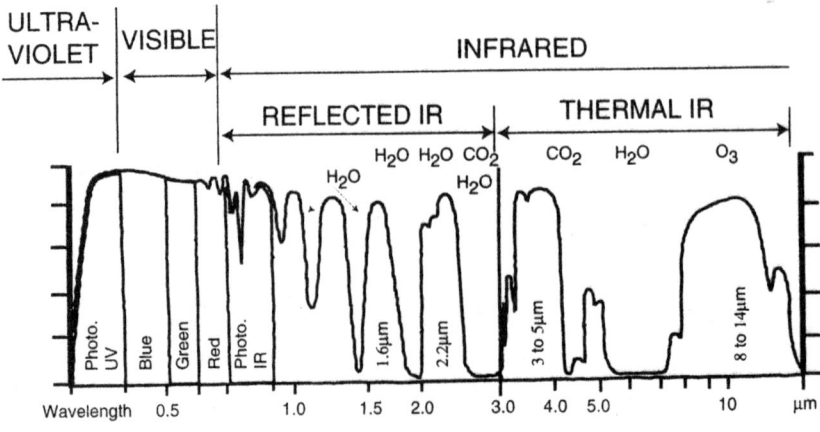

Figure 7–1. Electromagnetic spectrum in vicinity of visible and IR wavelengths.

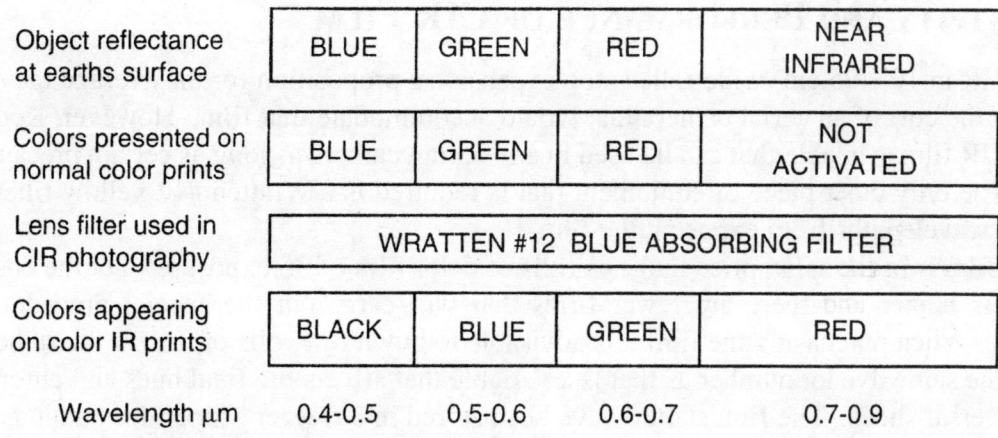

Object reflectance at earths surface	BLUE	GREEN	RED	NEAR INFRARED
Colors presented on normal color prints	BLUE	GREEN	RED	NOT ACTIVATED
Lens filter used in CIR photography	WRATTEN #12 BLUE ABSORBING FILTER			
Colors appearing on color IR prints	BLACK	BLUE	GREEN	RED
Wavelength μm	0.4-0.5	0.5-0.6	0.6-0.7	0.7-0.9

Figure 7–2. Color formation on normal color and color infrared films.

Since the mesophyll layer (internal tissue) of healthy plant life reflects strongly in the near IR region of the EMS, it is possible to use that signature characteristic in interpreting CIR photos. In normal color photography the chlorophyll layer absorbs blue and red wavelengths, yielding green reflectance and a green color for leaves and succulent stems. The near infrared energy penetrates the chlorophyll to reach the underlying mesophyll where much of it is reflected and most of the remainder is transmitted through the leaf. Various types of vegetation will reflect differing amounts of energy in the green and near IR bands, with the greatest variations in reflection occurring in the near IR.

CIR film records a greater range of reflectance from vegetation in the near IR band as well as varying amounts of reflectance from the green band for various types of vegetation. This means that CIR film will yield a greater range of colors and tones than will normal color film, which doesn't record the near IR, when vegetation is photographed. The end result of this situation is that CIR photography supplies considerable information to aid interpretation that is not readily available in normal color prints.

This fact is of the greatest significance and largely accounts for the rapid increase in CIR aerial photography around the world. As is shown in later chapters, the same wavelengths of the EMS may be recorded by imaging media other than film, e.g., Landsat MSS and TM, etc., and produce similar types of signature information about healthy vegetation. (See Chapters 9, 12, 13, and 14.)

HISTORY AND USES OF CIR PHOTOGRAPHY

Some of the earliest reports of IR-sensitive film date back to about 1931. Kodak's **Kodachrome CIR** was introduced in 1935 and refinements led to the creation of **Kodacolor Aero Reversal Film** in 1942. Much of the early work centered upon the military use of the film to detect camouflage. The film was modified and then adopted for use by the military in 1943 and a patent was issued to Kodak in 1946.

Initial use for camouflage detection eventually gave way to more important and long-lasting scientific application in the study of plants. This was pursued by the U.S. Forest Service and the Soil Conservation Service of the U.S. Department of Agriculture and the work has much wider applications today. A great deal of the knowledge gained in this area has been applied to sensors on satellites operating in the same wavelengths of the EMS.

SENSITIVITY AND PERFORMANCE OF CIR FILM

Aerial CIR in 10″ (25 cm) wide rolls is too expensive a proposition for the average interpreter to use as is the cost of an aerial camera that would accommodate that film. However, Kodak has a 35mm CIR film available that can be used in any 35mm camera as long as certain precautions are taken. The only other piece of equipment that is required is a Wratten #12 yellow filter to keep the blue wavelengths from exposing the film.

This film is in the same price range as rolls of color film of 36 exposures, but the cost of processing is higher and there are fewer firms that will carry out the special procedure that is required. When purchasing the film it is advisable to buy a few rolls of film at the same time all having the same dye lot number as that is a variable that affects the final hues and chroma in the color reversal slides. The film should have been stored in a freezer at the film outlet and should be kept in a freezer until used and then processed promptly. It is easy to see that this is a more sensitive film than most people ordinarily use. CIR is more sensitive to changes in amounts of light than is typical color film and so bracketing exposures above and below the value indicated by the exposure meter becomes more important. If one wanted to compare the seasonal differences in CIR response to the vegetative community at a particular locale, it is important to try to duplicate lighting conditions as well as use film from the same dye lot (marked on the package).

Black and White IR is also available in 35mm rolls and requires a Wratten #25 filter. This film produces unusual effects in black and white photographs due to tonal variations existing between panchromatic and IR reflectance. Nevertheless, it seems to be used more by photographers, artists, and architects than by scientists involved in airphoto interpretation.

NATIONAL HIGH ALTITUDE PHOTOGRAPHY (NHAP) PROGRAM

The Soil Conservation Service of the U.S. Department of Agriculture carried out the **NHAP** photography between 1980 and 1987. This was a cooperative effort by 12 separate agencies to acquire cartographic quality 9″ × 9″ (23 × 23 cm) image size CIR and black and white aerial photographs simultaneously. The flight lines ran north–south through the centers of USGS 7.5 minute topographic quadrangles and each CIR photo at RF 1:58,000 covered the area of one such quad. One full frame of CIR portrayed approximately 72 square miles (186 square km). The B&W photography was acquired at RF 1:80,000. See Figure 7–3 (Plate III) for an example of a NHAP CIR photo acquired at the beginning of the growing season.

NATIONAL AERIAL PHOTOGRAPHY PROGRAM (NAPP)

The requirements for higher resolution photography by many Federal agencies led to the replacement of the NHAP by the **NAPP** in 1987. This CIR coverage was flown at an altitude of 20,000 feet and the frames were centered on quarter sections of USGS quadrangles. The photography produced has a nominal scale of RF 1:40,000 and one full frame covers approximately 32 square miles (83 square km). The intent is to rephotograph areas every five years for the entire conterminous United States. See Figure 7–4 (Plate IV) for an NAPP example of the same locus and season as Figure 7–3.

Both the NHAP and NAPP flew some photography **leaf on** and some **leaf off.** Some types of uses for the photography, such as contour mapping, are better met if the ground surface can be seen and so obtaining photos at times when the leaves are off trees is more advantageous. If the need is to map the different types of vegetation, then photos that show trees in leaf will work better.

There are two **windows for acquiring CIR** aerial photography that best serve most of the needs for the coverage. The longest and therefore the easiest of these windows to use occurs in the spring season when the vegetative growth begins but is not to the full leaf out stage. In New England that period runs approximately from March 15 to April 30, but it varies somewhat based upon the temperature and precipitation conditions in a given year. The fall window is shorter in New England, from perhaps September 30 to October 31, and is harder to judge. This area of the United States has an average of about 40 days during the year when high quality aerial photography may be flown because of winds, clouds, and other atmospheric conditions occurring on other days.

Other parts of the nation will have windows of opportunity at different times based upon latitude, sun angle, temperature, precipitation, and local vegetative response to these conditions. It is easy to see that planning and flying a successful mission is affected by many variables. This is even more of a problem if CIR is being acquired rather than black and white photography, since the lighting conditions are more critical.

ESTABLISHING CIR SIGNATURES

It is apparent that color infrared film displays characteristics that are at variance with other types of films utilized in aerial photography. Despite these differences and the need to exercise care in the use of this film, it is a film medium that is rich in information about objects on the face of the Earth.

Figure 7–5 is a table that lists some of the general image characteristics of objects displayed on CIR photography and also on satellite images that portray reflectance values in the same combination of visible and near IR wavelengths. When using these general signature characteristics as a guide in interpretation of CIR reflectance information, it is important to remember several facts about CIR imagery reproduced from CIR film:

1. When the **dye lot of CIR film** is changed, the color values in the images are likely to vary. Each CIR photograph should be approached as though the specific colors are unique to that photo unless it is one of a number of exposures in a flight line that have already been interpreted and for which signatures have been established. (Even when that situation exists, such signature characteristics may only hold true for perhaps five airphotos in succession in a flight line.

2. The **reproduction of CIR prints or CIR film positives** is not always constant even by Federal agencies or private firms with strict quality control measures in force. This means that some levels of brightness or color values may be present due to the reproduction process and not to existing conditions when the photography was acquired.

3. CIR photography **cuts haze** conditions to some extent more than is possible with normal color photography. However, if the haze is excessive, it can produce a color shift on the resulting photos, e.g., growing vegetation appears orange rather than red.

4. If **clouds cast shadows** on the ground that is photographed there will be less photographic detail in the shadows on the CIR film than there would be with other film types.

5. If a person is acquiring CIR airphotos with a 35mm camera the **amount of light allowed onto the film** is critical. Making small changes in lens opening (f/stop) or shutter speed will lead to large changes in photo density and coloration. Experimentation and use of bracketing the "proper" exposure setting is most important so that a photo opportunity that may not be replicable is not wasted.

General Guide to Feature Characteristics
on CIR Photography and False Color CIR
Multispectral Satellite Imagery

Land Use Category	*Land Area Identified By:	Best Band/Dye Layer
Residential	Mottled gray surrounding commercial business district. Street patterns when visible.	Green Band Color Composite
Commercial and Services	Light tone in commercial business district, following major arteries: white/bluish in color composite.	Green Band Color Composite
Industrial	Regular boundaries; light tone; large buildings and much bare ground. Usually separate from center of urban area.	Green Band Color Composite
Extractive	Extensive light toned spoil areas; isolated road networks.	Green Band Color Composite
Utilities	Long, straight, fairly light features. Commonly do not conform to orientation of section lines.	Green Band Color Composite
Stripe and Clustered Development	Light tones or mottled areas along right-of-way; isolated from other urban areas.	Green Band Color Composite
Golf Course, Cultivated Grass	Light tones; "fingered" appearance within urban or forested areas.	Red Band
Cropland and Pasture	Light tone (pink to red on color composite). Extensive land areas occupied. Boundaries conform to section lines. Contoured boundaries adjacent to bottom lands.	Green Band Color Composite
Deciduous Forest	Bright red or red-yellow conforming to stream patterns in bottom land on color composite. Dark tone on B & W image.	Green Band Color Composite
Coniferous Forest	Brownish red or purple-red (magenta) in upland areas on color composite. Very dark appearance on B & W image. Usually separated from lowlands by deciduous forest.	Green Band Color Composite
All Water Areas	Very dark or even black tone on B & W image. On color composite, color is black to light blue depending upon amount of suspended sediments.	Red Band Color Composite
Beaches	White strip along coastline between water and vegetated areas. Surf may add to bright reflectance effect.	Green Band Color Composite

*Note! All signatures subject to variation due to seasonal and other changes.

Green Band on Landsat MSS = Band 5 on Landsat 1, 2, 3; = Band 2 on Landsat 4, 5;
Green Band on Landsat TM = Band 3

Color Composite refers to CIR photography, 3 dye layers (Blue, Green, and Red);
Landsat MSS, Bands 4, 5, 7 on Landsat 1, 2, 3 = standard color composite;
Landsat TM, Bands 2, 3, 4 = standard color composite.

Figure 7–5.

If the points listed above are observed, and if a person who desires to acquire good quality CIR photography experiments prior to the actual "shoot," then the results can be most rewarding. CIR photography and false color satellite imagery recording in the same wavelengths of the EMS have never been more popular than they are at present. Some of the applications of such imagery are explored in this manual.

CIR VIDEOGRAPHY

A relatively recent development in the field of Remote Sensing is **real time CIR video.** It is now possible to acquire imagery that produces the type of information discussed above relative to CIR photography, but through use of a video system. This allows an interpreter to view the CIR product as it is being recorded, or "real time" imagery. The impact of this development should be of major significance to many fields that study the Earth's surface condition. An example of such a video system will be discussed in Chapter 14.

Exercise 7 Information Extraction from CIR Images

The following material will deal with comparison of Figures 7–3 through 7–8, airphotos acquired at different times at varying scales and on different films. Figure 7–6 appears on Plate V and Figures 7–7 and 7–8 appear on the following pages. All of these aerials are centered over the same location on the ground, a section of the Ipswich River where it flows through Topsfield, MA and through a Massachusetts Audubon Society Wildlife Sanctuary (previously portrayed on maps, Figures 5–1 and 5–2). This approach should enable the interpreter to understand the value of CIR photos as a source of information, as well as revealing some of the pitfalls inherent in the use of this type of imagery.

Figure 7–6 (see Plate V) is a CIR airphoto acquired from the Skylab space station in the spring circa 1974. It is an enlargement made from a 4.5″ × 4.5″ film positive with a scale of RF 1:900,000. It was photographed from an altitude of approximately 270 miles (435 km) and shows the New England coast north of Boston in the vicinity of Cape Ann and the Merrimac River. The scale of the enlargement is RF 1:230,000.

Figure 7–7 is a black and white airphoto acquired on June 11 of 1971 at a nominal scale of RF 1:20,000, and Figure 7–8 is a 70mm black and white photo obtained on April 9, 1990 at a nominal scale of RF 1:6600. All of these airphotos fall within the spring window for CIR photography except the 1971 photo which displays full leaf out conditions.

1. Examine Figures 7–3 and 7–4. Note that the colors vary somewhat. Considering that these airphotos were taken in different years but within two weeks of the same date, what might account for the differences in color? Is the following information relevant to your answer?

   Figure 7–3 was photographed on 4/1/86 at a nominal scale of RF 1:58,000 at an altitude of 40,000 feet. Figure 7–4 was obtained on 4/13/92 at a nominal scale of RF 1:40,000 at an altitude of 20,000 feet.

2. Is it easier to isolate objects on Figure 7–3 or 7–4? Is scale a factor? What else might have a bearing on being able to see individual features and perhaps identify them?

3. In comparing these two airphotos, is there any evidence that atmospheric conditions might have been different on the days photography was acquired? Anything conclu-

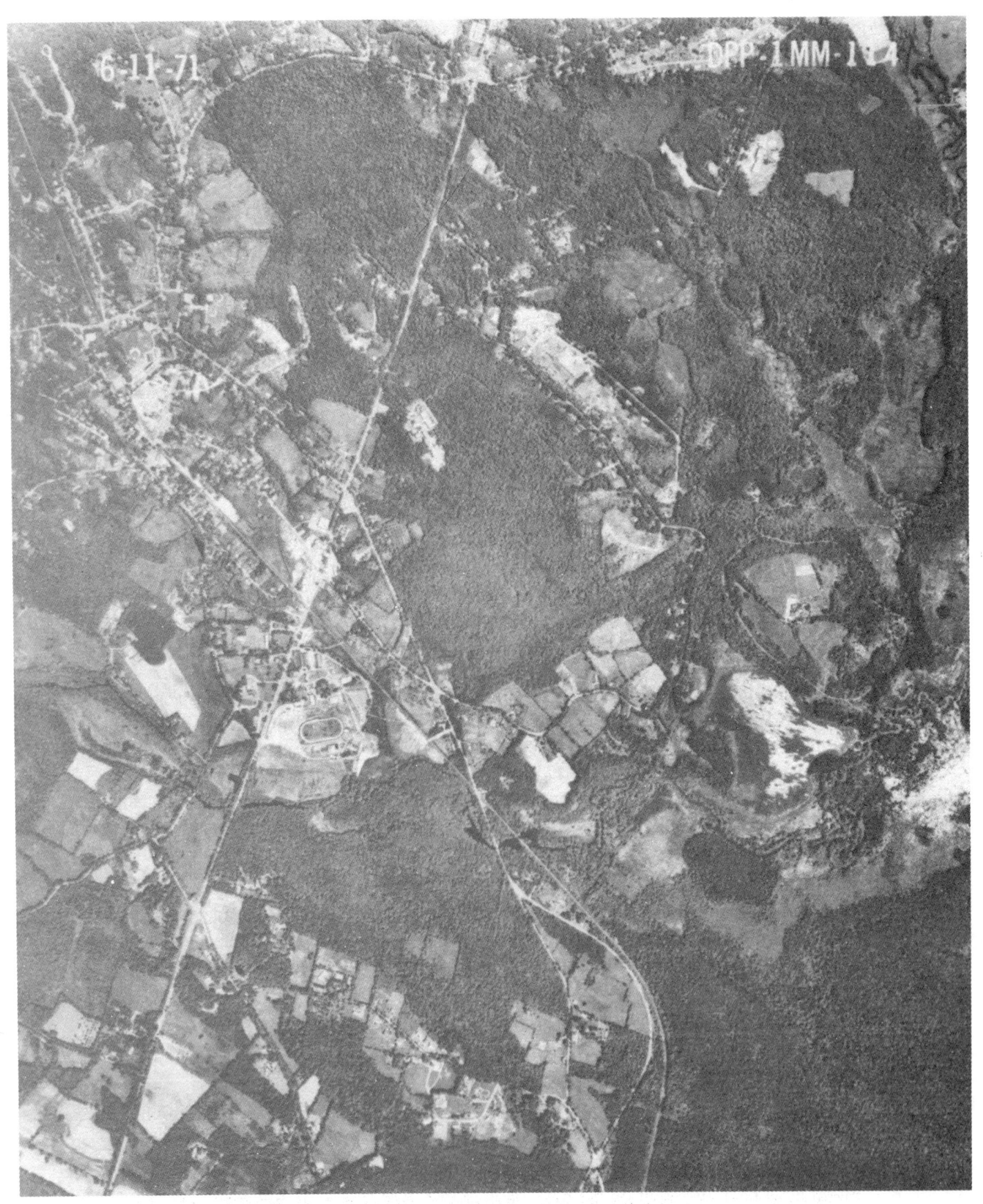

Figure 7–7. USDA airphoto #DPP-1mm-114. Topsfield, MA. Acquired 6/11/71.

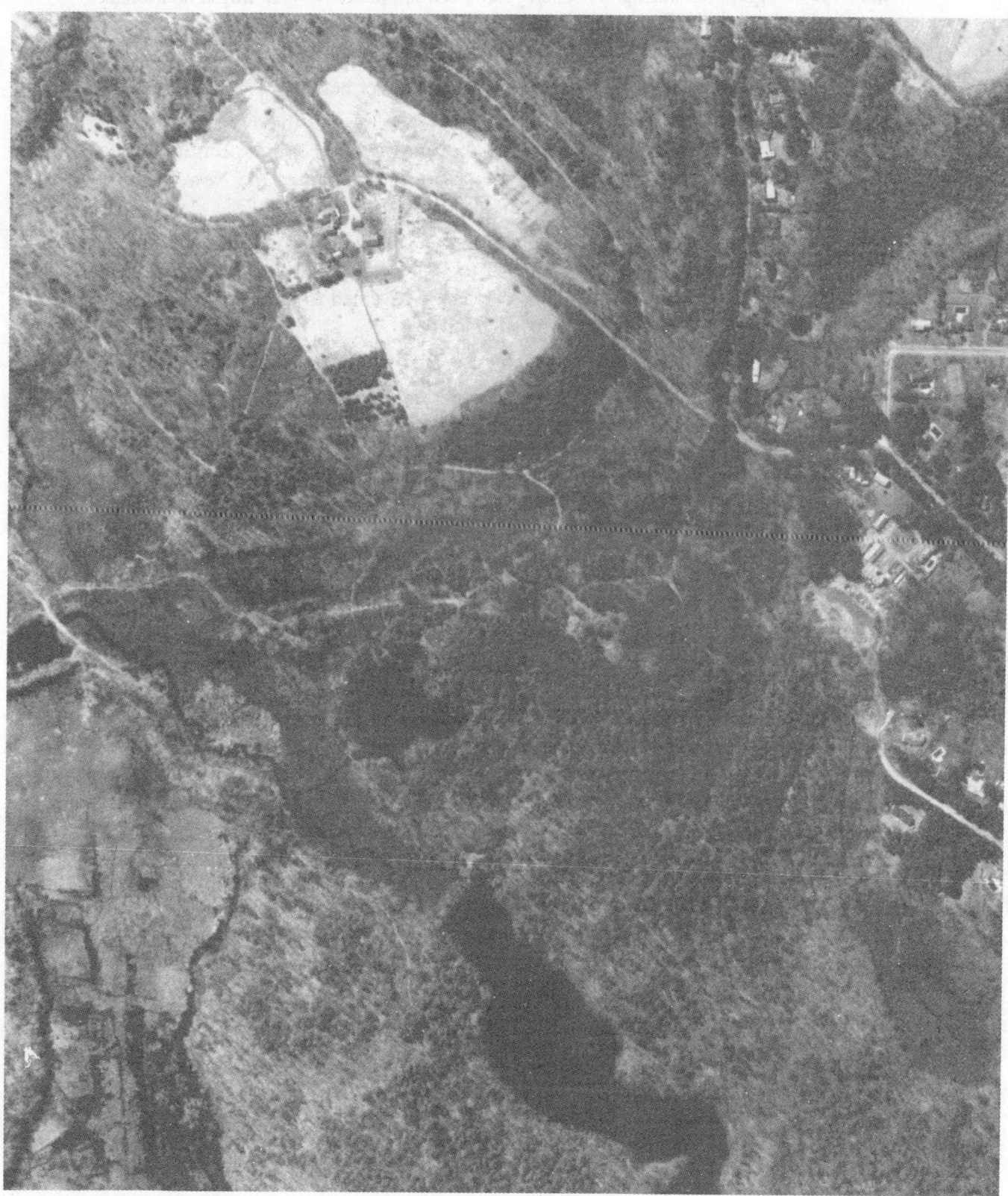

Figure 7–8. 70mm airphoto 4/9/90. Topsfield. MA. Courtesy Flight Survey & Mapping. Inc.

sive? If you compare the distribution of darker red areas on the two photos (primarily needleleaf, conifers) is the pattern similar? Has there been any large change in area of that type of vegetation during the seven years between photos? What might be human, economic, and natural factors that could lead to an increase or decrease in acreage of coniferous forest?

Do you see any noticeable change in land use/land cover between 1986 and 1992?

4. Now examine Figure 7–6, the Skylab CIR airphoto. Why does the overall coloration of this photograph seem to be shifted toward the blue direction of the spectrum? What might be the explanation for this? Was the sky over this part of the New England coast essentially free of cloud cover on the day the photo was taken?

5. Note that on Figure 7–6 much of the light pink area seems to be concentrated near the center of the photo and extends from B-2 to B-1 (refer to Chapter 4 for location method). On Figures 7–3 and 7–4 those areas are revealed to be primarily agricultural fields in close proximity to the Ipswich River.

Can you think of any historical reason why this land use pattern would exist?

6. On Figure 7–6 the dark red/purple areas are forested. Is it reasonable to assume that most of these areas represent presently undeveloped open space? Could you easily make a rough estimate of the area of such lands using a grid? If you were told that the coastal areas (inland of the beaches) in A-3 and B-3 were salt marsh/tidal flats, could you make a similar estimate of the area of such wetlands using the grid method?

7. In Figure 7–6 along the Merrimac River (bottom of A-1 to A-3) several cities can be seen in blue/gray tones. Also in the coastal area (C-1, C-2) there are several urban areas in blue/gray tones. Are there some distinguishing characteristics of these areas that help to identify them as different than the coastal wetlands referred to above? How might a map of the local area aid in this process?

8. Now study Figures 7–7 and 7–8. The area centered on B-3 in Figure 7–7 (6/11/71 photo) is essentially the same as most of the area of the photo shown in Figure 7–8 (4/9/90). Can you identify any differences in gray tones that might be accounted for by differences in vegetative cover or height of the water table? Are these variations likely related to seasonal changes? Note in particular the hill (drumlin) in C-3 area of Figure 7–8 that is almost covered with pasture. Also note the wet areas in A-2, B-3 on the same photo (darkest areas are open water, see also streams and adjacent area of marsh). Compare these areas with the same locations on the June photo.

9. Now review all of the photographs in this exercise. What kinds of information are present on the CIR photos that are totally missing or difficult to interpret on the black and white photos? Can you list some types of land use/land cover that are more readily identified on the CIR?

10. If you were to use the maps in Chapter 5 (Figures 5–1 and 5–2) in conjunction with these photos, how would that assist you in identifying features on the Earth's surface? What other maps or materials related to this area might be useful in helping to interpret the airphotos?

chapter 8

INTERPRETATION MAPPING WITH MULTIPLE IMAGES

TERMINOLOGY

mosaic	elimination key
orthophoto mosaic	dichotomous key
controlled mosaic	selective key
ground control	change through time
uncontrolled mosaic	modification of physical
photo index mosaic	and cultural landscapes
drafting tape	

INTRODUCTION

Most of the discussion to this point has dealt with the interpretation of one photograph or a stereopair of photographs. It is often the case that the interpretation involved in solving a particular problem demands the investigation of multiple images, simultaneously or in some sequence. The focus of this chapter is how to deal with the problems that arise when multiple airphotos are employed as a mosaic or in time sequence.

AIRPHOTO MOSAICS

As was discussed in Chapter 4 aerial photography is most often acquired in a series of flight lines in such a fashion that the photos obtained will overlap within and between flight lines. This produces a **mosaic** of airphotos that covers the area of study and allows for stereoviewing of successive photos within a flight line. There are several different types of mosaic.

In Chapter 5 the orthophoto was mentioned. These photo products may consist of one or several vertical airphotos from which the parallax has been removed so that they may be aligned with points on the Earth's surface. In this manner **orthophoto mosaics** may be used with the same degree of accuracy as maps in measuring compass angles and distances between points.

Another form of mosaic is the **controlled mosaic.** This form requires that the central portions of each vertical airphoto (the part with the least image displacement) be cut out and matched to the central portions of adjacent aerials. Then the positions of these image parts are aligned with templates that depict **ground control** (survey reference points) for the area. Approximate measurements of distance and direction may be made with confidence on such products.

A third category is termed the **uncontrolled mosaic.** This may be created by cutting out the central parts of vertical airphotos and aligning them with images on the photos, or it may be produced by aligning uncut aerials using image features. Obviously this latter approach would be less accurate, but neither form of uncontrolled mosaic is the most desirable platform to use in making measurements of distance and direction. All of these mosaic forms provide an overall view not available from an individual, large scale photo.

Figure 8–1 is a **photo index mosaic** which is the result of arranging the airphotos in sequence within flight lines as they were photographed. This is done with the identifying numbers displayed and then the mosaic itself is photographed. This allows a prospective user to select the photo or photos needed for a specific project. Many Federal and private agencies that sell airphotos create photo mosaics to aid the customer in photo selection. See Appendix E for a selected list of such sources.

GUIDELINES FOR UNCONTROLLED MOSAIC LAYOUT

The focus here will be on the uncontrolled mosaic in which the airphotos are not cut. The approach is similar to that wherein only the central portions of photos are used, but this method allows the airphotos to be disassembled for later use in other projects.

The aim is to acquire the skill in creating an overlay map for a mosaic of airphotos that could be used for a variety of planning purposes. Such an overlay could also provide an initial look at an area for which more detailed and precise interpretive mapping must be done at a later date. This would be appropriate for regional planning efforts, highway engineering layout, power line right of way investigation, and the like.

1. The airphotos should be arranged by adjacent flight lines as well as by consecutive exposure number within flight lines. The number of photos to deal with may be reduced by eliminating every other photo in stereo coverage and this will still allow a forward overlap within flight lines of 25 to 30 percent. Complete coverage of the ground will be acquired, but matching major features such as roads, coastlines, etc., becomes more difficult because of increasing parallax. Choose the approach that best suits your needs.

2. In assembling the photos of the mosaic, the overall representation of the area will be closer to reality if important, linear features such as interstate highways, large rivers, or coastlines are aligned as continuous rather than broken or offset lines. This is especially true if these large linear elements appear closer to the center of the individual prints than at the edges, because of parallax displacement. See Figure 8–2 in which the coastline of Cape Ann, MA was the guiding characteristic.

3. Once the mosaic has been assembled with these specific features aligned as best as possible, the photos should be taped together using **drafting tape.** (It is most important to use drafting tape rather than masking or transparent tapes because of the amount of adhesive present on these other varieties.) All such taping should be carried out on the back of the photos, never on the image surface. The emulsion is very thin and once it is damaged the airphoto is ruined. Note that if the photos are stacked for storage, remnants of adhesive on the back of one photo may be transferred to the front, emulsion surface of another photo.

4. Once the mosaic has been assembled and taped together, an overlay material should be placed over the mosaic with the frosted or prepared side up and the shiny side down. There are many brands of mylar drafting film available on the market that are suitable for overlay mapping. At least one side of the film used should be prepared for pencil and ink and the film should be at least 3 mils (0.003″) thick. A 3H lead

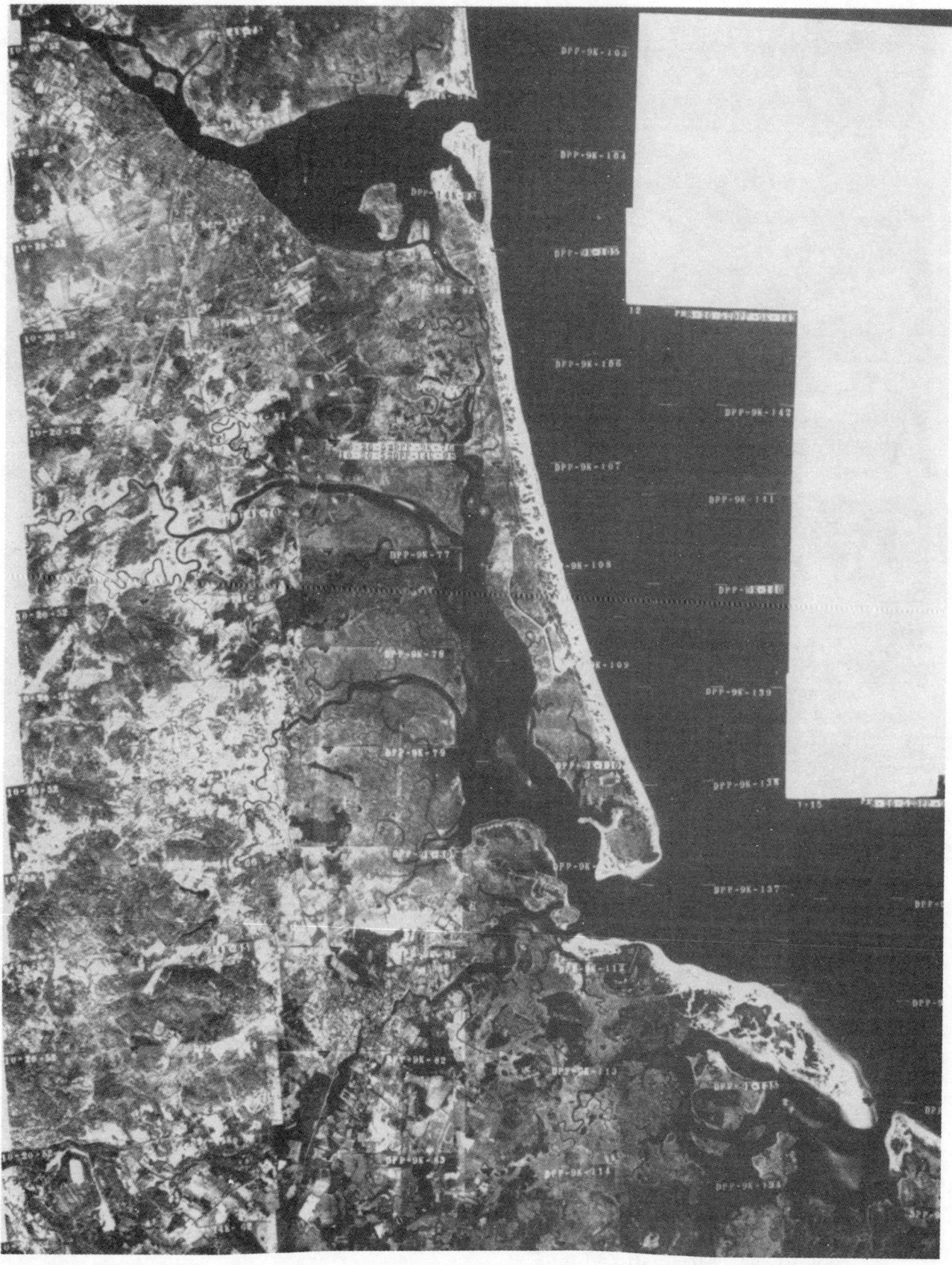

Figure 8–1. Photo index mosaic. Essex County, MA (part) USDA, ASCS, 1952.

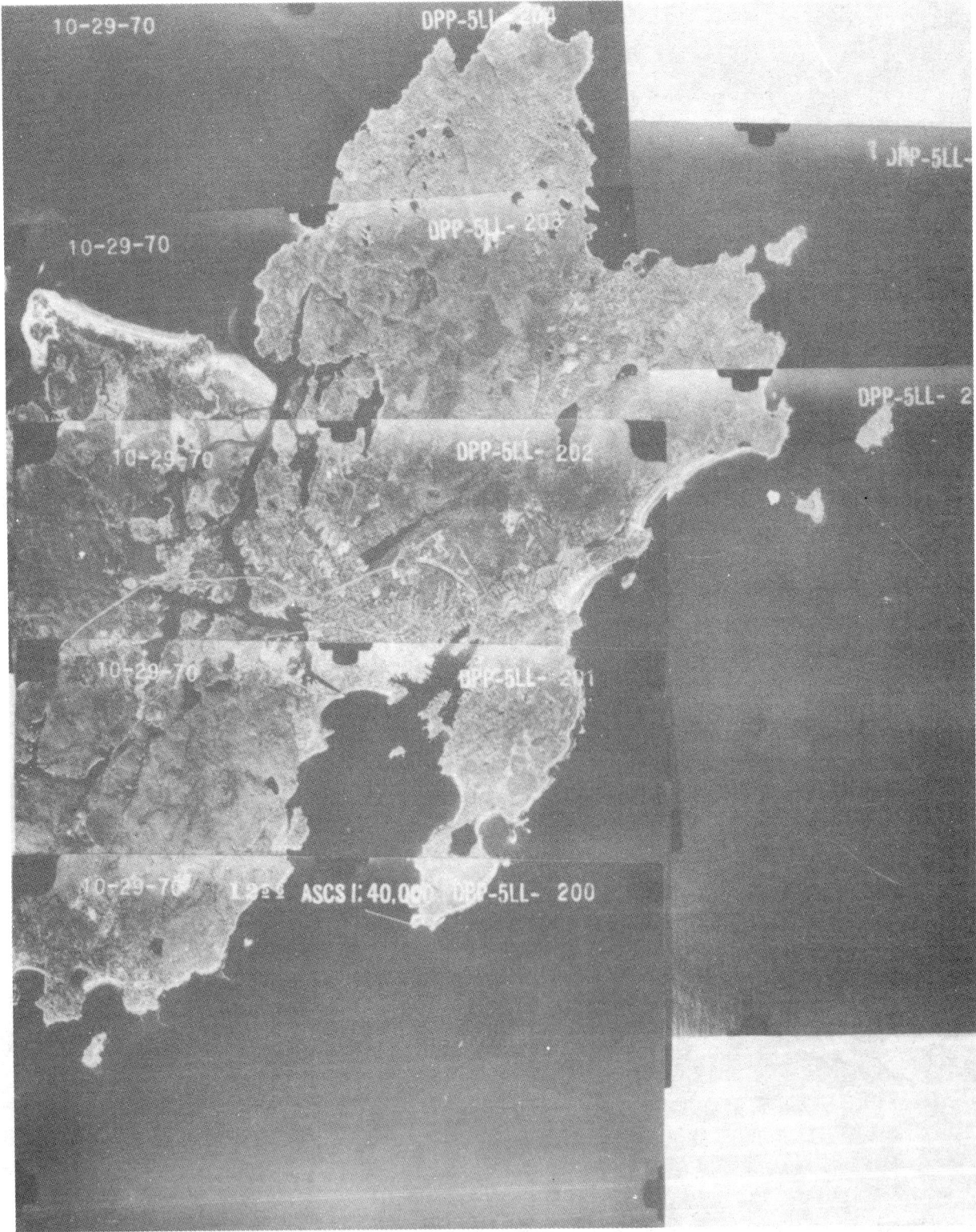

Figure 8–2. Uncontrolled mosaic. Cape Ann, MA USDA, ASCS 10/29/70.

pencil is hard enough to maintain a sharp point, but not so hard as to damage the air-photos beneath the drafting film.

5. The size of the mylar sheet used should be large enough to cover the complete mosaic and still leave room for a title of up to four lines, and also for the key that

explains symbols used in the overlay. It is better to have a larger piece than is needed, as it can always be trimmed when work is complete.

6. Using the 3H pencil, lightly mark the location of all of the corners of the photos in the mosaic. This will facilitate removing and replacing of individual photos for stereo viewing or detailed work. These corner marks can be eliminated after the interpretation and mapping have been completed.

INTERPRETATION AND MAPPING PROCEDURES

At this point the actual interpretation and mapping of objects on the photos can begin. The individual interpreter needs to select a methodology to carry out consistently so that the results are reliable. There are a number of paths that may be followed and some of these were alluded to in Chapter 5.

In Chapter 5 several land use/land cover classification schemes were discussed and examples of such mapping were shown. The soils mapping of the Soil Conservation Service, the vegetative mapping of McConnell, and the classification system of the U.S. Geological Survey were examined. One of these methods may be suitable for an individual interpreter's needs, one of those in Appendix D, or a new scheme may need to be created. The specific classification used is dependent upon several variables. What type features are of primary interest? How detailed is the classification needed to be and how detailed a scheme will the resolution of the photography allow?

1. **Constructing a classification key.** When faced with the task of creating a classification system for interpreting imagery of a particular portion of the Earth's surface, there are several points to consider. Do the results of the mapping have to be compatible with other work completed or planned? Is there an existing classification system that is recommended? Does the skill level of the interpreter require more or less guidance in decision-making regarding what image features should be assigned to which classification class? There will likely be other considerations involved for specific interpretation projects.

   For certain fields of investigation, **elimination keys** have been created that present an interpreter with making a choice between two or more classes for an observed feature. Many of these keys are **dichotomous** so that it is an either-or decision. Such phenomena as agricultural crops, forest types, and soils are adaptable to this approach as are the identification of industrial activities and certain military targets. This approach is sometimes constraining and may not allow for unusual or anomalous objects to be classified.

   Another form is the **selective key** where the interpreter compares the ground feature to known examples. The identified class representative may be in the form of diagrams of tree shape or building style, as a stereophoto log, or as written description. While this approach is more open-ended, it follows the same approach as most human learning and probably leads to developing a more skilled, independent image interpreter.

   Some classification scheme must be created that meets all of the requirements mentioned above. If one is developing a key of the land use/land cover types present on an image, it is better to have more classes that can be combined rather than start with a few classes and add more as you go. This latter approach would require that the interpretation of areas on the overlay be reexamined, whereas the former technique calls for combining identified areas on the overlay. This type of problem is not unusual if the interpreter finds that he or she cannot consistently identify a particular feature and must drop back to the more general level of classification.

2. **Mapping procedures.** There is a temptation when interpreting an airphoto or satellite image to identify and delineate on an overlay all of the occurrences of a particular feature, e.g., all the single family residential areas or all agricultural land, and then go on to the next most easily identified feature. While this is perhaps confidence building, it is not efficient. Each area that is examined on an image with the aim of drawing a line on the overlay that separates it from unlike areas, calls for the interpretation of the delineated parcel and all contiguous areas.

If one type of use or class is concentrated upon, then the same identification decisions will be made several times as one comes back to a portion of the image to look at adjacent uses.

A better approach to mapping is to divide the whole image or mosaic into units by utilizing major linear features such as rivers, roads, or railroads as boundaries. This allows the image interpreter to concentrate on discrete areas and identify and classify all features there before moving on to the next area. Those features that are not positively identified should be noted for location and returned to after everything else is classified. It is often the case that a feature that cannot be classified at one location may be identified elsewhere because of slight changes in appearance or form.

Figure 8–2 on page 62 is a photographically reduced version of a mosaic of airphotos of Cape Ann, MA. The mosaic was initially subdivided by utilizing the tidal channel running north–south and the main highway running east–west to divide up the area of study. Figure 8–3 below is a diagram illustrating the manner in which the areas were blocked out. Figure 8–4 is the final interpretive overlay produced, and Figure 8–5 presents the accompanying title and key that were utilized. This key was based upon observed features and symbols were derived subjectively. Classes were created of land uses that were roughly equivalent in importance and then these were subdivided where possible and where it seemed logical. Note that Figures 8–4 and 8–5 have been photographically reduced from the scale at which the interpretation was done.

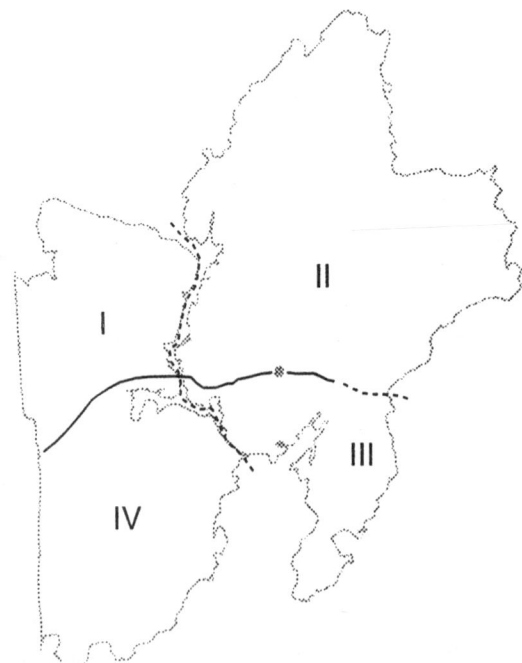

Figure 8–3. Blocking out areas for interpretation.

LAND USE/LAND COVER
GLOUCESTER/ROCKPORT, MA
1970

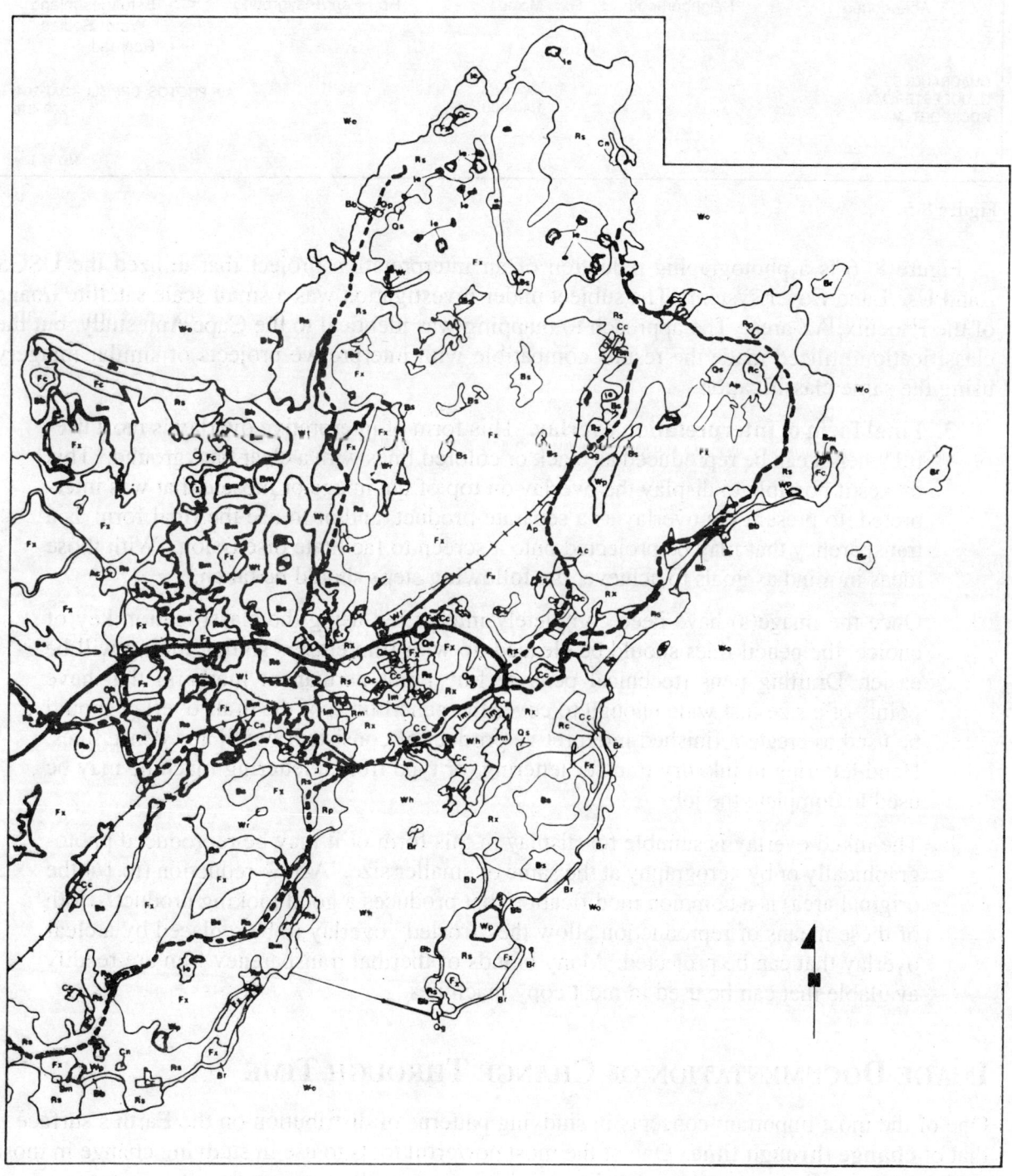

Figure 8–4. LU/LC interpretation, Cape Ann, MA. Courtesy James T. Parker, 1981.

WATER	BARREN LAND	WOODLAND	INDUSTRY	QUASI-PUBLIC
Wh Harbor	Bb Beach	Fc Coniferous	Ie Extractive	Qc Cemetery
Wo Ocean	Bm Tidal Marsh	Fd Deciduous	Im Manufacture	Qg Government/admin.
Wp Pond	Br Rock Outcrop	Fx Mixed		Qh Hospital
Wr Reservoir	Bs Swamp			Qs School

AGRICULTURE	COMMERCIAL	RESIDENTIAL	RECREATIONAL	SYMBOLS
Ac Cropped Land	Cc CBD/Strip	Rs Single Family	Rc Country Club/	—— Primary Road
Ap Pasture and/or	Cn General/	Rm Multi Family	Gulf Course	- - - Secondary Road
Abandoned	Neighborhood	Rx Mixed	Rp Park/Playground	⬭ Estuaries/Inland Water Bodies
				↦ Railroad

QUADRANGLES:
GLOUCESTER, MA
ROCKPORT, MA

1:40,000

AIR PHOTOS: DPP 5LL - 200–204;
- 209–210

J.T. Parker
May 11, 1981

Figure 8-5

Figure 8–6 is a photographic reduction of an interpretation project that utilized the USGS Land Use/Land Cover system. The subject under investigation was a small scale satellite image of the Phoenix, AZ area. The approach to mapping was identical to the Cape Ann study, but the classification utilized made the results compatible with interpretive projects of similar imagery using the same classification.

3. **Final form of interpretation overlay.** This form of interpretive overlay is most useful when it can be reproduced as black or colored lines with a clear background. This makes it possible to display the overlay on top of the image or mosaic that was interpreted, to present the overlay as a separate product, and/or to use the final form as a transparency that may be projected onto a screen to facilitate discussion. With those ideas in mind as goals to achieve, the following steps should be taken.

Once the image(s) have been completely interpreted using the classification key of choice, the pencil lines should be cleaned up with film erasers so that inking will be easier. Drafting pens (technical pens or fine point permanent markers) that have points of a size just wide enough to cover the pencil lines of the draft overlay should be used to create a finished interpretive product of consistent width and dense lines. Hand lettering in ink, dry transfer lettering, or type from a lettering machine may be used to complete the job.

The inked overlay is suitable for display in this form or it may be reproduced photographically or by xerography at the same or smaller size. A 50% reduction (to 1/4 the original area) is a common modification that produces a good looking product. Both of these means of reproduction allow the "frosted" overlay to be replaced by a clear overlay that can be projected. Many brands of thermal transparency film are readily available that can be used in most copy machines.

IMAGE DOCUMENTATION OF CHANGE THROUGH TIME

One of the most important concepts in studying patterns of distribution on the Earth's surface is that of **change through time.** One of the most powerful tools to use in studying change in those surface patterns is the record provided by airphotos and satellite images.

Modification of the physical landscape. The field of Geomorphology deals with the processes that bring about change in landforms on the Earth's surface. Many of the changes take

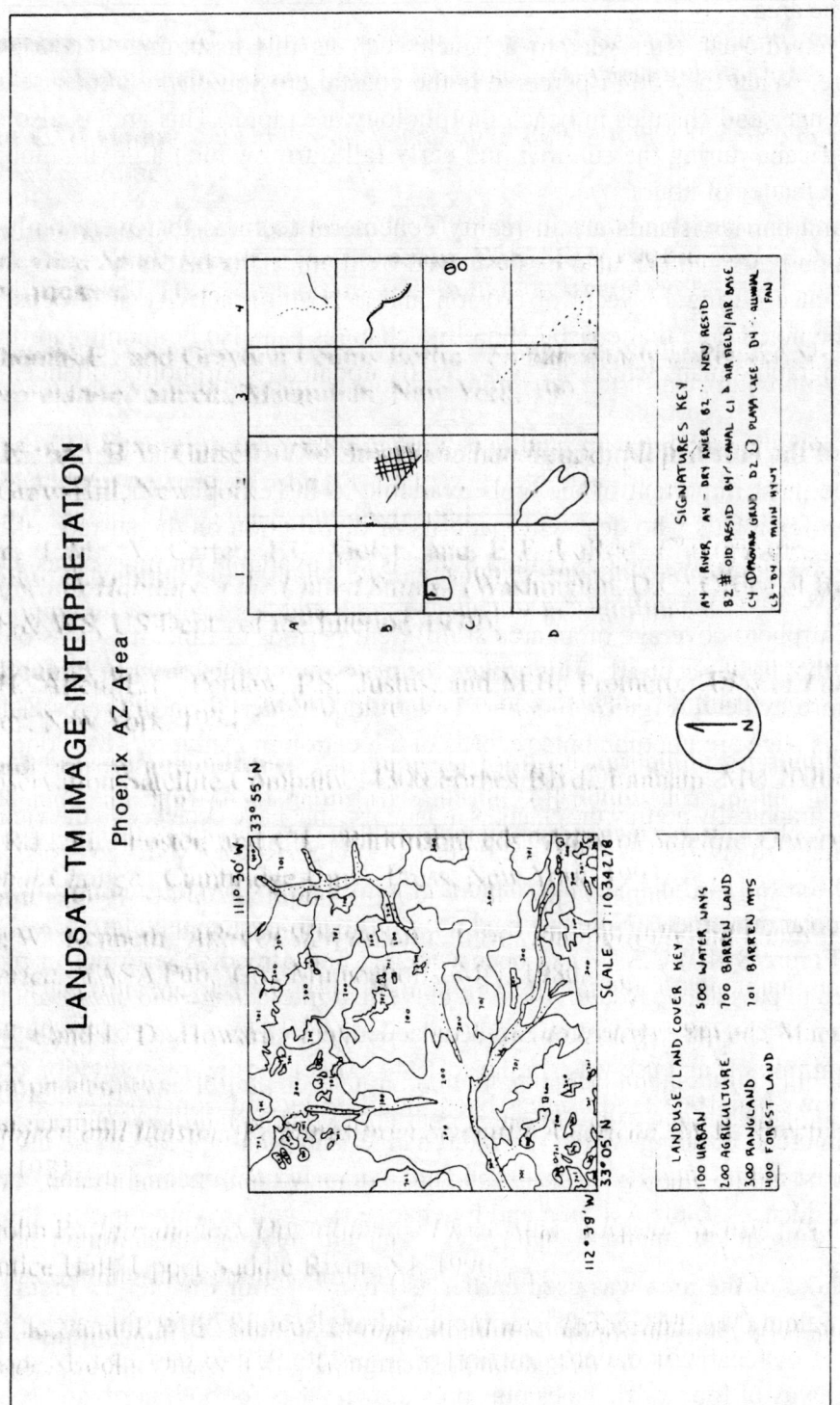

LANDSAT TM IMAGE INTERPRETATION

Phoenix AZ Area

MANNY SANTOS 12-14-93
REMOTE SENSING GGR 343

SIGNATURE'S KEY

A1 RIVER, AN - DRY RIVER B1 ' NEW RESID
B2 # OLD RESID B4 ' CANAL C1 ▲ AIR FIELD/AIR BASE
C4 ⬡PHOENIX GRID D2 ◯ PLAYA LAKE D4 ≋ ALLUVIAL
L5 - BY ≈ MAIN HWY   FAN

SCALE 1:1034276

LANDUSE/LAND COVER KEY
100 URBAN   500 WATER WAYS
200 AGRICULTURE   700 BARREN LAND
300 RANGELAND   701 BARREN MTS
400 FOREST LAND

PHOENIX AZ LSTM8678 5050917335 FRAME 1511 7-23-85
USGS TOPOS 1:250000 PHOENIX 1969, MESA 1978, AJO 1982

place over extremely long periods of time, thousands or millions of years. However, other modifications to Earth landforms occur within the average life span of a human and so are susceptible to being recorded by airphotos or satellite imagery several times during a period of years to document the changes.

Figures 8–7 and 8–8 show two aerial views of a portion of the Texas gulf coast in Matagorda County southwest of Galveston and Freeport. These stereograms of a section of barrier island were photographed in 1943 and 1960. One has to examine the photos carefully to determine that they cover the same area.

Many people return year after year to a beach such as this assuming it is a permanent, unchanging feature. What they don't perceive is the coastal erosion of the winter season when wave action is stronger and changes in beach morphology are rapid. This area is also subject to the occasional hurricane during the summer and early fall during which a great amount of erosion can occur in a matter of hours.

Coastal spits and barrier islands are in reality ephemeral features that were built by wind, waves, and ocean currents and can also be destroyed by them. This fact is corroborated by the photographic evidence during 17 years of erosion and deposition activity at Greens Bayou in Texas. It should be noted here that coastal shoreline changes can also be monitored through the use of USGS Topographic maps which are periodically updated by airphoto evidence.

Modification of the cultural landscape. Since the late 1920s black and white aerial photography has been the most important of the tools available to the city planner, highway engineer, and many other professionals who deal with patterns of distribution on the surface of the Earth. Photography covering the conterminous United States is still available for most areas dating back to at least the 1930s. (See Appendix E.)

By comparing airphoto coverage of an area at different periods of time, it is possible to document the changes that have occurred. This makes the economic forces that are largely responsible for change more evident. Figures 1–3 and 1–4 from Chapter 1, and Figures 8–9 through 8–12 following this page are photographic records of a location in Danvers, MA along with map information from the USGS quadrangle photorevised in 1979.

These airphotos graphically portray the changes in land use that have occurred in the vicinity of the junction of some major roadways. In 1944 (see Figure 8–9) U.S. Route 1 (NE–SW road at left side of photo) and state Route 114 (NW–SE road diagonally across photo) met at grade with a traffic signal to control vehicular movement. Note that land use is primarily agricultural, forest, and wetland.

By 1954 (see Figure 8–10) U.S. 1 had been widened and a median strip put in place. That roadway was raised to pass above Route 114 and a cloverleaf interchange had been built to accommodate increasing traffic. Some structures had been added along U.S. 1 and on or close to Route 114, but the general mix of land use was similar to 1944 except for the transportation changes.

Figure 8–11 shows that 1963 land use has been modified by additional trailer park use north on U.S. 1, a commercial building within the interchange, and what appears to be industrial use south of 114 and east of U.S. 1. The agricultural land is somewhat impacted by the construction of single family residences in the A-1 area and by change to a golf driving range in the B-1, C-1 part of the photo.

The 1980 airphoto of the area was used earlier as Figure 1–3 in Chapter 1. Figure 1–4 displays part of the Salem, MA USGS Topographic Quadrangle of 1979 for this area. The major change that is most noticeable is the construction of Interstate 95 just east of U.S. 1. This is a limited access highway of four traffic lanes plus breakdown lanes for both north and south movement. By this date the agricultural land has been greatly impacted by the construction of I–95 and also by many new residential structures.

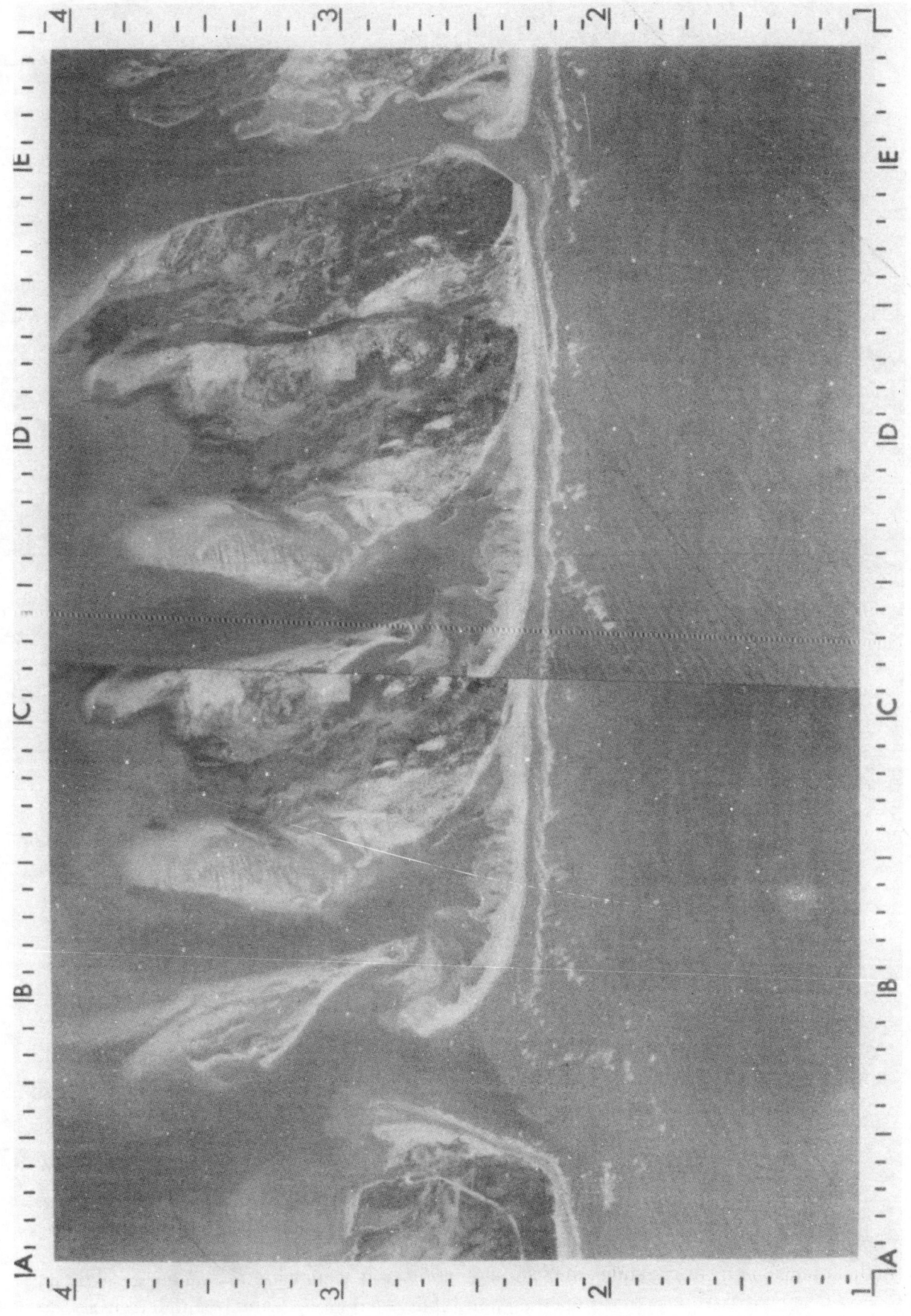

Figure 8–7. Greens Bayou (1943) Matagorda County, TX. Stereogram No. 605.
Courtesy University of Illinois, GIS Laboratory.

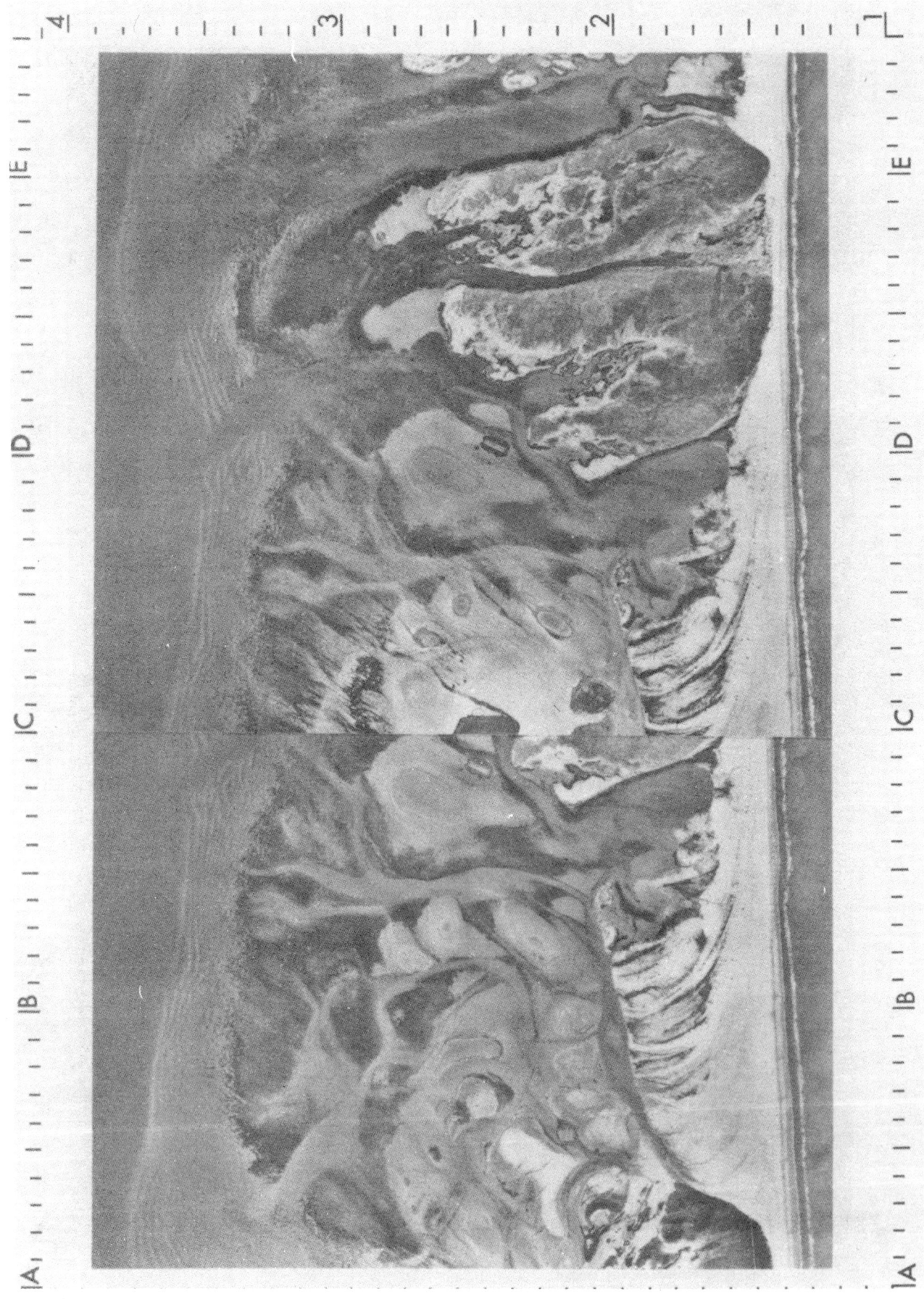

Figure 8–8. Greens Bayou (1960) Matagorda County, TX. Stereogram No. 607.
Courtesy University of Illinois, GIS Laboratory.

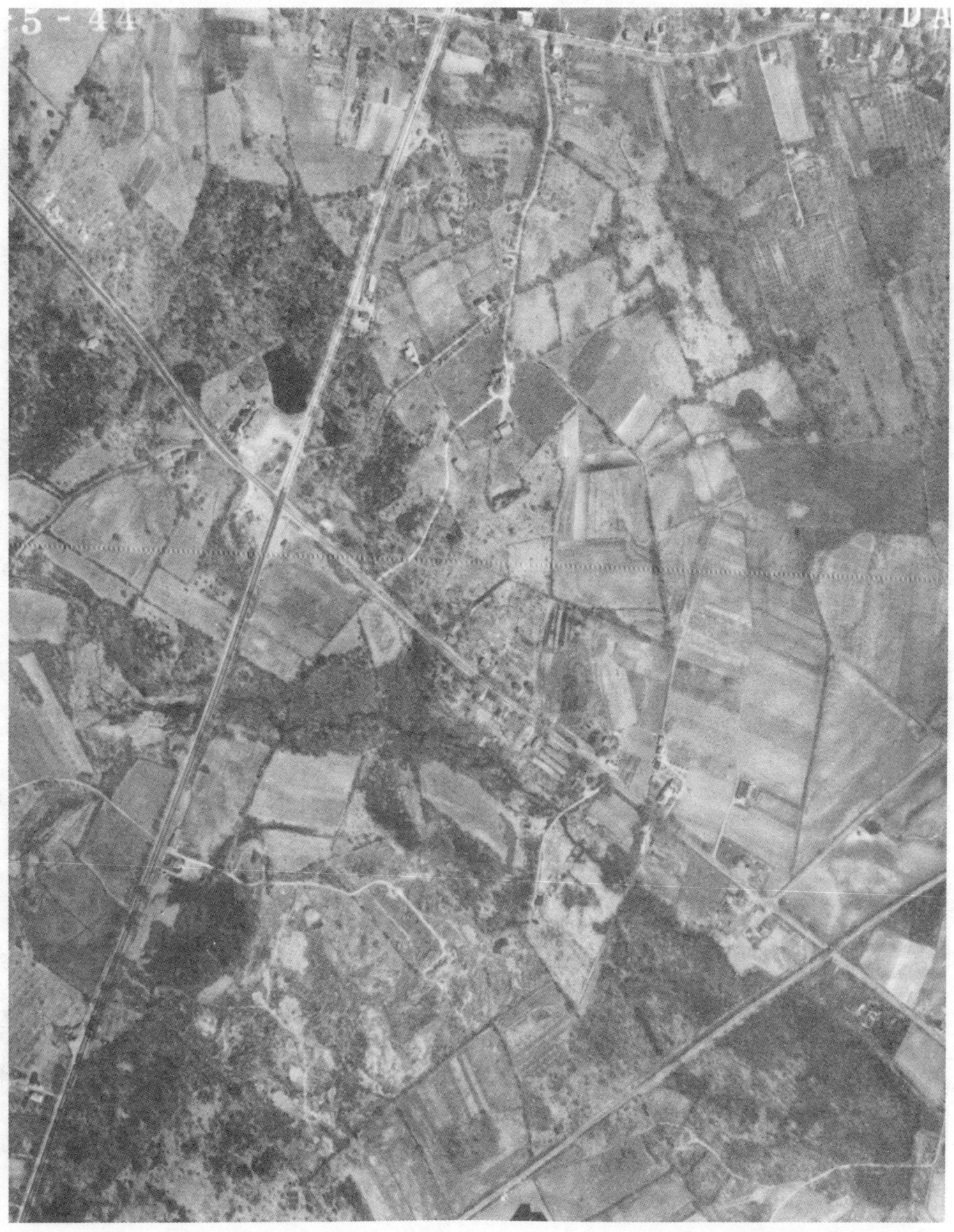

Figure 8–9. Danvers, MA 5/5/44. Exposure #DAN-21. Courtesy Town of Danvers.

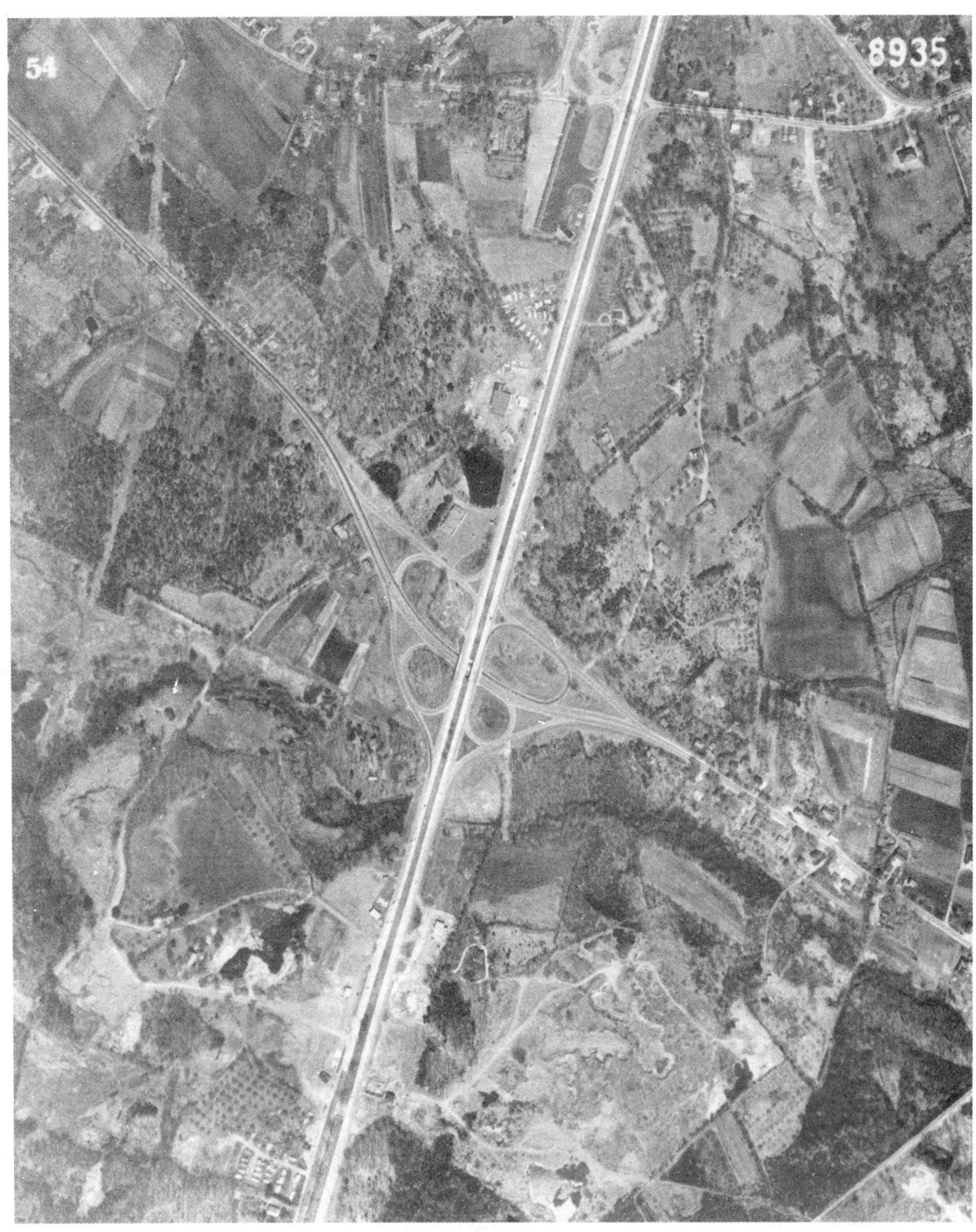

Figure 8–10. Danvers, MA 4/29/54. #8935 3711 . Courtesy Fairchild Aerial Surveys.

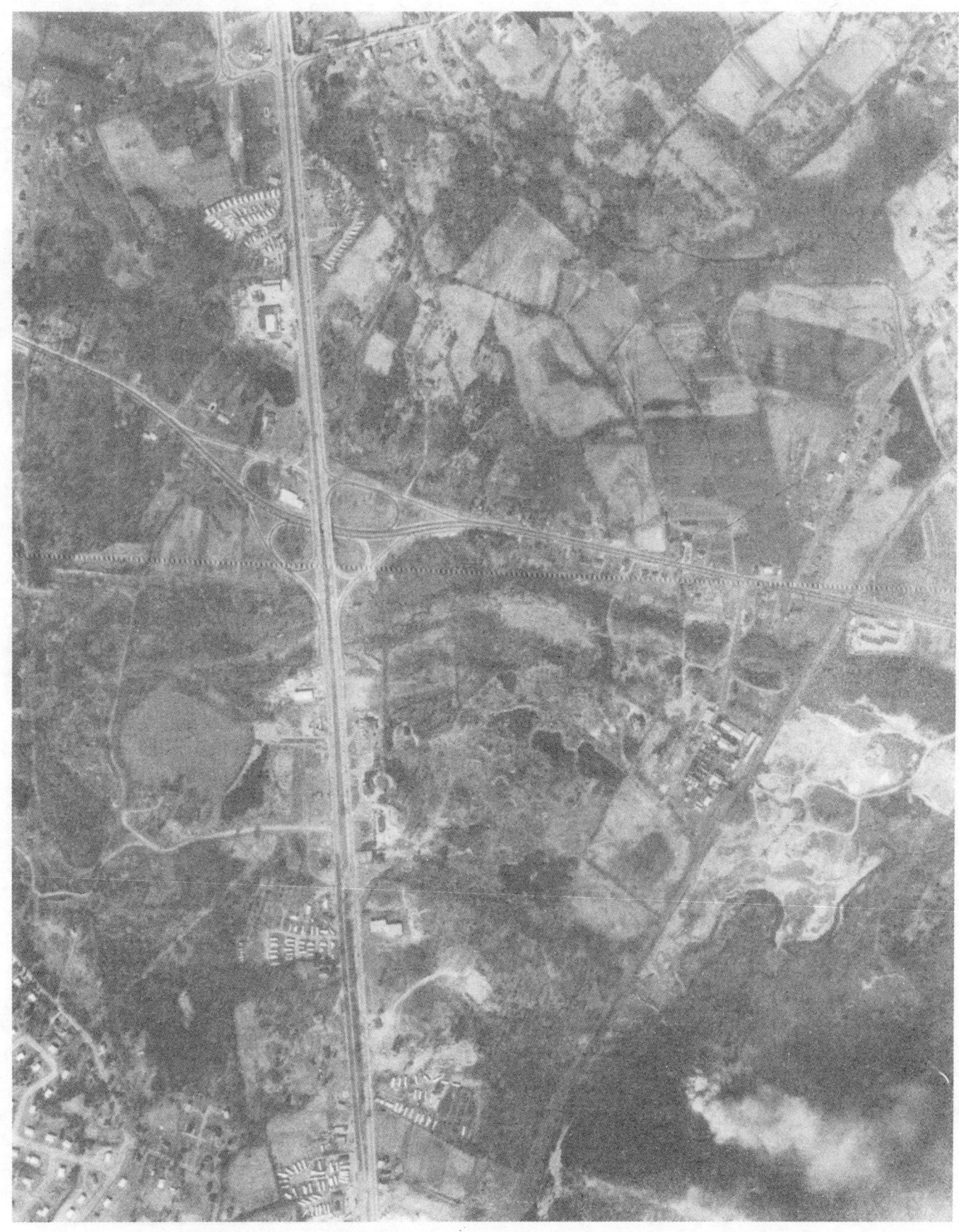

Figure 8–11. Danvers, MA 4/9/63. #4638 01 08. Courtesy Col-East, Inc.

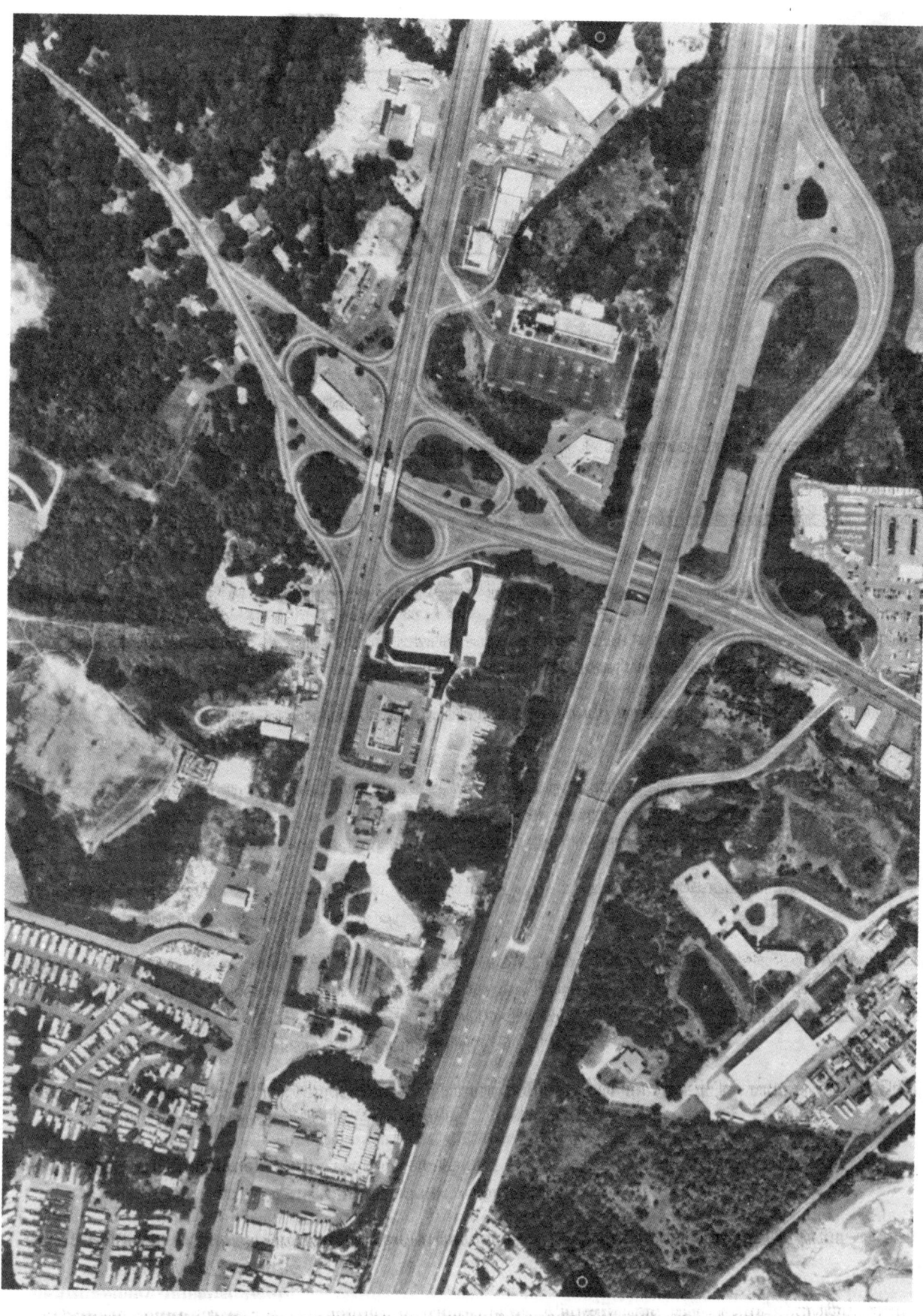

Figure 8–12. Danvers, MA 6/12/88, #2750-9. Courtesy Flight Survey & Mapping, Inc.

By 1988 (see Figure 8–12) the undeveloped areas between I–95 and U.S. 1 had rapidly begun to fill in. In this area north of Route 114 an office building and strip mall had replaced a poorly vegetated area, and south of 114 a chain restaurant had been constructed and a motel-type use was being built. The agricultural land in B-3 was giving way to commercial development north of 114 and south of that road the commercial development was increasing. All of these areas of development represent primarily a change from undeveloped land or agricultural land to the higher intensity use.

Exercise 8   Interpretive Mapping of Multiple Images

1. It is desirable for those interested in mastering image interpretation techniques to apply the "Guidelines for Uncontrolled Mosaic Layout" discussed earlier. Due to the constraints of space, it is not possible to supply contact airphotos in their complete 10″ × 10″ size here. For those wishing to experience working with an uncontrolled mosaic, there are several possible approaches.

   a. Work with airphotos on hand, or try to obtain "surplus" aerials from a local firm or agency.

   b. Visit a nearby college or university or a consulting firm to use some of their aerials at their site.

   c. Visit your county ASCS office (see white pages of telephone directory—"United States Government, Agriculture Dept., ASCS") for assistance in ordering photographs. You will require at least two overlapping exposures in two adjacent flight lines to create a small mosaic.

2. Examine Figure 8–2. What are some of the obvious limitations in trying to interpret images in this form? What is the reason that airphotos will not overlap precisely in regard to matching image features? Do you think this effect is more of a problem with large scale or small scale airphotos?

3. See Figures 8–3 and 8–4. If the student interpreter of this work had used the USGS classification system mentioned in Chapter 5 (and Appendix D), what numbers would replace the letter symbols shown? Can you find a number for every category? What is the significance of your finding?

4. Examine Figures 8–7 and 8–8. What evidence is present that indicates the material of which these coastal features are made? Does there appear to be a difference in materials in the same locations where there are varying albedo values? What might account for these differences?

   Considering the origin of these features, what is the likely highest elevation to be found here? A stereoscope may help you answer the question. What forces are at work along this coastal area that could bring about changes in shapes and locations of inlets, etc., in such a relatively short period of time?

5. Now examine Figures 1–3, 1–4, and 8–9 through 8–12. If the aim of investigation of these photos was to document the type and amount of change in land use/land cover, how might one go about this task? What parameters of such a study conceivably could affect consistency and reliability of data produced?

6. Study Figure 8–13 on the next page. Note that I–95 is now connected to Route 128, a major artery encircling the Boston metropolitan area. Note also that the Town of

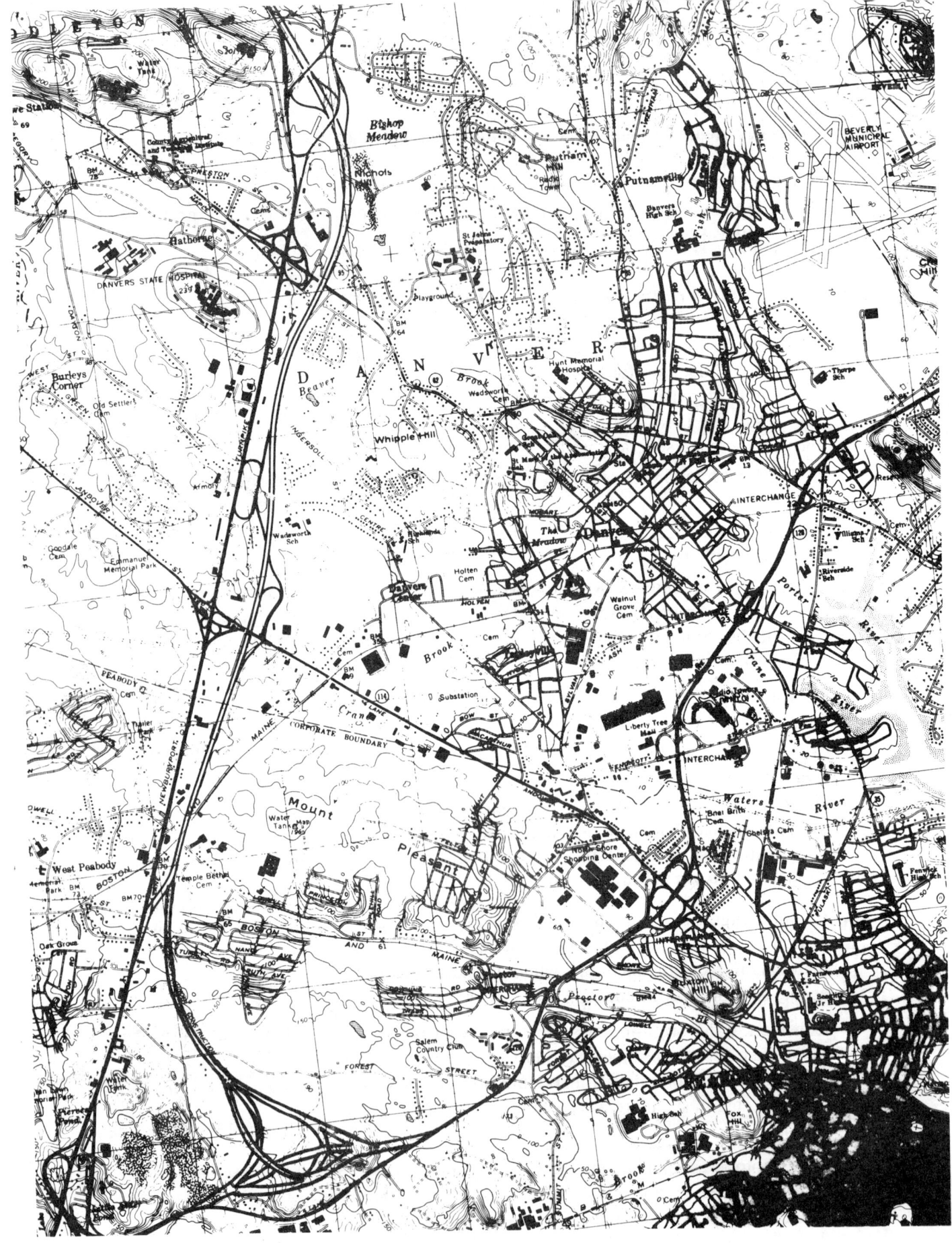

Figure 8–13. USGS topographic map. Salem, MA (part) 1985 revised.

Danvers is traversed by I–95, U.S. 1, Route 128, and Route 114 which is a connector of the other highways. Does the proximity of these roads to each other and the Town of Danvers help to explain why these photos show so much change in land use? Explain.

7. Examine the series of airphotos and maps listed in #5 and devise a methodology that will enable you to document the changing land use by amount and location. This will entail map and airphoto interpretation and overlay mapping. Will you use one overlay with different colors or symbols or will you use several overlays to illustrate land use distribution for different periods of time or for different uses? If you were given the task of publishing your completed interpretive overlays, how might that impact your methodology?

8. Use the map and airphotos referred to and produce a final interpretive overlay in ink that portrays the change over time for one or more types of land uses. Create a classification key that utilizes the USGS classification system (see Appendix D) with most symbols placed at Level III.

chapter 9

SATELLITE MAPPING OF EARTH RESOURCES

TERMINOLOGY

Landsat
ERTS
multispectral scanner (MSS)
geographic registration
pixel
IBM precision processing
EROS Data Center
Return Beam Vidicon
Thematic Mapper (TM)
EOSAT
sunsynchronous

high resolution visible (HRV)
Space Transportation System
large format camera
PE&RS journal
hand-held photography
panchromatic
false color composite
Skylab
Apollo, Mercury, Gemini
SPOT program
standard RGB array

HISTORICAL PERSPECTIVE

Beginning about 1960 the world has seen numerous satellites lofted into space by a few countries. Some of the vehicles have been sent on exploratory missions to other planets in our solar system and beyond. Others have been devoted to short term experimental journeys that dealt with measuring some characteristic of the Earth's atmosphere, magnetic field, shape, or other interest.

It is not possible in the space available here to deal with all of these satellites or those that were employed prior to the 1960s dating back to the late 1800s. Also, many satellites flown by the military or intelligence agencies and whose products are classified cannot be discussed here. Rather the intent is to present basic information about those satellite systems that are the most prolific in acquiring images and those whose image products have the greatest dissemination and use.

THE LANDSAT SYSTEM OF THE UNITED STATES

On July 23, 1972 the first Landsat vehicle was launched into orbit around the Earth (see Figure 9–1). Because of changes in the status of the Landsat program, this vehicle has been referred to as "ERTS 1," "ERTS A," or "Landsat 1." **ERTS = Earth Resources Technology Satellite.** In each case the name refers to the same vehicle and the same image products, so a word of caution is in order when studying older image materials.

The program was renamed **Landsat** in 1975 and maintains that designation today. This system of satellites was designed to follow a specific orbit that would bring it back over the same locus on the Earth's surface, thereby providing repetitive coverage to monitor change through

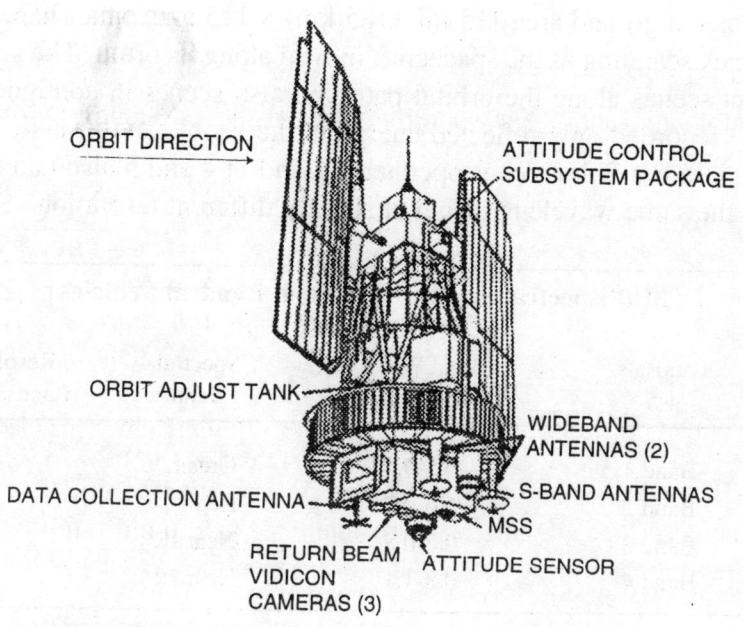

Figure 9–1. Landsat 1, 2, 3 vehicle.

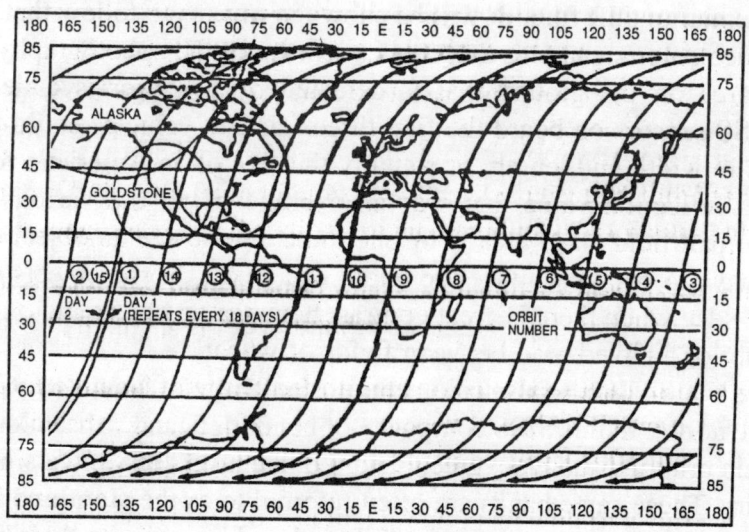

Figure 9–2. Landsat 1, 2, 3 orbits.

time. Figure 9–2 illustrates the polar orbital nature of the spacecraft's path which is almost north–south in its heading. As the Earth rotated under the path of the satellite, it would return to a specific location every eighteen days.

Landsat MSS. The Landsat program was created to provide data for scientists to monitor the Earth's resources. The primary type of imagery intended to fill that need was the product of a **multispectral scanning system (MSS).** Such a system recorded reflected and/or radiated energy from the Earth simultaneously in four contiguous bands of the EMS. This would provide multiple data sets to produce **geographically registered images,** one for each band being recorded. This would allow an interpreter to compare the energy values recorded in each band for the same place on the Earth's surface and so increase the likelihood of identification.

Each scene covered a ground area 115 mi. (185 km) × 115 mi. in the shape of a parallelogram due to the across track scanning as the spacecraft moved along its orbit. The scenes thus produced overlapped adjacent scenes along the orbital path and also scenes in contiguous orbits. Landsat vehicles 1, 2, and 3 recorded only reflected energy in the visible and near IR bands (one thermal band was on board Landsat 3, but never operated). Landsat 4 and 5 also had the MSS system on board operating in the same wavelengths but at slightly different resolution. See Figure 9–3.

Multispectral Scanner Bands on Landsat Vehicles						
Landsats 1-3	Landsats 4-5	Band Width micrometers	Spectral Band	Resolution in meters Landsat 1-3		4-5
Band 4	Band 1	0.5-0.6	Green		79 m	82 m
Band 5	Band 2	0.6-0.7	Red		79 m	82 m
Band 6	Band 3	0.7-0.8	Near IR		79 m	82 m
Band 7	Band 4	0.8-1.1	Near IR		79 m	82 m

Figure 9–3.

Note that comparing Figure 9–3 with Figures 3–1 and 7–1 and the text concerning CIR film in Chapter 7, reveals that bands 4, 5, and 7 on Landsats 1-3 and 1, 2, and 4 on Landsats 4-5 record reflected energy similar to CIR film. It will be shown in images to follow that satellite multiband imagery can produce color signatures that replicate those of CIR photography.

The smallest area on the ground that is detected by a sensor is called a **pixel** (picture element). The MSS sensors on board the first three Landsat vehicles could detect areas on the ground 79 meters in width and length, or roughly the size of an American football field. Each sensor recorded a reflection value that was an average of all the various reflectance levels present in that football field-sized area. For one to be able to see an object on a MSS image it would normally have to be about twice the width of a pixel. Some objects closer to pixel size can sometimes be seen when there is great contrast between an object and its surroundings, e.g., a bridge over water or a paved road between fields of wheat.

Figure 9–4 (see Plate VI) is a **false color composite** image of New York City and the Hudson River Valley to the north. In this MSS composite of bands 4, 5, and 7 the colors that are produced in the photographic print use the blue, green, and red dye layers to yield a similar color response to that of CIR film. These same bands are often referred to as the **standard RGB array.**

It is possible to see the interstate routes as light colored lines due to their high reflectivity and considerable width. If one follows some of the north–south roads that lead out of the peninsula containing Manhattan (C-2 area of image), the road disappears and then farther north reappears.

This is especially true of the Taconic Parkway, the visible road closest to the Hudson River. In NYC the road is wide, but to the north it passes through rugged topography and narrows to only two lanes of pavement without a breakdown lane in some places. In these areas the adjacent grass and trees are more dominant than the reflectivity of the pavement and so the road's signature is replaced by that of growing vegetation (red). Farther north the road is up to five lanes of pavement plus a breakdown lane with a narrow median strip and then another 6 lane width of pavement. The parkway is very easy to see in this location. This example illustrates the importance of pixel size as a limiting factor in interpretation.

Figure 9–4 is an example of **Digital Precision Processing** by **IBM** Corp. This image was created by photographically exposing film positives of bands 4, 5, and 7 onto color film through blue, green, and red filters respectively. See Figures 9–5, 9–6, and 9–7 for black and white images of those three bands. IBM precision processing of Landsat MSS images in the 1970s produced a much broader range of color tones and chroma than the standard processing by the then government-run **EROS Data Center** in Sioux Falls, SD. The process was eventually sold to the U.S. Government to use in custom image production.

Different band combinations with varying color filters will produce other false color results. This point will be discussed further in Chapter 13. The false color rendition in Figure 9–4 is a result that replicates the type of reflectance information that is present in color infrared photography. Because many scientists around the world had been using and interpreting CIR photographic products, this type of false color satellite image became and has remained the most popular.

Return Beam Vidicon. Also on board Landsats 1, 2, and 3 was another remote sensing system called **Return Beam Vidicon (RBV).** This imaging system was very similar to that of a video camera. It was used briefly on Landsat 1 as a multispectral system, was not activated on Landsat 2, and on Landsat 3 was used as a higher resolution panchromatic single band system with a 30 meter ground resolution. This latter RBV recorded about the same levels of reflectance as Bands 4, 5, and 6 in the MSS system. Reference to Figure 3–1 will assist in placing all of these band locations in perspective.

Figure 9–8 is an RBV image of the New York City area that was acquired on 04/06/80. The ground resolution for RBV images was 30 meters and so was an improvement over the 79 meter resolution of the MSS bands. Because the RBV images provided more detail of the Earth's surface, they were sometimes merged with MSS products and the result yielded multiband reflectance information with added detail. This concept will be pursued in Chapter 13.

Landsat TM. In 1982 Landsat 4 (also sometimes referred to as Landsat D in the literature) was placed in orbit by NASA with a new Earth resource mapping system on board, **Thematic Mapper (TM).** This was an imaging system that was intended to operate alongside the MSS on Landsat 4 and later on Landsat 5. (See Figure 9–9.)

The TM system offered increased ground resolution, a greater number of spectral bands, and faster return time to image a particular Earth surface location than Landsats 1-3, 16 days vs. 18. The spectral bands were modified so that identification of soil, rock, and vegetation types could be improved. (See Figure 9–10.) Ground resolution was improved to 30 meter pixel width.

The orbital path of Landsats 4 and 5 is similar to that of the first three Landsat vehicles. The altitude of the satellites averages 438 miles (705 km) above the ground which is less than that of earlier Landsats, but the angle of view of Landsats 4 and 5 has been adjusted so that a full scene covers the same path on the Earth, 115 miles (185 km) wide.

Figure 9–11 (see Plate VII) is a subscene of the Charleston, SC area that is a small part of a full scene that was acquired by the Landsat 4 Thematic Mapper on November 9, 1982. This image was selected for this manual because it is the same area shown in an EOSAT foldout poster which has been widely disseminated and has good educational value. **Earth Observation Satellite Company** is the quasi-public organization that distributes satellite images and data for the Federal Government; see Appendix E for address. For those needing a display of the basic information concerning TM or for printed examples of some false color combinations, the EOSAT poster entitled **"Landsat Thematic Mapper Imagery"** is a useful piece and is available free.

The false color image displayed in Figure 9–11 is the result of photographically combining the reflectance values of bands 2, 3, and 4 using blue, green, and red filters as was the case with the MSS imagery. This is the most commonly produced false color image as it replicates the type of information provided by color infrared photography. Figures 9–12 through 9–18 are

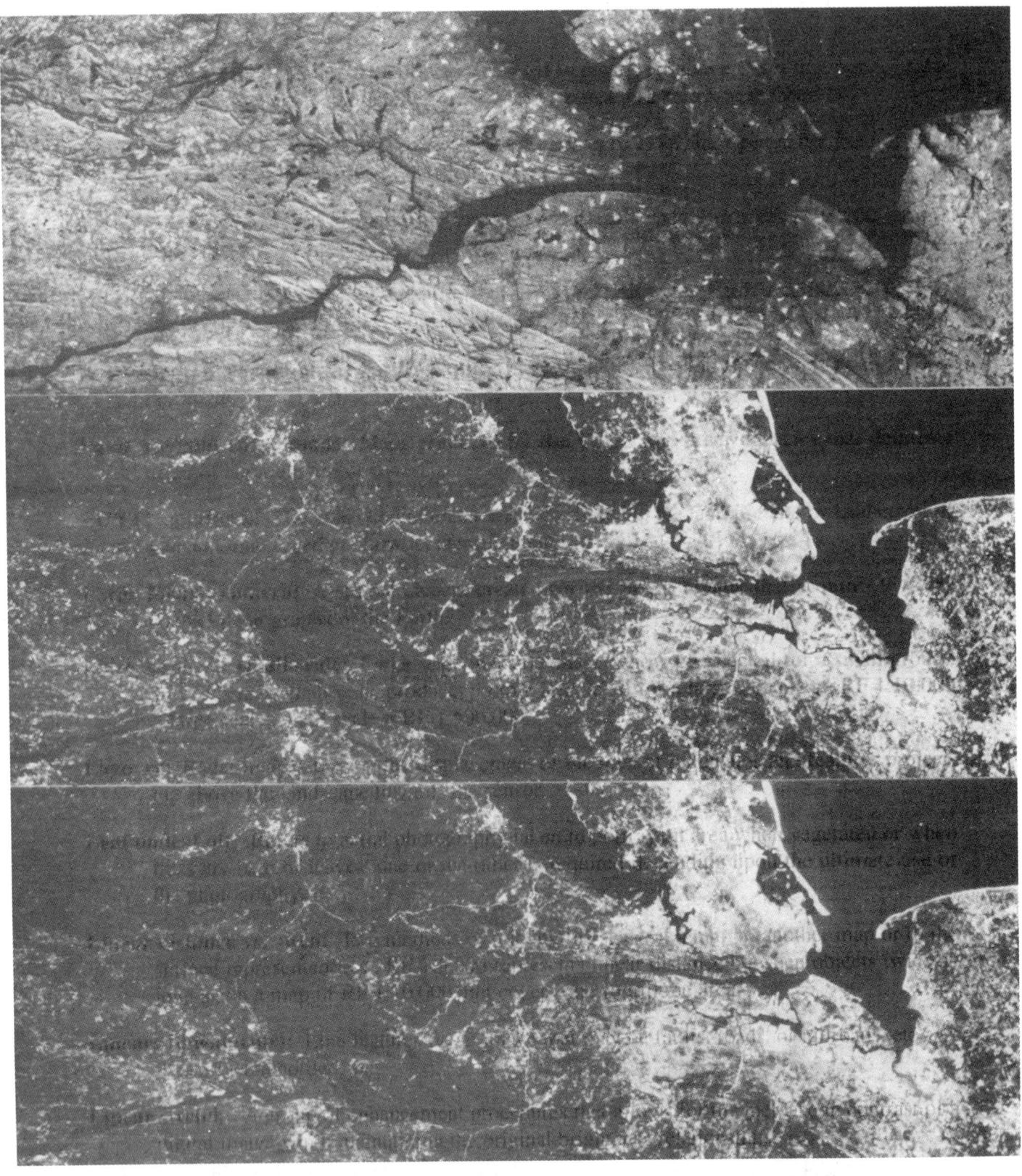

Figures 9–5, 9–6, and **9–7.** NASA ERTS1 Bands 4,5,7. Scenes 1079-15124,15131, and 15133. IBM Digital Precision Processing of Mosaic. Imaged 10/10/72. Courtesy IBM Corp.

black and white reproductions of the seven bands recorded by the Thematic Mapper. Note that Figure 9–17 is a coarser appearing image than the other bands. This is band 6 which is the thermal image and it is characterized by 120 meter resolution whereas all the other TM bands display 30 meter resolution.

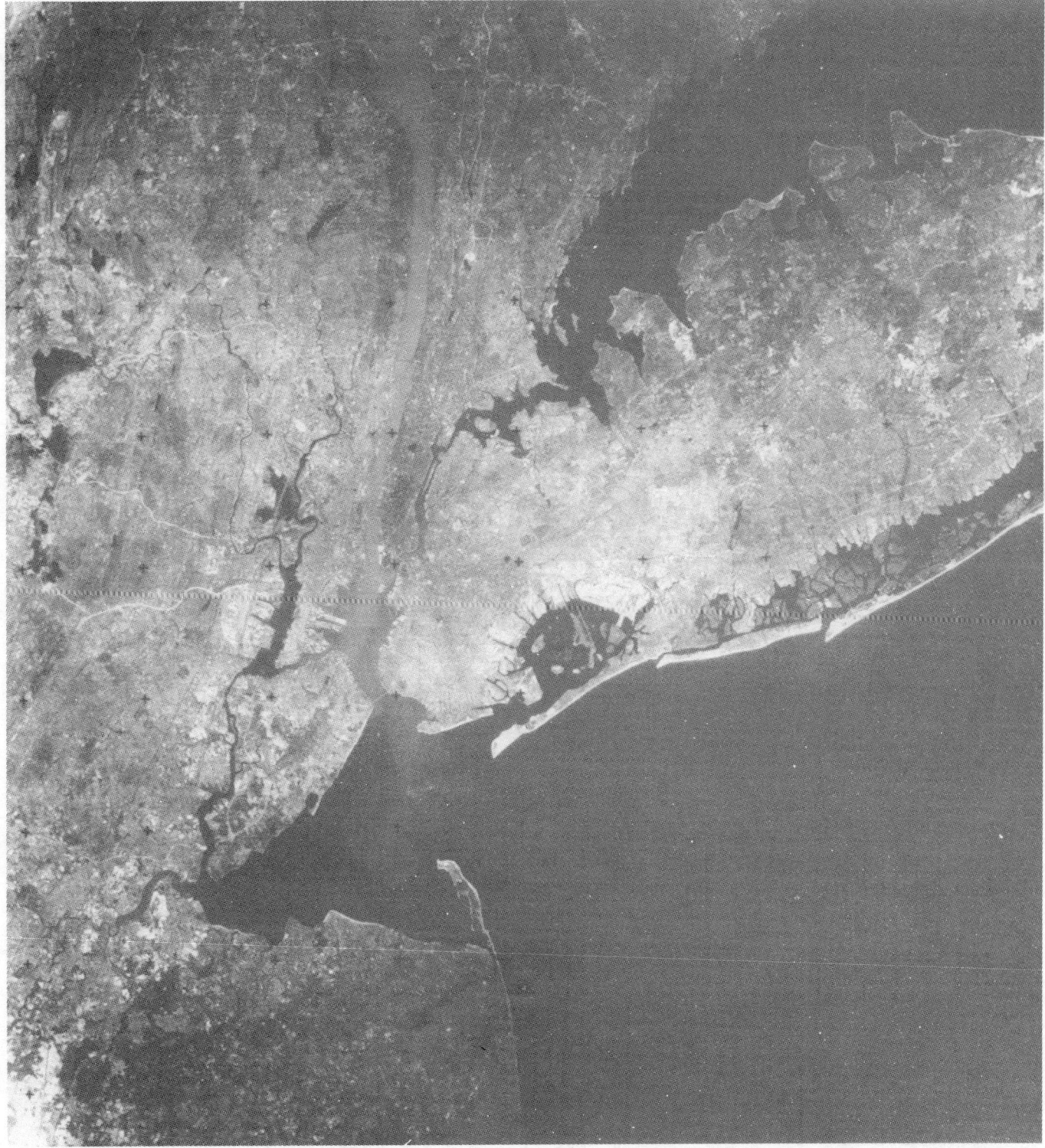

Figure 9–8. NASA Landsat RBV subscene, NYC area image ID E-30763-14500-A. Acquired 4/6/80.

THE SPOT PROGRAM

In 1977 the French government, in cooperation with Belgium and Sweden, initiated the **SPOT (Systeme Pour l'Observation de la Terre)** remote sensing program. SPOT 1 was launched on February 21 in 1986 and began commercial operation on May 6th of that year. It got off to a well publicized start during the testing period when on May 1st it acquired an image of the Chernobyl nuclear power plant in the Soviet Union shortly after the disastrous accident.

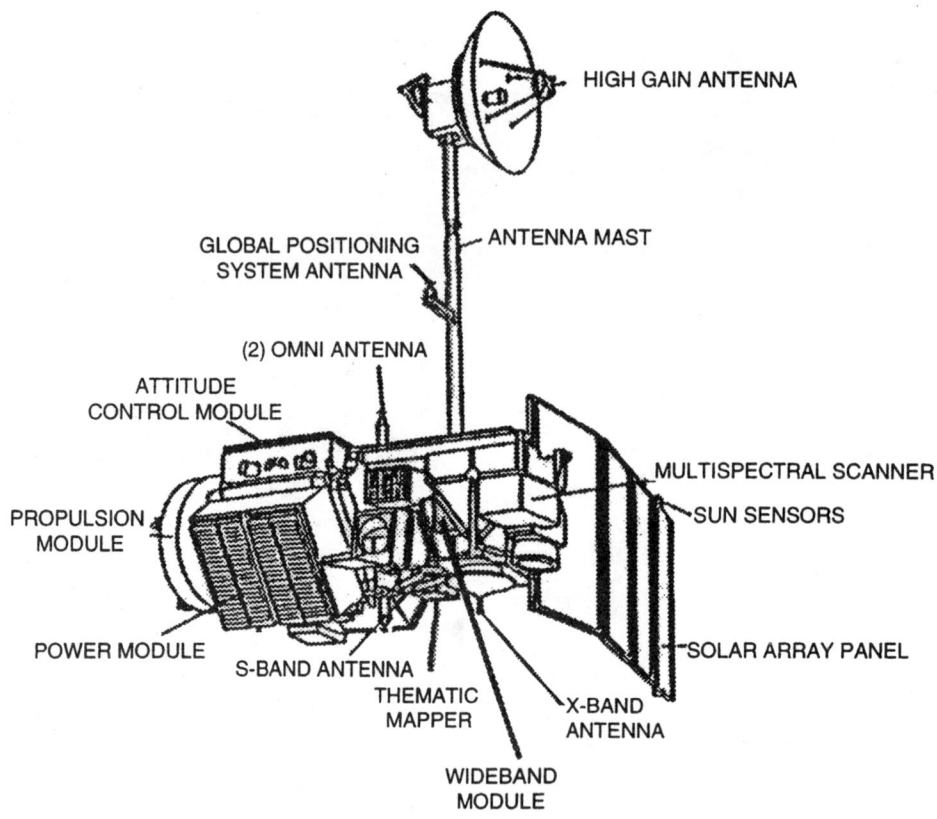

Figure 9–9. Landsat 4, 5 vehicle.

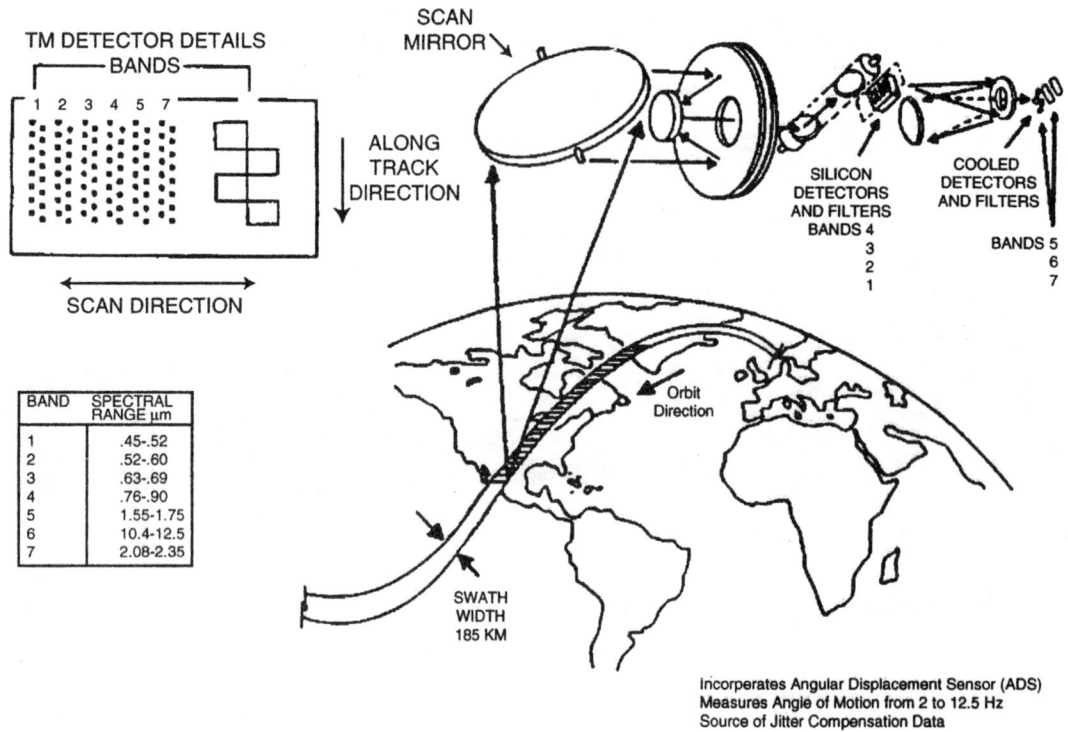

BAND	SPECTRAL RANGE µm
1	.45-.52
2	.52-.60
3	.63-.69
4	.76-.90
5	1.55-1.75
6	10.4-12.5
7	2.08-2.35

Incorporates Angular Displacement Sensor (ADS)
Measures Angle of Motion from 2 to 12.5 Hz
Source of Jitter Compensation Data

Figure 9–10. Thematic mapper system.

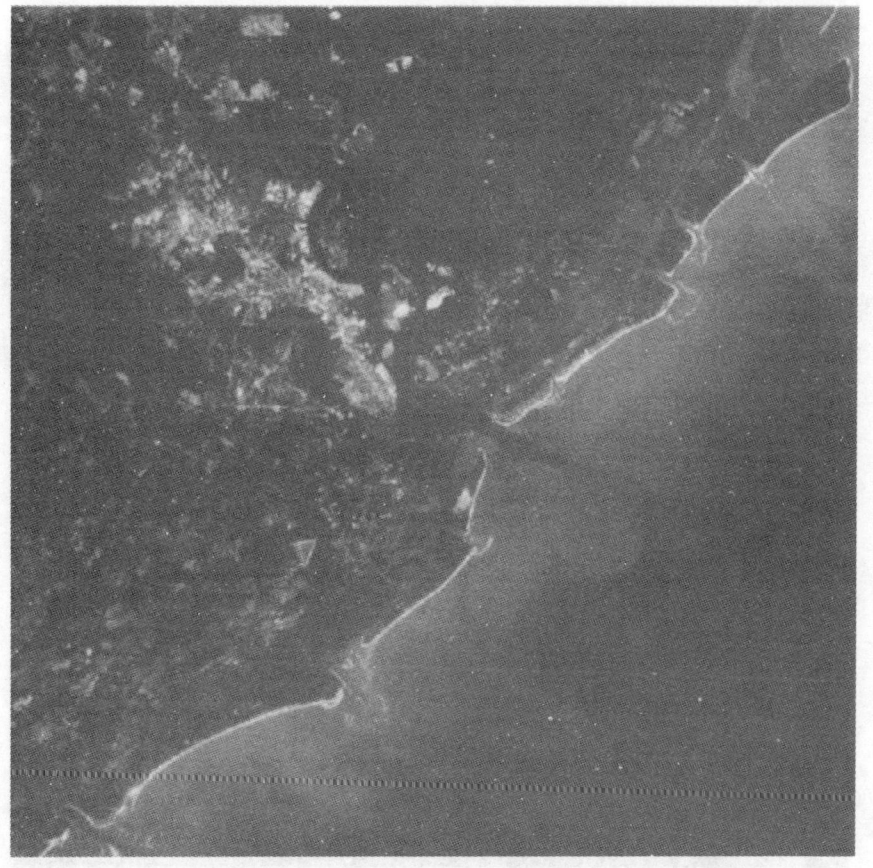

Figure 9–12. TM Band 1
Charleston, SC.
Courtesy NASA/GSFC.

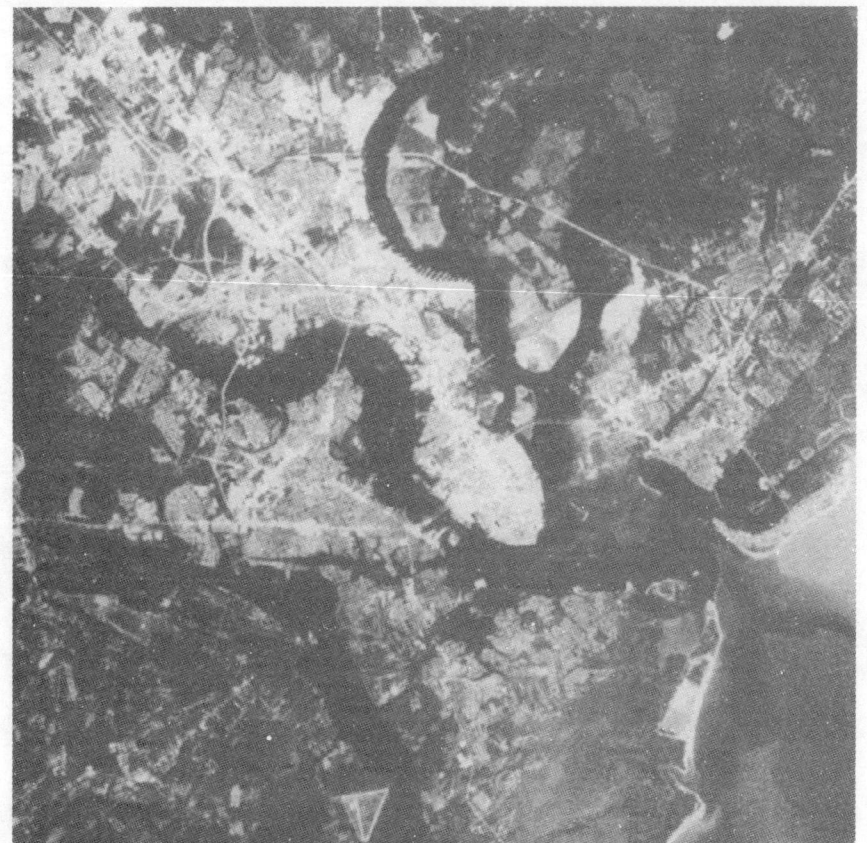

Figure 9–13. TM Band 2
Charleston, SC.
Courtesy NASA/GSFC.

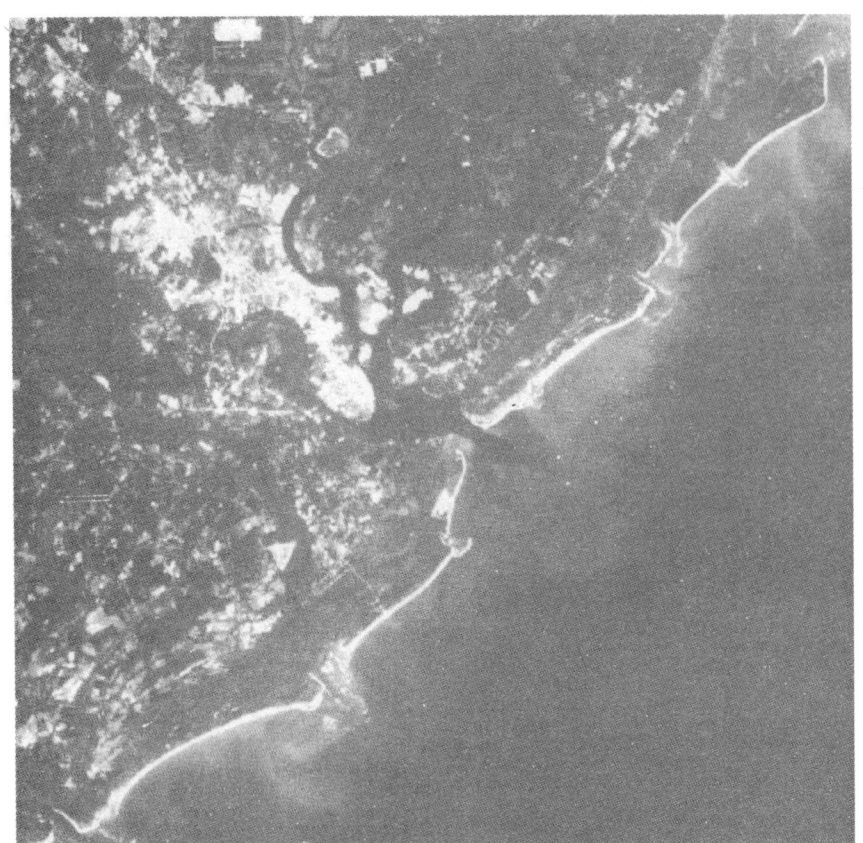

Figure 9–14. TM Band 3
Charleston, SC.
Courtesy NASA/GSFC.

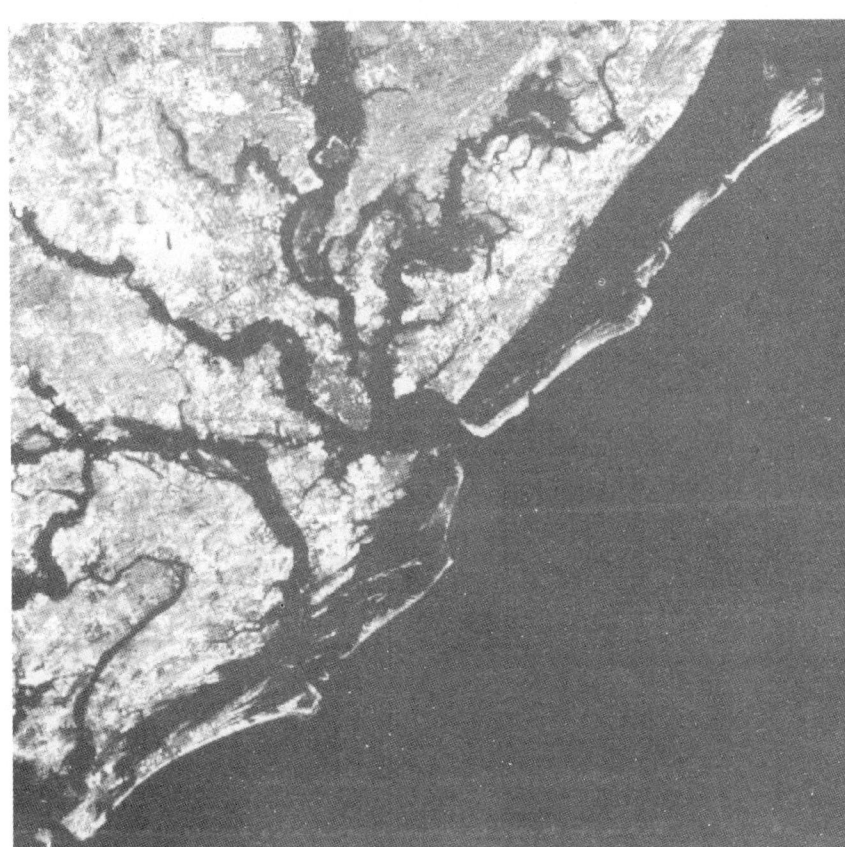

Figure 9–15. TM Band 4
Charleston, SC.
Courtesy NASA/GSFC.

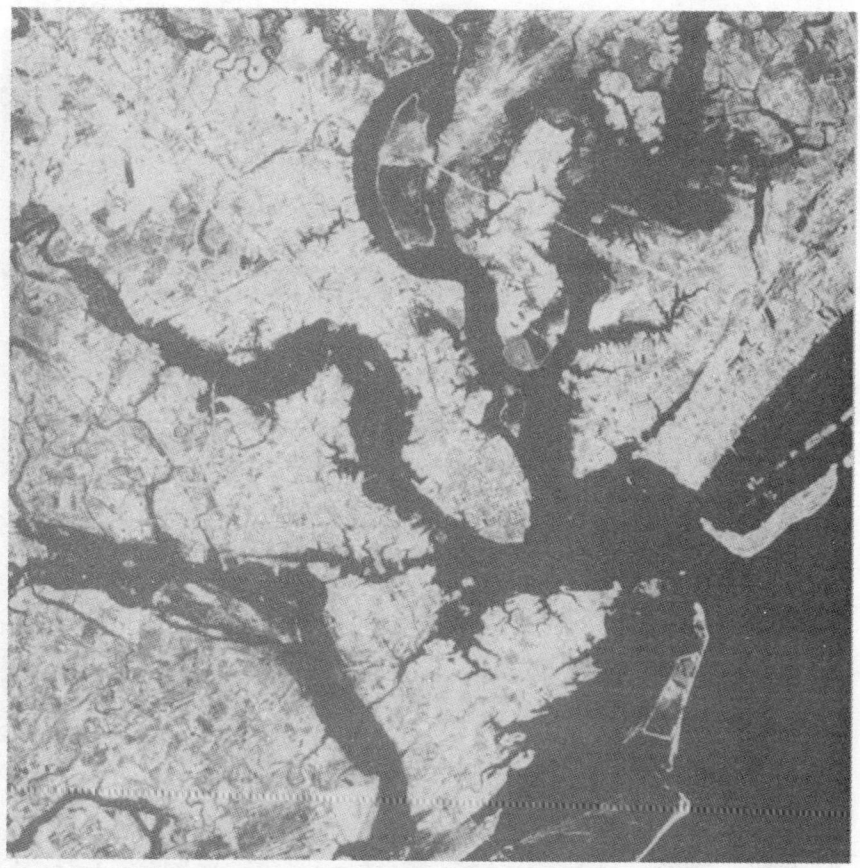

Figure 9–16. TM Band 5
Charleston, SC.
Courtesy NASA/GSFC.

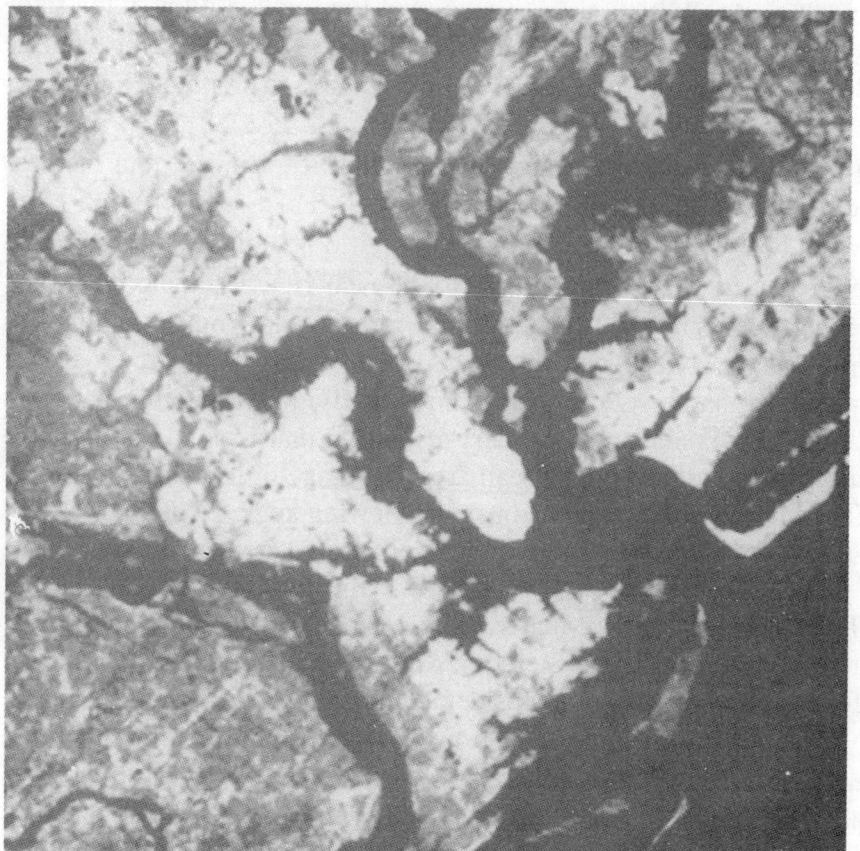

Figure 9–17. TM Band 6
Charleston, SC.

Chapter 9   Satellite Mapping of Earth Resources   **87**

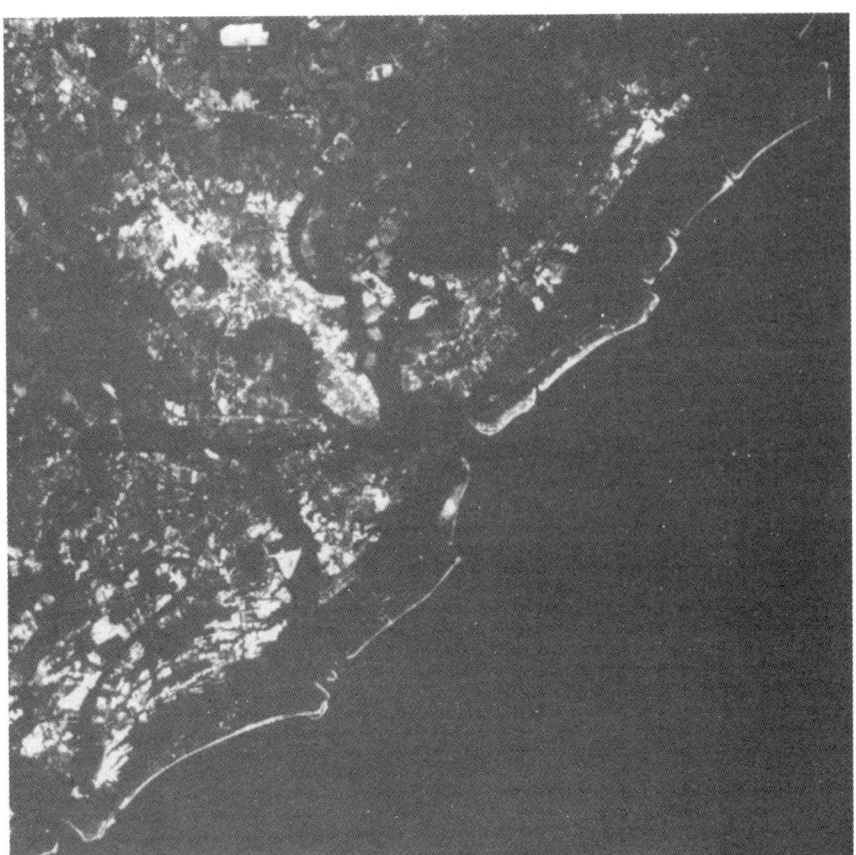

Figure 9–18. TM Band 7 Charleston, SC. Courtesy NASA/GSFC.

The SPOT satellites follow an orbital path around the Earth in a fashion similar to the Landsat vehicles, i. e., near-polar, **sunsynchronous** (passing over the same latitude at the same local time), and crossing the Equator at 10:30 AM local solar time on each orbit. The satellite maintains an average altitude of 516 miles (830 km) and requires 26 days to complete Earth coverage and come back above the same Earth location.

SPOT images have higher resolution characteristics than do those of Landsat, but the area covered in one scene is much smaller and the band coverage is less. There are two **high resolution visible (HRV)** sensors on board, one being panchromatic (.5 to .73 micrometers) and the other multispectral (.5 to .59 green, .61 to .68 red, and .79 to .89 near IR). Utilizing the green, red, and near IR bands with blue, green, and red filters produces a false color composite that displays the CIR color signatures discussed earlier.

The panchromatic sensor has a ground resolution (pixel size) of 10 m × 10 m while the multispectral pixel is 20 m × 20 m. This is currently the best resolution that is available to those without security clearance in imagery that provides broad areal coverage of the Earth's surface.

Another characteristic of the SPOT satellites is an adjustable viewing angle 27 degrees east or west of the orbital path. This allows for stereopairs of imagery to be created on successive orbital passes over an area. This type of product has been produced using Landsat imagery, but the procedure with the fixed, near-vertical view of that system is expensive and time consuming and so has been carried out relatively few times. Stereoviewing is especially useful for larger scale SPOT subscenes and for images of areas with considerable relief.

Figure 9–19 is a SPOT 1 panchromatic full scene 60 km x 60 km (37 x 37 miles) of the Boston, MA metropolitan area acquired on March 23, 1988. Boston Harbor, Logan International Airport, and the central city area are all visible in the bottom portion of the A-2,

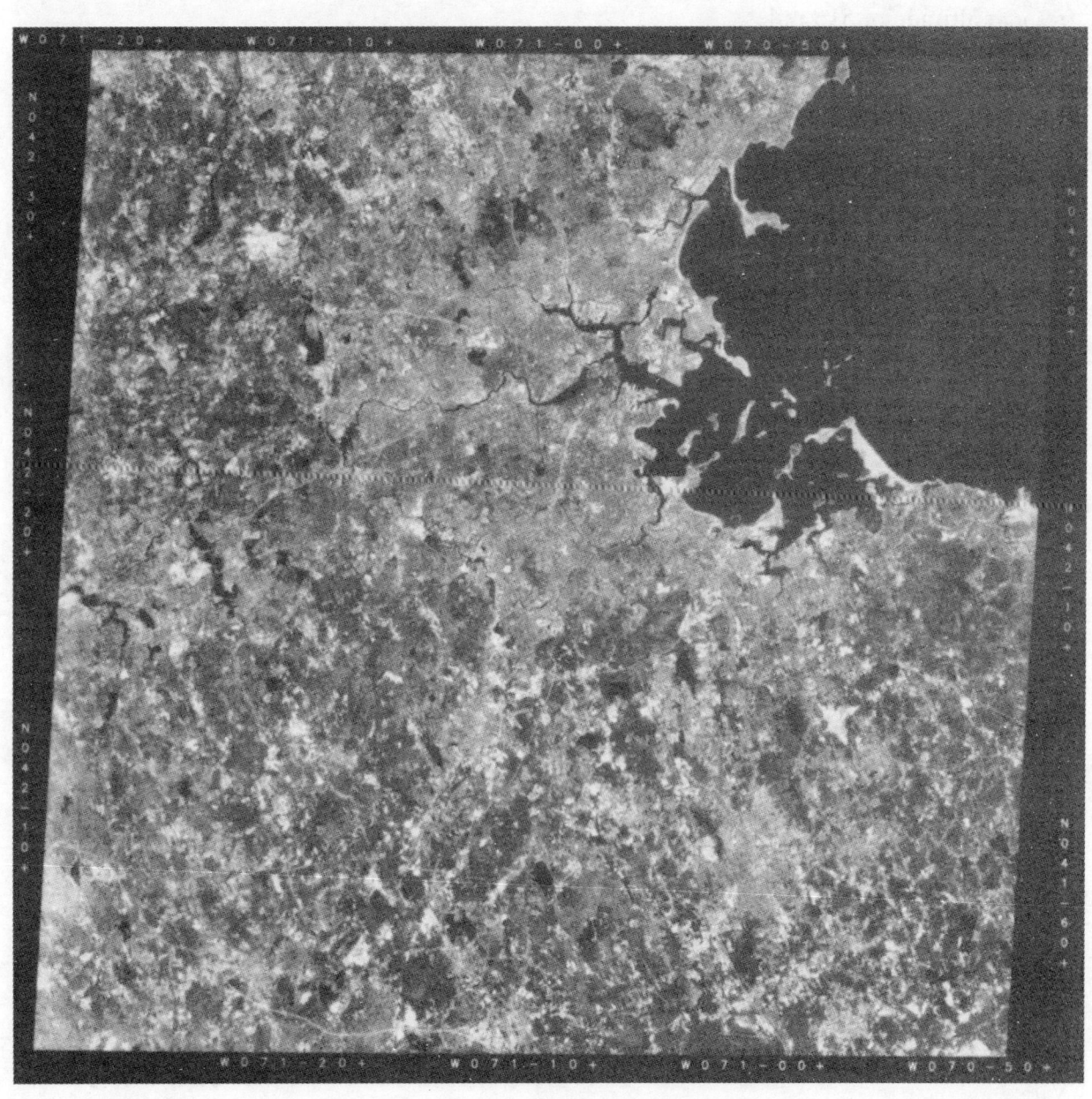

Figure 9–19. SPOT 1 panchromatic band. Boston, MA, 632-634 acquired 3/23/88. C 1993 CNES Provided by SPOT Image Corp.

A-3 section of the image. Compare this image with Figure 9–20 (see Plate VIII) which is a photographic enlargement of a portion of a multispectral SPOT image obtained on November 7, 1988 that covers the inner harbor, Logan Airport, and Boston's central area.

The former scene would allow for considerable enlargement and increased detail from the scale shown here of RF 1:400,000 (1″ = 6.3 miles). The original full scene of Figure 9–20 covered the same size area as 9–19 and this part of it has been enlarged photographically to RF 1:72,000 (1″ = 1.14 miles). Note the considerable detail present in this image scanned from an altitude of 516 miles.

Figure 9–21 is a black and white version of a SPOT multispectral image of the Los Angeles basin and the part of the coast that includes the Long Beach Naval Base has been enlarged approximately 450 percent in Figure 9–22. The detail in the enlargement is quite impressive.

SPACE TRANSPORTATION SYSTEM (STS) PHOTOGRAPHY

The more commonly used name of the **STS** is the **Space Shuttle.** There have been many Shuttle missions over the years and taken as a whole they have been characterized by diversity in both mission objectives and by the hardware carried into space. The focus here is on photographs acquired on various Shuttle missions.

Large Format Camera (LFC). The LFC was conceived as a concept as long ago as 1965 by a group working within NASA on the **Apollo** program. A recommendation was made to include a large format terrain mapping camera along with a pair of cameras to be used for establishing the system's location in space and the exact direction of the Earth nadir for the LFC. The concept came to fruition in 1977 when NASA provided funding to construct the system, not for the Apollo program but for the Space Shuttle. All of the cameras were built by Itek Optical Systems, a division of Litton Industries, Lexington, MA.

The LFC was delivered in 1980, and the reference cameras in 1983. The system was created as a "precision, wide-angle, cartographic instrument capable of producing high resolution stereo photography of great fidelity" as stated in a press release by Litton. The focal length of the LFC is 30 cm and the photographs are 23 × 46 cm (9 × 18 inches). Photographs are acquired from an average altitude of 225 km (140 miles) and are taken with overlap of ground detail to provide stereo coverage. Overlap was increased in mountainous terrain and decreased in flatter areas to make adjustments for changing parallax.

The outside cover of this manual is almost all of frame number 0664 that was taken at 12:56 EST on October 7, 1984 from an altitude of 237 km (147 miles) on Shuttle Mission STS-41G. The area covered was 178 × 356 km or 63,368 square km (111 × 221 miles or 24,500+ square miles) and photo scale was RF 1:778,000 (1″ = 12.3 miles on the original photo).

The photo extends from Cape Cod, MA, Martha's Vineyard and Nantucket Islands, to Boston, Providence, and Worcester cities, and westward to Quabbin Reservoir and the Connecticut River Valley.

The inside manual cover photograph was a 15.6× enlargement of the Boston and Logan Airport area from the outside cover. The scale of this photo was RF 1:50,000 when published as the cover of the *Photogrammetric Engineering & Remote Sensing* journal of the American Society of Photogrammetry (name changed since then) in February, 1985. The quality of these two photographs attests to the success of the system construction and operation on STS-41G.

Shuttle hand-held photography. Many thousands of photographs have been taken from space from **Mercury, Gemini, Apollo, Skylab,** and space shuttle vehicles. A large number of

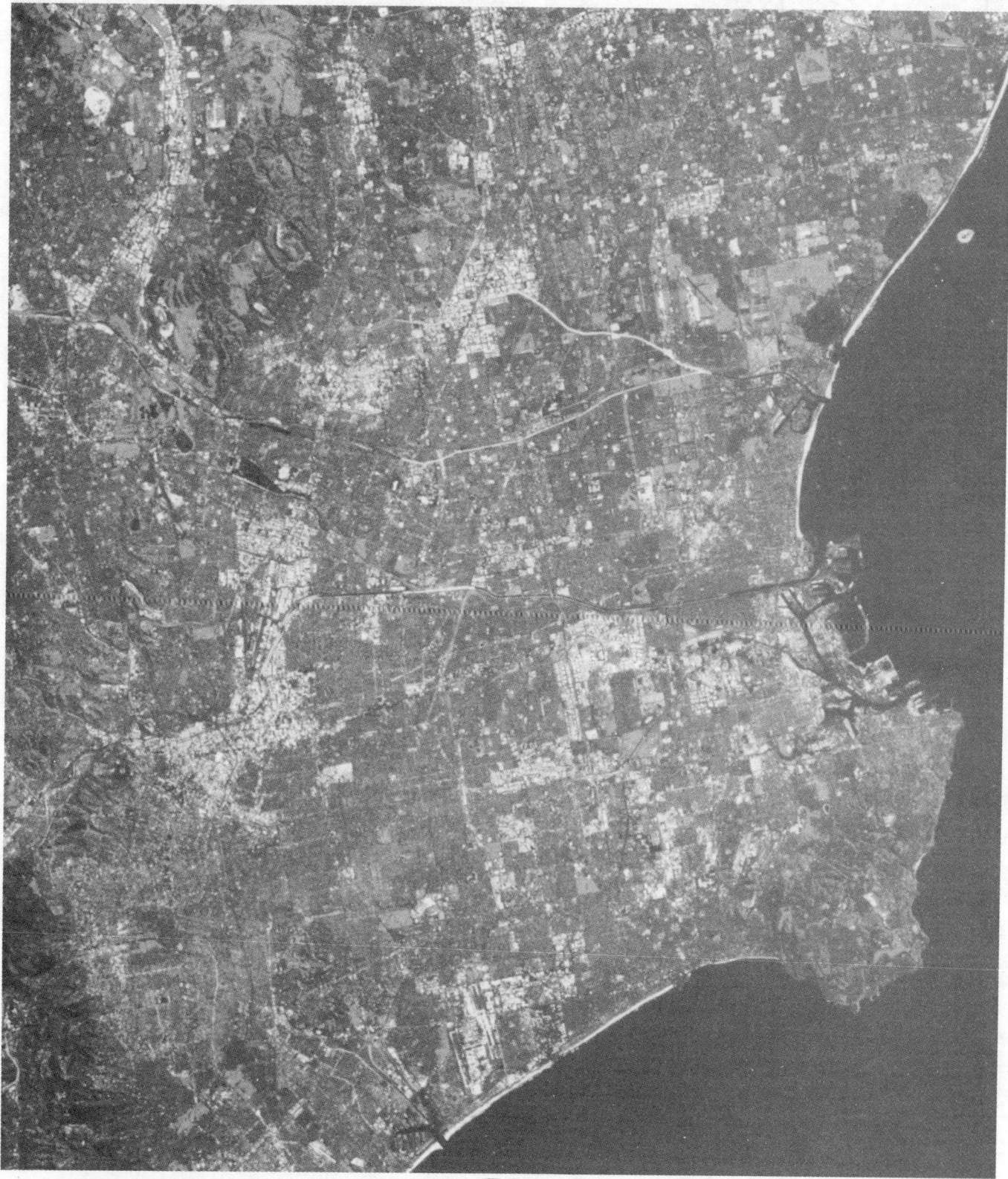

Figure 9–21. SPOT multispectral image (B&W here) Los Angeles, CA, acquired circa 1990. C 1993 CNES Provided by SPOT Image Corp.

these were acquired by astronauts using hand-held cameras. The themes or primary objectives of the various missions covered a wide range of topics that impacted upon what types of photographs were taken. Suffice to say that the photographs archived at EROS Data Center (EDC), Johnson Space Center (JSC), Goddard Space Flight Center (GSFC), and elsewhere (see Appendix E) represent a rich resource for any person interested in the Earth and its environment.

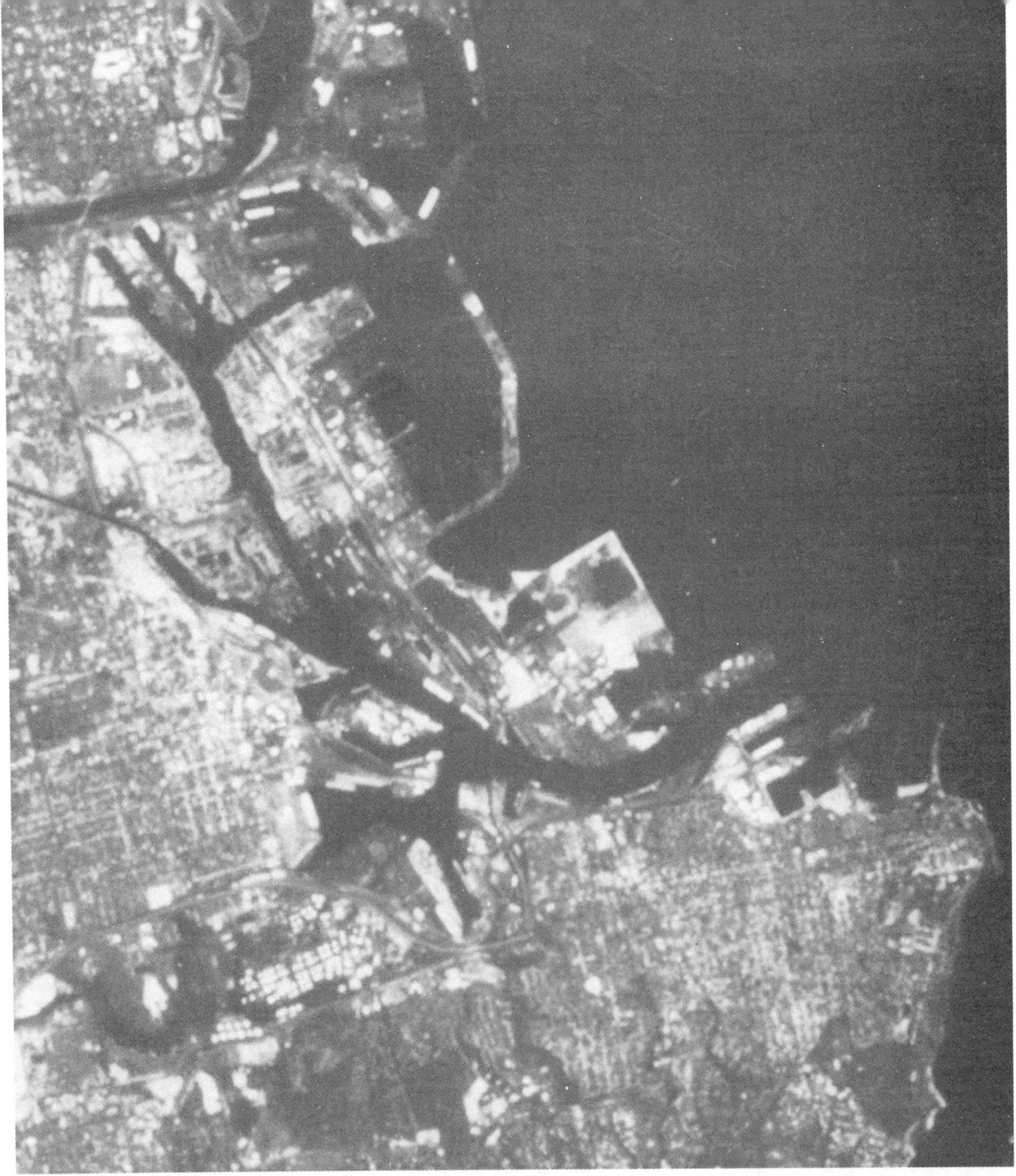

Figure 9–22. 450% enlargement of Figure 9–21. Longbeach area C 1993 CNES original Provided by SPOT Image Corporation.

Photographs portraying meteorological phenomena, hydrologic features, geomorphic landforms, and the patterns of human habitation on the Earth are available in wide variety. Many photographs were acquired that dealt with natural or man-made disasters, e.g. volcanic eruptions of Mt. St. Helens and Mt. Pinatubo, and the destruction and burning of the Kuwaiti oil fields by the Iraqi army.

Two hand-held photographs acquired from Space Shuttle missions have been selected to illustrate some of the diversity and wealth of information that can be obtained. Figure 9–23, in black and white rather than the original color, is a **low oblique** view (close to but not vertical) of the

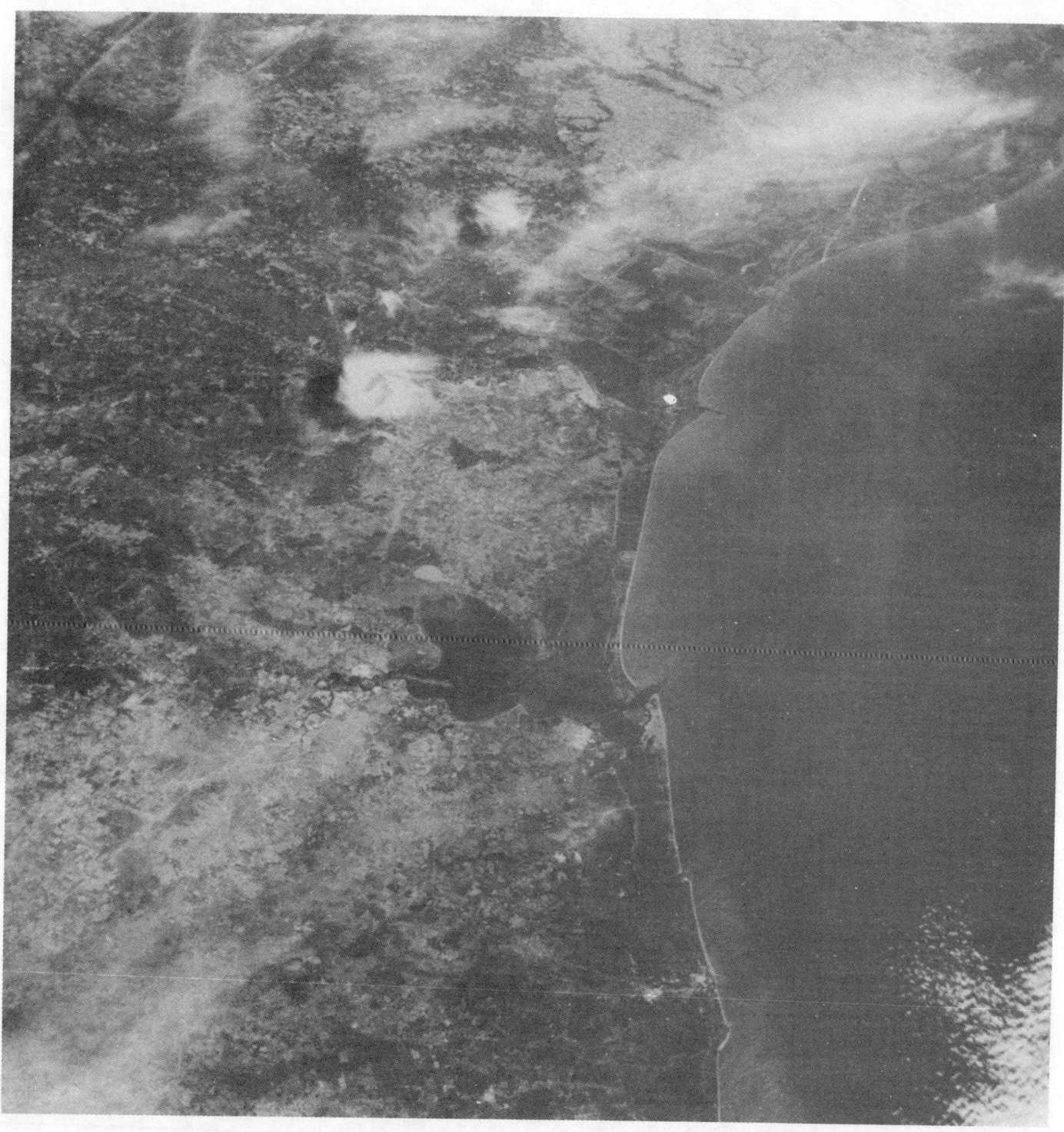

Figure 9–23. NASA handheld shuttle photo. Texas-Louisiana coast. ID #51C-143-0013.
Courtesy Johnson Space Center.

Texas–Louisiana Gulf coast as it appeared from Space Shuttle mission STS 51C. Figure 9–24 is a black and white version of a color photo that is vertical or very nearly so, acquired over Houston, TX from Space Shuttle mission STS 61A.

Figure 9–23 covers the area of the Gulf of Mexico coastline from Freeport, TX to Atchafalaya Bay in Louisiana. Some of the main features to be noted: two convectional cells in the A-2 area

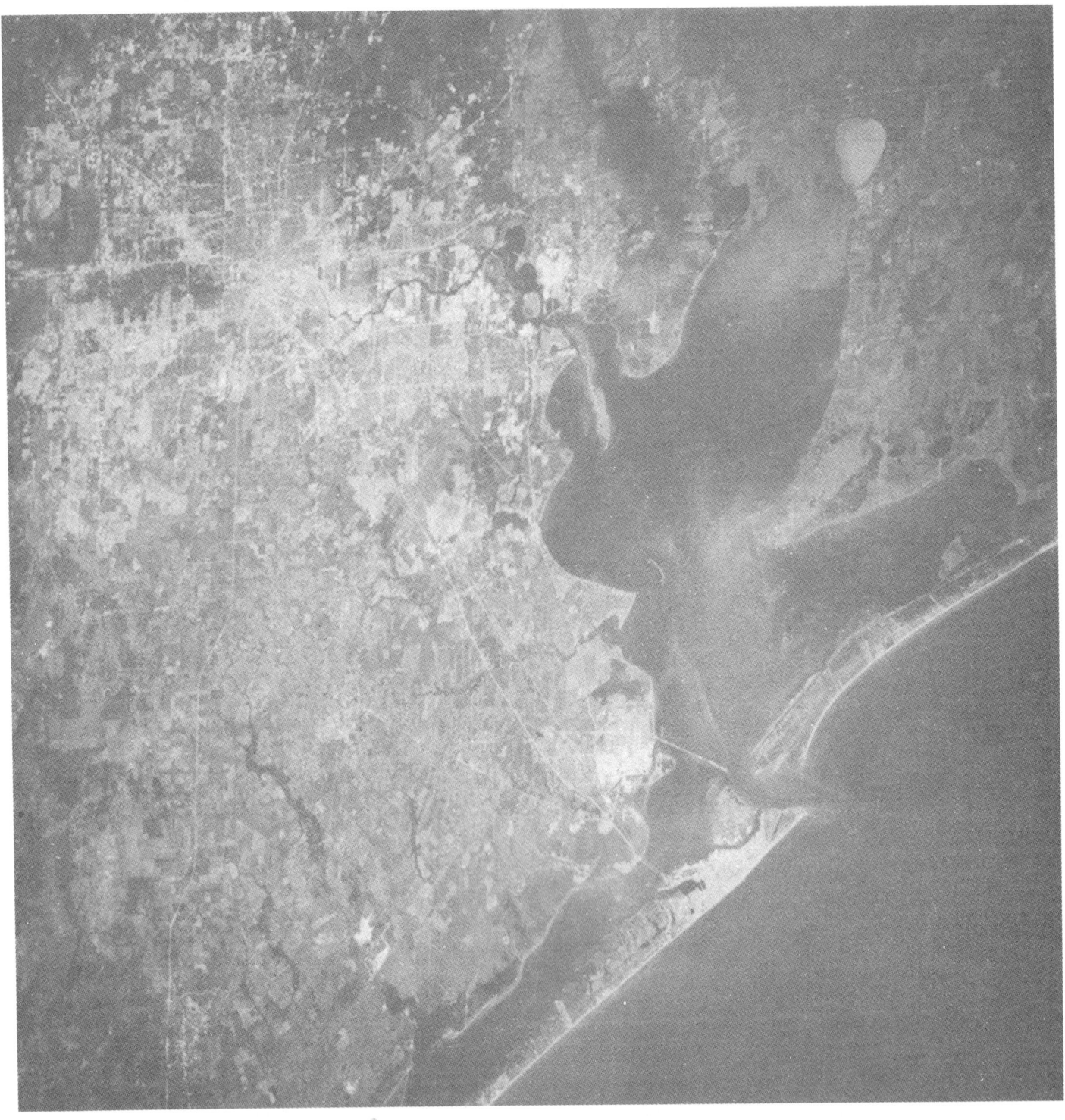

Figure 9–24. NASA handheld shuttle photo. Houston, TX. ID #61A-042-0046. Courtesy Johnson Space Center.

above the center, barrier islands and sand spits along the coast, the city of Galveston at the upper end (east) of the barrier island at the entrance to Galveston Bay (largest bay), and the city of Houston below and left of that bay (northwest).

Figure 9–24 is centered on the west side of Galveston Bay so that Galveston is prominent in C-3 area of photo and Houston covers most of the A-1 area. Obviously the scale of this photograph is larger than that of Figure 9–23 and the city features are better defined.

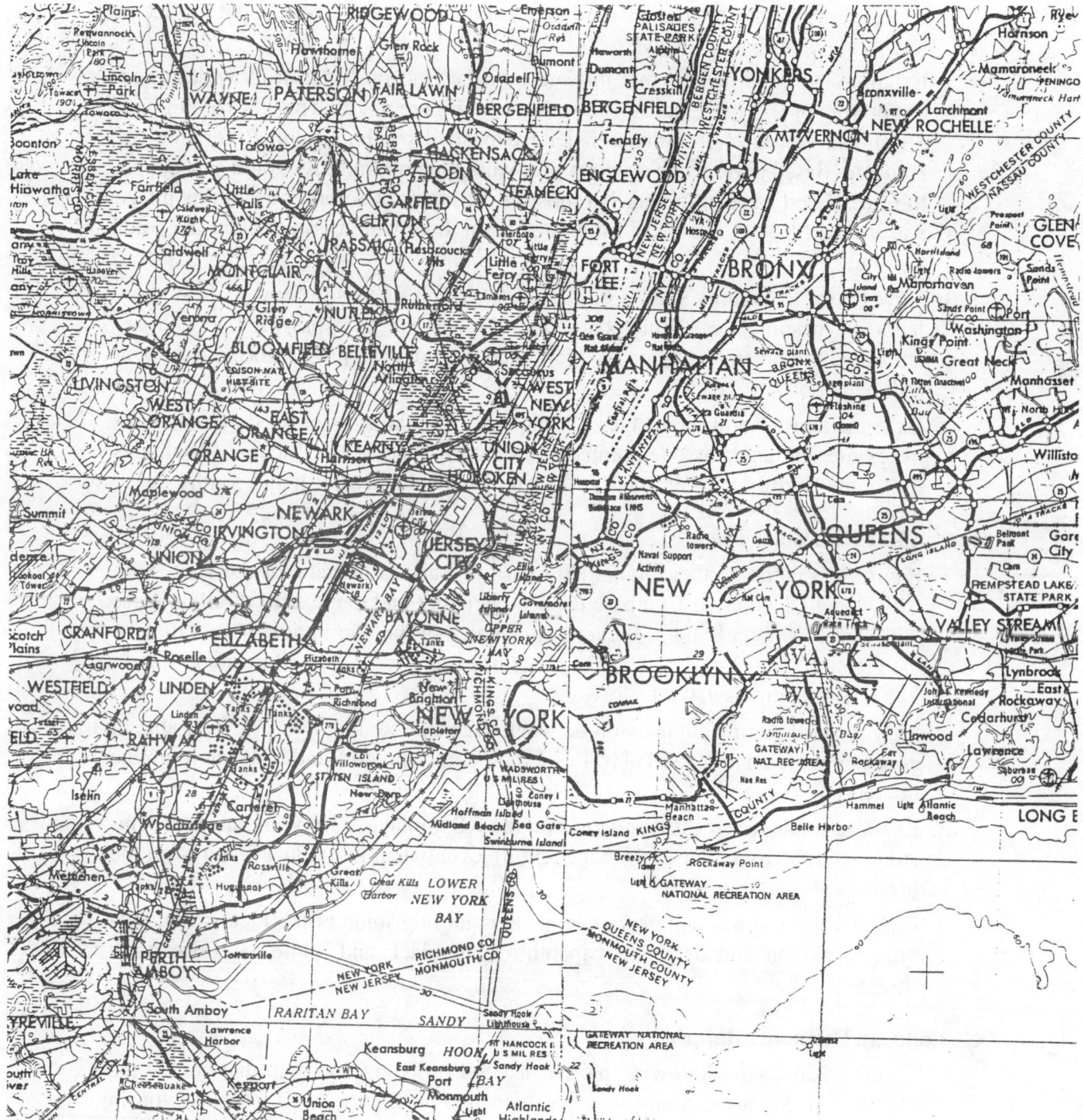

Figure 9–25. USGS topographic map, NYC area.

Exercise 9   Satellite Mapping of Earth Resources

A. **Landsat MSS imagery.**

Before answering the following questions, study Figure 9–25, part of a USGS Topographic map of the New York City area.

1. Study Figure 9–4, the false color composite of bands 4, 5, and 7 (CIR replication) IBM precision processed image of the NYC area. Note that urbanized areas have a color signature ranging from light blue to dark gray. As one moves from a heav-

ily developed urban area to a suburban location, what would you expect to see in regard to color change? Why?

2. Examine the point of land that juts out into the Hudson River at its widest area. (At bottom of B-3 in Figure 9–4; not shown in Figure 9–25) This is Croton Point, a remnant of a glacial meltwater delta deposited by the Croton River into the Hudson. Based upon its color signature, what type of land use would you expect to find? If you were informed that most of the area of Croton Point in this image was a land fill with relatively little vegetation, how would you explain the color signature?

3. Now return to Figures 9–5, 9–6, and 9–7. Which of these seems to best show land–water boundaries? Which band does the best job in displaying the "culture" attributed to human habitation? Which band provides the most information about sedimentation?

4. Examine all four of these images. What factors might account for the differences in texture and color as one compares the appearance of the Taconic Mountains (cross the Hudson R. SW to NE) with areas immediately north and south?

B. **Landsat RBV subscene.**

1. Figure 9–8 is the northeast quadrant of a full Landsat 3 scene of the New York City–Long Island area. Compare the road network and the texture of urban areas with that shown on Band 5, Figure 9–6. Which of the two images seems to have the greatest amount of detail? If the pixel size in Figure 9–8 is 30 m and in Figure 9–6 is 79 m, how is that a factor?

2. Now make the same comparison between Figures 9–8 and 9–4. Are you able to ascertain more on the false color image than you could on Band 5? What might be the reason?

3. Examine a number of different areas on the RBV image. Can you draw a conclusion about the importance of image texture even when individual features on the ground are hard to see?

4. Can you make some summary statements regarding some of the factors that provide resolution and detection capabilities on black and white vs. false color images?

C. **Landsat Thematic Mapper.**

Figure 9–26 on the following page is a USGS topographic map that covers the Charleston, SC area. Study this map and review Figure 9–10 to assist you in answering the following questions. Also examine closely the black and white images of TM bands 1-7 (Figures 9–12 to 9–18) for specific types of surface materials that are displayed band by band.

1. Examine Figure 9–11 (see Plate VII), a false color composite image using bands 2, 3, and 4 with blue, green, and red filters to produce simulated CIR photographic signatures. This is a subscene of a TM image of the Charleston, SC area. If this image was acquired on November 9, 1982, why is there so much red color indicating growing vegetation? Are there areas on this image that you would expect to show red shades if this were a June acquisition? Explain why the red signature is not present in November in those areas.

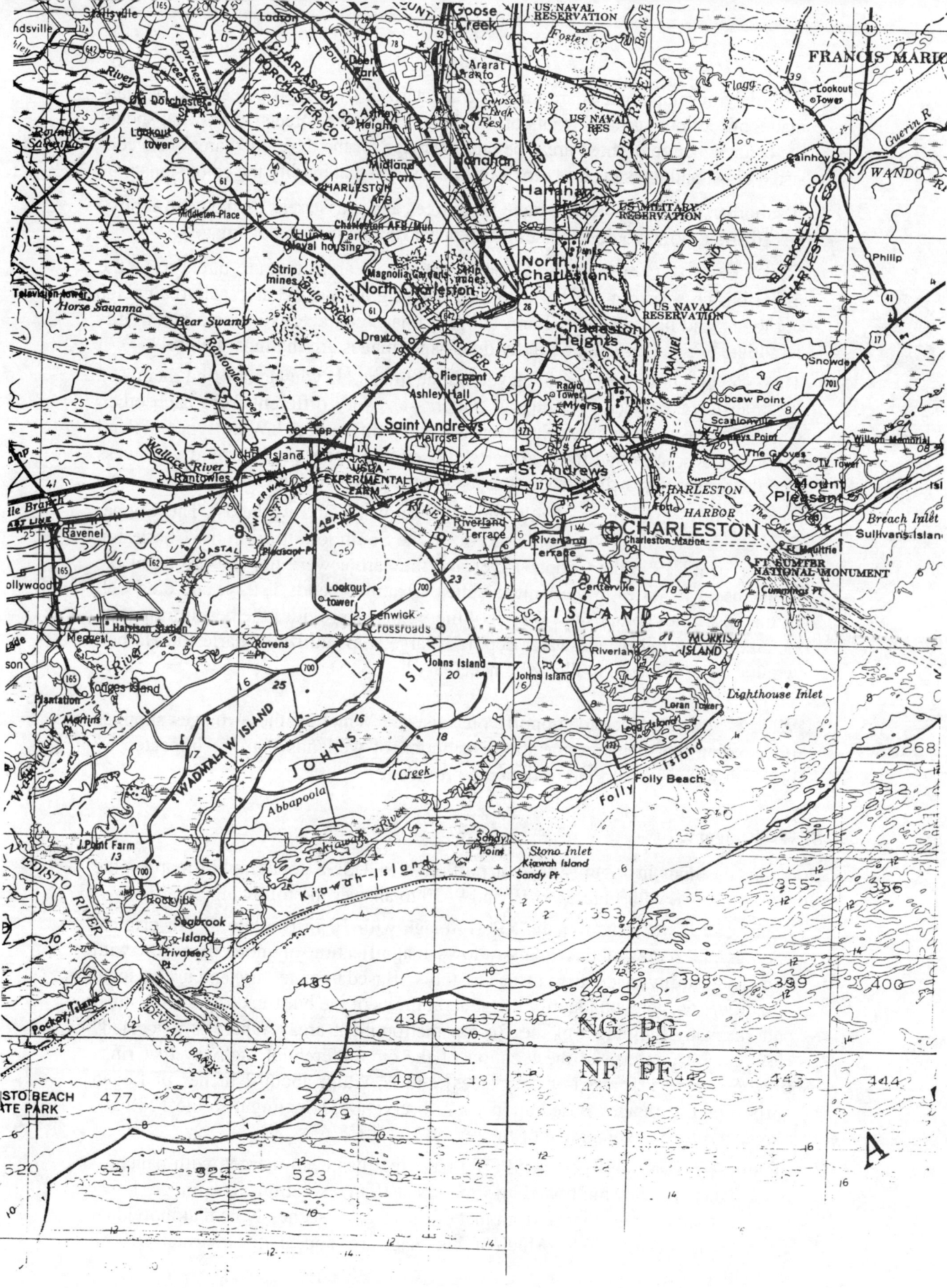

Figure 9–26. USGS topographic map, Charleston, SC.

2. If heavily developed areas are largely characterized by a high percentage of man-made, impervious surfaces (roofs of buildings, sidewalks, streets, etc.), what should the color signature of those areas be on this image? If the developed areas also consisted of tall buildings in close proximity, how might that modify the color signature?

3. If suburban areas are likely to have more trees and grass areas than a central business district, how would such suburbs appear on this false color image?

4. Locate by A, B, C method some good examples of areas that are likely to be the major business district of Charleston, an outlying business center, suburban neighborhoods, and new residential areas. What image characteristics help you to differentiate these areas? Which individual band images are the most helpful?

5. There are two large airports displayed on the image. If you cannot find them, use the map for assistance. The layout patterns are very different from each other. Can you give a possible explanation for these differences? What types of land use/land cover surround these airports? Based upon your study would you guess that the airports or their surrounding uses were present on the land first?

6. The city of Charleston is greatly impacted by water and wetlands. The Cooper River to the North and the Ashley River to the South join to form the Charleston Harbor. Until 1994 the Cooper River and the harbor were the site of one of the major naval facilities in the United States. Examine bands 1, 2, 4, and 6 as well as the false color composite. Which of these images provides the best information on siltation, turbidity, and water depth? If large ships utilized Charleston Harbor, would frequent dredging likely be necessary?

What is your evidence to support your answer? Which of these images supplies information that helps define the distribution of wetlands that are impacted by brackish water? What kind of data are provided?

D. SPOT Imagery.

Compare the map in Figure 9–27, a portion of a USGS topographic map of the Boston, MA area, with Figures 9–19 and 9–20 to answer the following.

1. Figure 9–19 is a panchromatic SPOT image with 10 m resolution. An original photographic copy of this image allows magnification of more than 10× with increasing ability to identify surface features. Based upon texture and use of some magnification, can you identify the general limits of the built-up area (dense street pattern, many buildings) of the Boston metropolitan area? Where is it located? How does it conform to the location of the circumferential highway that rings Boston (Route 128 and I–95 in places) and is within about 2 inches of Logan Airport on this image? Based upon your observation, how might Route 128 have played a role in development of the region?

2. Figure 9–20 is a multispectral SPOT image that has been merged with panchromatic data to provide multiband information at 10 m resolution. The image shown here has been enlarged 556% from that of the original scene. Note that individual structures are easily visible. Much of this image area appears to be virtually 100% developed.

What image characteristics and map information would you use to delineate the areas of most intense development? Explain.

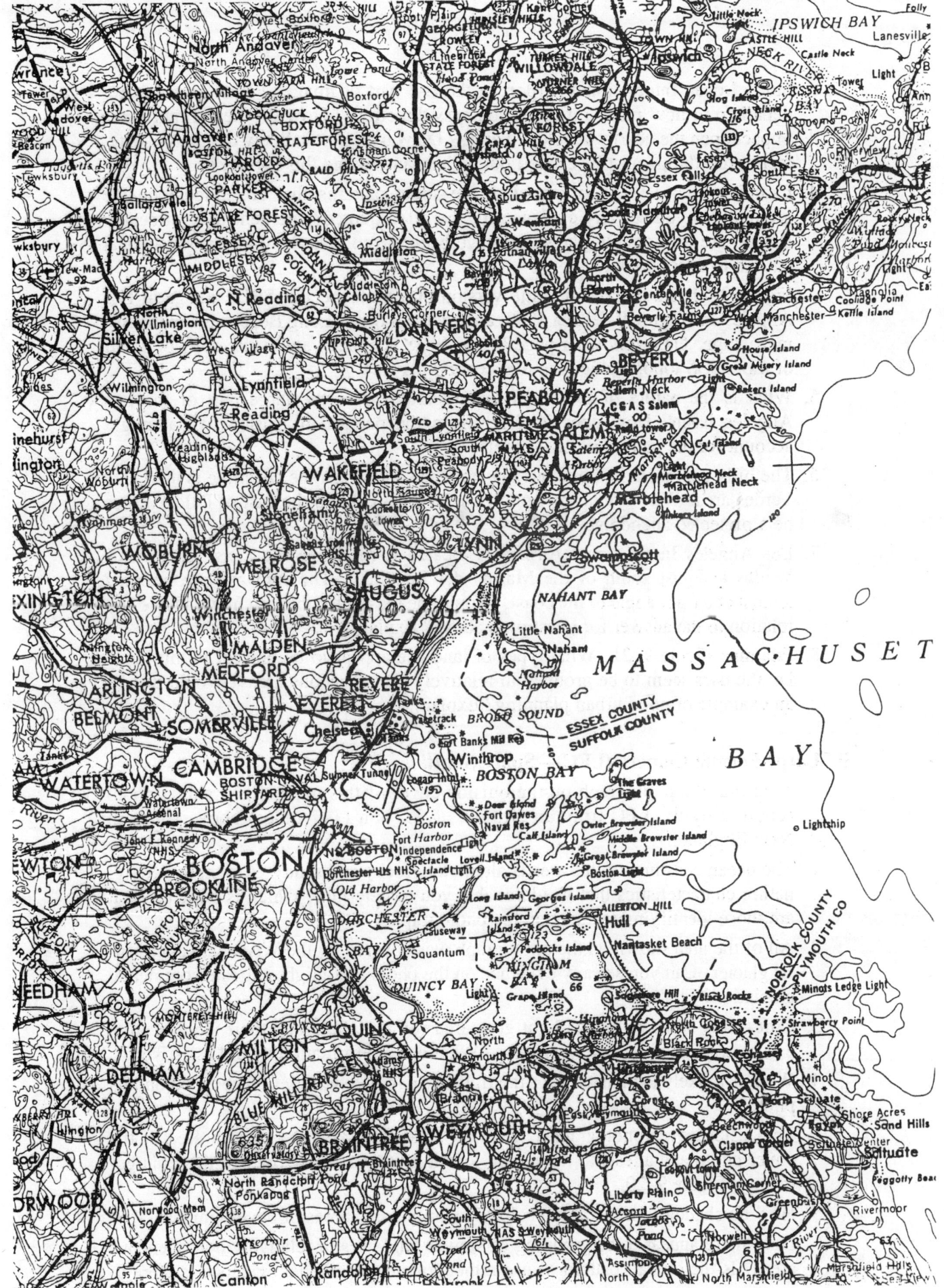

Figure 9–27. USGS topographic map, Boston, MA.

3. Study Logan International Airport in the B-3 part of image. Planes may land or depart from how many different compass directions? How do these flight paths impact upon the nearby developed areas? Do you think it is likely that noise abatement procedures must be carried out by airplanes using this facility?

4. Examine Figure 9–20 in regard to the street patterns that are evident on the ground. Do you see any evidence for the often quoted statement that "streets in Boston are former cow paths that have been paved"? Is there an overall logical pattern to street layout?

Figure 9–21 is a black and white version of a multispectral SPOT image of Los Angeles and Figure 9–22 is a 450% enlargement of the Long Beach area in the former image. On a following page Figure 9–28 is a black and white copy of a part of the USGS topographic map that covers this area.

5. What is the nature of the street layout/pattern over the large part of the Los Angeles basin? How does it differ from the pattern of Boston? What might account for the pattern in Los Angeles?

6. The hilly peninsula that juts out towards the south from the coastline is the Palos Verdes area. Why would the Long Beach Naval Base to the east of Palos Verdes be a protected anchorage?

7. Los Angeles International Airport is located north along the coast from Palos Verdes and just south of the Marina del Rey inlet. The runways are oriented almost at a right angle to the coast. Would the reason for this orientation have any relation to the answer for question 6 above? Explain.

8. Examine Figure 9–21. What types of land use appear to be present in this area? Do the uses seem to be grouped in relatively homogeneous areas? Would that be an example of good urban planning? Explain.

E. Large Format Camera (LFC)—Space Shuttle.

Note that the photograph on the outside cover of this manual is a single photo acquired by the LFC from an altitude of 237 km (147 miles) above southern New England.

1. The ocean coastline of this area is alternately rocky and sandy. Are there any signature characteristics that would enable you to determine which is which? What are those identifying features on the photograph?

2. All of the area shown on the photograph was covered by the Wisconsin continental glacier. Can you see any evidence on the photo that might indicate continental glaciation? Explain.

3. Cities may be identified as light gray areas on the photo and their identification can be confirmed by other evidence. What other information would assist you?

4. What is a likely explanation for the light lines in the vicinity of Cape Cod on this photo?

The inside cover photograph is a 13.6× enlargement of the Boston area of the cover photo and attests to the quality of the LFC system. Compare this enlargement to Figure 9–20.

5. How do shadows on the photo help to identify some areas of development?

6. Does there appear to be grouping of similar land uses in relatively homogeneous areas? Is the overall pattern of use rather fragmented? What evidence is there to support your conclusion?

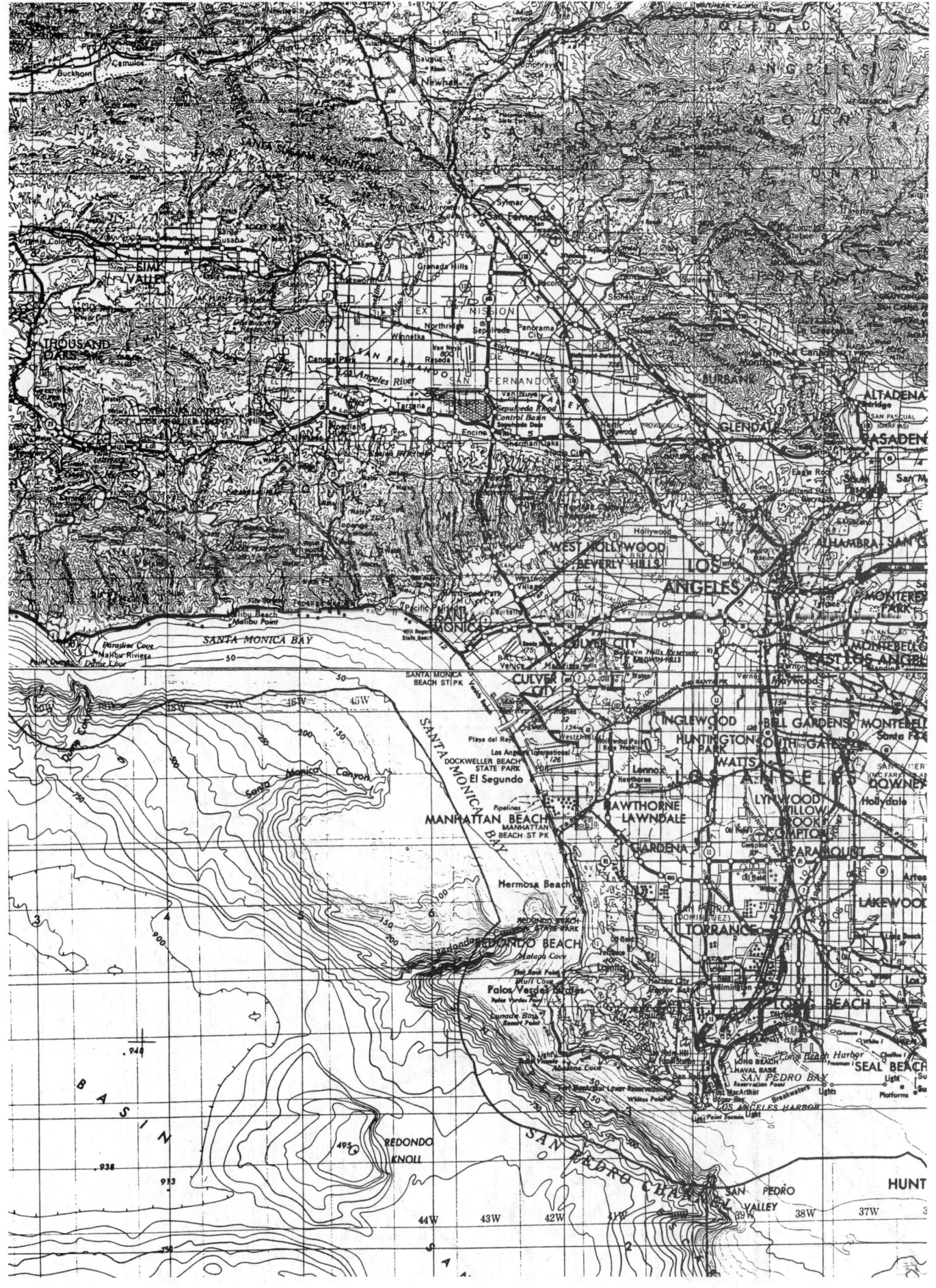

Figure 9–28. USGS topographic map, Los Angeles, CA.

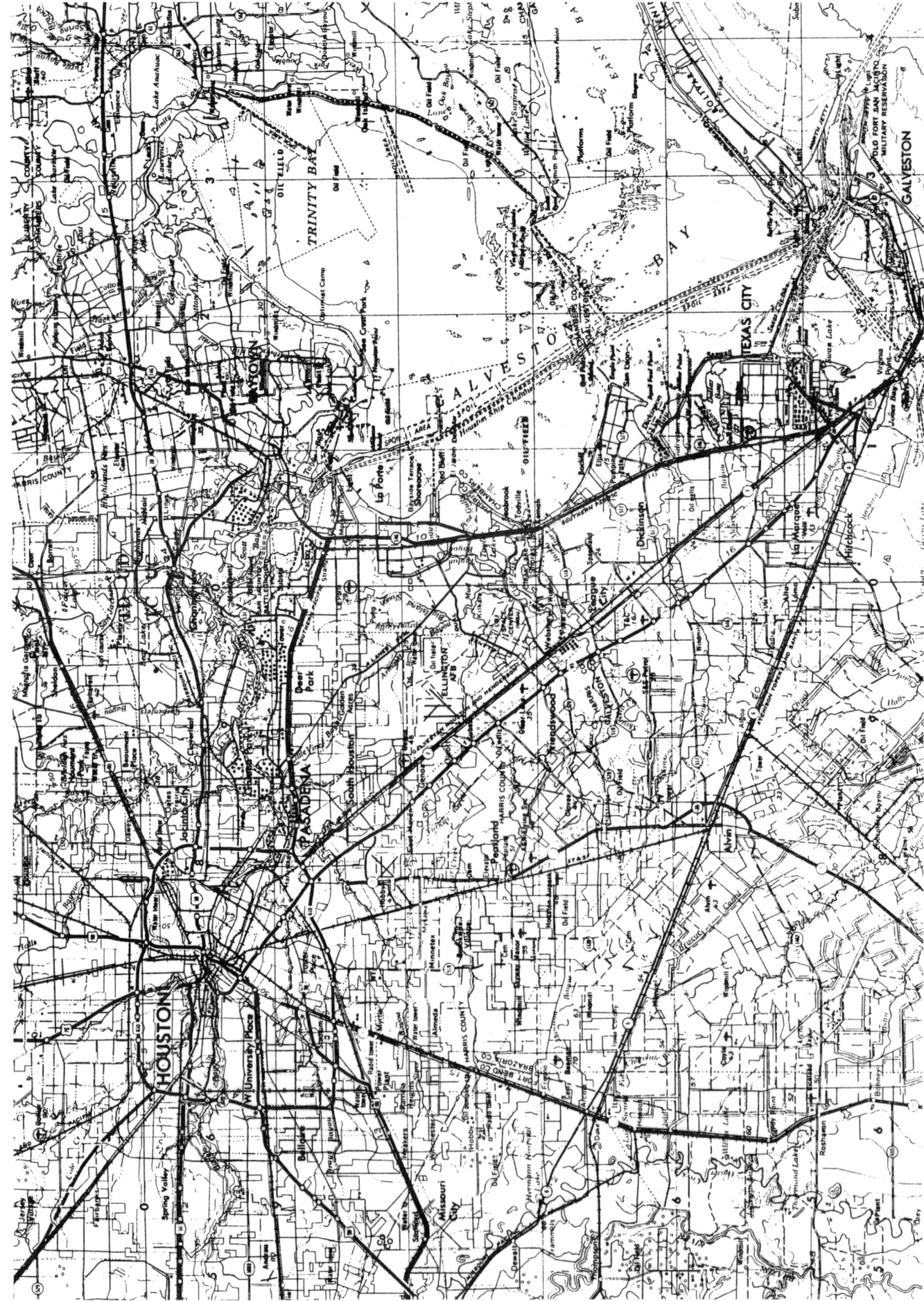

Figure 9–29. USGS topographic map, Houston, TX.

F. **Hand-held photography—Space Shuttle.**

Figures 9–23 and 9–24 are photographs taken over the Gulf of Mexico, Texas coast, by astronauts on two Space Shuttle missions. Figure 9–29 on page 102 is a reduction of part of the Houston USGS topographic map in black and white.

1. Which of these two photographs seems to represent better weather conditions? Evidence?

2. In what compass direction does the wind seem to be coming from in Figure 9–23? Evidence?

3. Is there any evidence of sedimentation on either photo? Give details.

4. Determine the scale (RF) of each photo at its center.

5. Using Figure 9–24 what signature features might be used to define the built-up (developed) area of Houston?

6. Can you find indications that Houston is a seaport?

7. The city of Galveston is in the C-3 area of Figure 9–24. Is there evidence in the photo of the city's vulnerability to hurricane destruction? Explain. In 1900, 6000 people lost their lives to a hurricane.

8. The city of Galveston lies on a barrier island. Are there similar features elsewhere along this coast? Study both of the photos. What is the likely origin of these features? Is there evidence on the photos that might indicate how they came into existence?

chapter *10*

SATELLITE MAPPING OF EARTH'S ATMOSPHERE AND WATER RESOURCES

TERMINOLOGY

spectral bands utilized	TIROS, TIROS-N
NASA, NOAA	TOVS
polar orbiters	AVHRR
geostationary	SEM
GOES	ARGOS DCS
VISSR	SARSAT
full disc visible image	NIMBUS
nephanalysis	Coastal Zone Color Scanner
full disc infrared image	Mission to Planet Earth
polarity of gray tones	Global Change Research Program
thermal infrared images	TOMS
water vapor imagery	Earth Observing System
standard IR	DMSP Visible-IR
Daily Weather Maps	Operational Line-scan System
NOAA, NESDIS	hand-held photography

BACKGROUND

In Chapter 9 the image products of a number of satellite systems used to survey the Earth's land surface resources were examined. The emphasis in this chapter will focus upon the resources of the Earth that are found in the atmosphere and large water bodies.

Satellites have been used to monitor and record a variety of atmospheric and oceanic conditions that hold great significance for their impact upon human habitation of the Earth. During the last thirty-five years the variety and sophistication of satellite technology has greatly contributed to the knowledge of the planet Earth.

Figure 10–1 is a table that summarizes the general thrust of satellite use for environmental investigation over that period by the United States. Several of these areas of investigation will be portrayed by examining the energy levels utilized and the nature of imagery produced.

The table in Figure 10–2 lists many of the instruments carried on board these satellites. It also indicates the wavelengths of energy being monitored and the types of data studied.

U.S. Environmental Satellites 1960–1994

Satellite	Launch Date or Planned	Avg. km.	Altitude miles	Ceased Operation	Remarks (see also Figure 10–2)
TIROS 1	04/01/60	720	447	06/19/60	First weather satellite providing cloud cover photography
TIROS 2	11/23/60	672	417	02/01/61	
TIROS 3	07/12/61	760	472	10/30/61	
TIROS 4	02/08/62	773	480	06/12/62	
TIROS 5	06/19/62	694	431	10/11/63	
TIROS 6	09/18/62	694	431	10/11/63	
TIROS 7	06/19/63	645	401	02/03/66	
TIROS 8	12/21/63	749	465	01/22/66	First APT satellite
Nimbus 1	08/28/64	677	420	09/23/64	Carried AVCS, APT, and High Resolution Infrared Radiometer for night pictures
TIROS 9	01/22/65	1630	1012	02/15/67	First TIROS satellite in Sun-synchronous orbit
TIROS 10	07/01/65	792	492	07/03/66	
ESSA 1	02/03/66	765	475	05/08/67	First satellite in the operational system; carried 2 wide-angle TV cameras
ESSA 2	02/28/66	1376	854	10/16/70	Carried APT cameras. APT carried on all even numbered ESSA satellites
Nimbus 2	05/15/66	1136	705	01/18/69	
ESSA 3	10/02/66	1427	886	10/09/68	Carried first AVCS cameras. ABCS carried on all odd numbered ESSA satellites
ATS 1	12/06/66	35,765	22,210	10/16/72	WEFAX discontinued (pictures) 12/31/78
ESSA 4	01/26/67	1373	853	10/06/67	
ATS 2	04/05/67	—	—		Unstable altitude. Data not useful
ESSA 5	04/20/67	1379	856	02/20/70	
ATS 3	11/05/67	35,815	22,241	10/30/75	WEFAX discontinued (pictures) 12/31/78
ESSA 6	11/10/67	1437	892	11/04/69	
ESSA 7	08/16/68	1440	894	07/19/69	
ESSA 8	12/15/68	1429	887	03/12/76	
ESSA 9	02/26/69	1456	904	11/15/73	
Nimbus 3	04/14/69	1100	683	01/22/72	Provided first vertical temperature profile data of the atmosphere on a global basis
ITOS 1	01/23/70	1456	904	06/17/71	2nd generation prototype
Nimbus 4	04/08/70	1108	688	09/30/80	
NOAA 1	12/11/70	1438	893	08/19/71	First NOAA funded 2nd generation satellite
Landsat 1	07/23/72	918	570	01/16/78	
NOAA 2	10/15/72	1460	907	01/30/75	First operational satellite to carry all scanning radiometer
Nimbus 5	12/12/72	1110	689	03/29/83	
NOAA 3	11/06/73	1510	938	08/31/76	First operational satellite to permit direct broadcast of VTPR data
SMS 1	05/17/74	35,788	22,224	01/29/81	Deactivated. Boosted out of geosynchronous orbit
NOAA 4	11/15/74	1460	907	11/17/78	Deactivated
Landsat 2	01/22/75	918	570	03/31/83	On standby
SMS 2	02/06/75	35,800	22,232	08/05/82	Deactivated. Boosted out of geosynchronous orbit
Nimbus 6	06/12/75	1110	689	03/29/83	
GOES 1	10/16/75	35,796	22,229		First NOAA operational geostationary satellite; 130° west

Figure 10–1.

Satellite	Launch Date or Planned	Avg. km.	Altitude miles	Ceased Operation	Remarks (see also Figure 10–2)
NOAA 5	07/29/76	1511	938	07/16/79	Deactivated
GOES 2	06/16/77	35,787	22,224		Second NOAA operational geostationary satellite; 113° west supporting central WEFAX
Landsat 3	03/05/78	918	570	03/31/83	First Landsat with IR capability
GOES 3	06/16/78	35,784	22,222		
TIROS-N	10/13/78	850	528	02/27/81	Deactivated
Nimbus 7	10/24/78	954	592	mid 1986	Coastal Zone Color Scanner
NOAA 6	06/27/79	807	501		First NOAA funded TIROS-N system satellite
GOES 4	09/09/80	35,782	22,221		First geostationary satellite to carry VISSR Atmospheric Sounder (VAS), now failed; at 139° west. Provides west WEFAX and DCS.
GOES 5	05/22/81	35,785	22,222		At 75° west; also carried VAS, now failed. Provides east DCS, WEFAX, and relay of GOES 6 imagery
NOAA 7	06/23/81	847	526		2nd NOAA funded TIROS-N
Landsat 4	07/16/82	700	435		Carries MSS and TM
NOAA 8	03/28/83	815	506	06/12/84	Had search and rescue capability
GOES 6	04/28/83	35,791	22,226	01/21/88	Alternates between 98° west and 108° west. Only spacecraft with operating VAS
NOAA 9	12/12/84	815	506		Search and rescue capability and sensors for ozone and Earth radiation budget
Landsat 5	1985	700	435		MSS and TM
NOAA 10	9/17/86	833	517		AVHRR 4 channels
NOAA 11	9/24/88	833	517		AVHRR 5 ch. add 11.5–12.5
NOAA 12	5/14/91	833	517		Add channel 5
NOAA 13	8/9/93				communications lost 8/21/93
Landsat 6	10/5/93				Did not achieve orbit. Enhanced Thematic Mapper
GOES 7	1987	36,000	22,356		Operational as "Central" GOES DMSP (Defense Meteorological Satellite Program of U.S. Air Force has made some data available to civilian users since April, 1973. Launch dates not known. Variety of meteorological data are acquired. See Figures 10–24 and 10–26.)
GOES 8	4/13/94	35,789	22,225		Undergoing checkout; will be "EAST" 3-axis stabilization replaces spin-stabilized satellites to allow for continuous observation of Earth as opposed to 5% of time previously.
DMSP F-12	8/29/94	847	526		Meteorological satellite for DoD

Explanation of abbreviations in table:

APT	Automatic Picture Transmission	Nimbus	2nd of NOAA's polar ortbiter satellite series
ATS	Applications Technology Satellite	NOAA	National Oceanic and Atmospheric Administration's designation of sun-synchronous satelites
AVCS	Advanced Vidicon Camera System		
AVHRR	Advanced Very High Resolution Radiometer	SMS	Synchronous Meteorological Satellite
DCS	Data Collection System	TIROS	Television and Infrared Operational Satellite
ESSA	Environmental Science Services Administration	TM	Thematic Mapper
GOES	Geostationary Operational Environmental Satellite	VAS	Visible and IR Spin-Scan Radiometer Atmospheric Sounder
Landsat	Land Satellite (also known as ERTS, Earth Resources Technology Satellite)	VTPR	Vertical Temperature Profile Radiometer
MSS	Multispectral Scanner Sensors	WEFAX	Weather Facsimile

Figure 10–1. (continued)

Some Sensor Systems Employed on U.S. Environmental Satellites

Instrument Name	Vehicle	Wavelengths in Micrometers	Types of Data Acquired
AVHRR	NOAA	5 bands range of 0.58–12.5	temperature of clouds, sea surface, and land; visible and IR; index of vegetation
TOVS	TIROS-N NOAA-6-11	20 channels 0.7–15	temperature profiles, total ozone, outgoing long-wave radiation, water vapor content; visible
VISSR	GOES 4-7	0.55–7.0 vis 10.5–12.5 IR	day and night weather observations, visible and IR, temperature profiles, water vapor, sea surface temperature
CZCS	Nimbus 7	0.43–0.80 10.5–12.5	chlorophyll, phytoplankton, surface vegetation, sea surface temps.
OLS	DMSP	0.4–1.1 10.5–12.5	visible-near IR scanner obtains day and night imagery; thermal data
MSS	Landsat 1-5	0.5–1.1	green, red, and 2 near IR bands; replicate CIR data
TM	Landsat 4, 5	0.45–1.75 2.08–2.35 10.4–12.5	visible, near IR, mid IR, & TIR; visible, CIR, & TIR imagery

Explanation of abbreviations in table: (see also explanations in Figure 10–1.)

CZCS	Coastal Zone Color Scanner	TOVS	TIROS Operational Vertical Sounder
OLS	Operational Line Scanner	VISSR	Visible & IR Spin-Scan Radiometer

Figure 10–2.

SPECTRAL BANDS UTILIZED

As can be observed in Figure 10–3 much of the data gathered by the instruments on board the environmental satellites was concentrated in several bands of the EMS. Not all of the instrument packages used in space have been listed here and the satellites and sensors of other countries have not been covered. Of those selected and displayed here, it is obvious that much attention has been focused upon the visible, reflected IR, and thermal IR portions of the electromagnetic spectrum.

Figure 10–3 illustrates that there are transmission windows in the areas most frequently sampled, and Figure 10–2 indicates that there are useful data available in those bands. Longer wavelengths of the microwave part of the EMS will be discussed in the following chapter.

NASA (National Aeronautics and Space Administration) and **NOAA** (National Oceanic and Atmospheric Administration) have worked closely for many years in developing **research** and **operational satellites.** There are two types of civil operational satellites in use by these agencies, **polar orbiters** and **geostationary** vehicles. As discussed earlier, polar-orbiting satellites follow a path roughly parallel to the Earth's axis from pole to pole and as the Earth rotates past the orbital path additional swaths of Earth surface are imaged. Geostationary satellites are characterized by an orbit and speed that positions them above any point on the Earth where they maintain position constantly. See Figure 10–4 for a comparison of orbital relationships.

Polar orbiters record a wide variety of Earth environmental parameters with relatively higher resolution as they are closer to the Earth. Geostationary satellites provide the constant monitoring of the Earth's atmosphere which is the single most valuable tool for meteorologists.

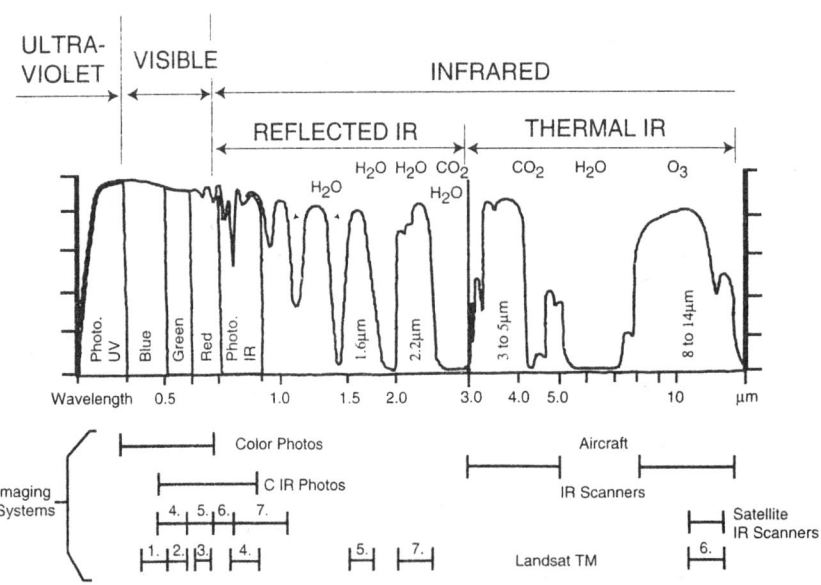

Figure 10–3. Bands of EMS commonly utilized.

GEOSTATIONARY WEATHER SATELLITES

GOES (Geostationary Operational Environmental Satellite) was first launched in 1975 (see Figure 10–5). To date seven more GOES satellites have been orbited. The objective of the program is to station GOES East above the equator at 75° west longitude and GOES West above the equator at 135° west as part of an international network of satellites constantly monitoring the Earth's weather. The two satellite coverage of North America was interrupted when GOES West (GOES 6) failed on January 21, 1988 and GOES East (GOES 8) was moved over the center of the United States. The coverage of these two approaches is illustrated in Figure 10–6. Currently GOES 8 is serving as the lone satellite for North America and GOES 9 is being tested for service as GOES West.

These satellites have a day-night imaging capability as they have on board a **VISSR** (Visible-Infrared Spin-Scan Radiometer) that records energy in one visible band (0.55–0.70 microns) and in the thermal IR (10.5–12.5 microns). Figure 10–7 is an example of a **full disc visible image** of part of the Earth and its cloud cover centered over the Equator at 75° west longitude. All clouds at any elevation above the surface show up in this image. This allows **nephanalysis**—the study of cloud patterns and formations, to be carried out. This approach to meteorology has yielded a number of quantifiable relationships between cloud size, the definition of a cloud mass, atmospheric pressure, and wind speed, especially in relation to tropical cyclones and hurricanes.

Figure 10–8 is a **full disc infrared image**—here centered over the Equator at 115° west longitude. Note that each of the GOES images has a geographic grid of meridians and parallels, all continental outlines, and country/state boundaries superimposed on the image. This greatly aids locations when surface features are difficult to see on an image.

The full disc IR image portrays thermal values for land, water, and clouds. The **polarity of gray tones** is presented with the warmer temperatures as dark and the colder temperatures as light. This approach emphasizes the coldest, and therefore highest, of the clouds as white, while the lower level, warmer clouds blend into the background of the Earth's surface. Therefore, the clouds that are a result of the greatest lifting and that may represent well developed and unstable systems are more easily identified on the images.

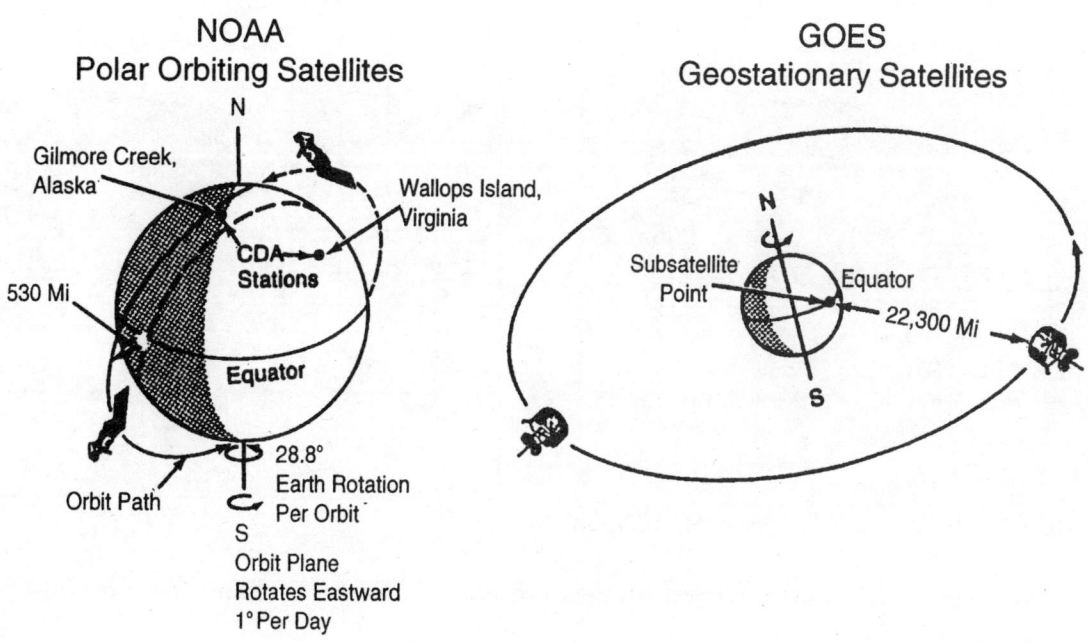

NOAA
Polar Orbiting Satellites

Gilmore Creek, Alaska

Wallops Island, Virginia

CDA Stations

530 Mi

Equator

Orbit Path

28.8°
Earth Rotation Per Orbit

Orbit Plane Rotates Eastward 1° Per Day

GOES
Geostationary Satellites

Subsatellite Point

Equator

22,300 Mi

Figure 10–4.

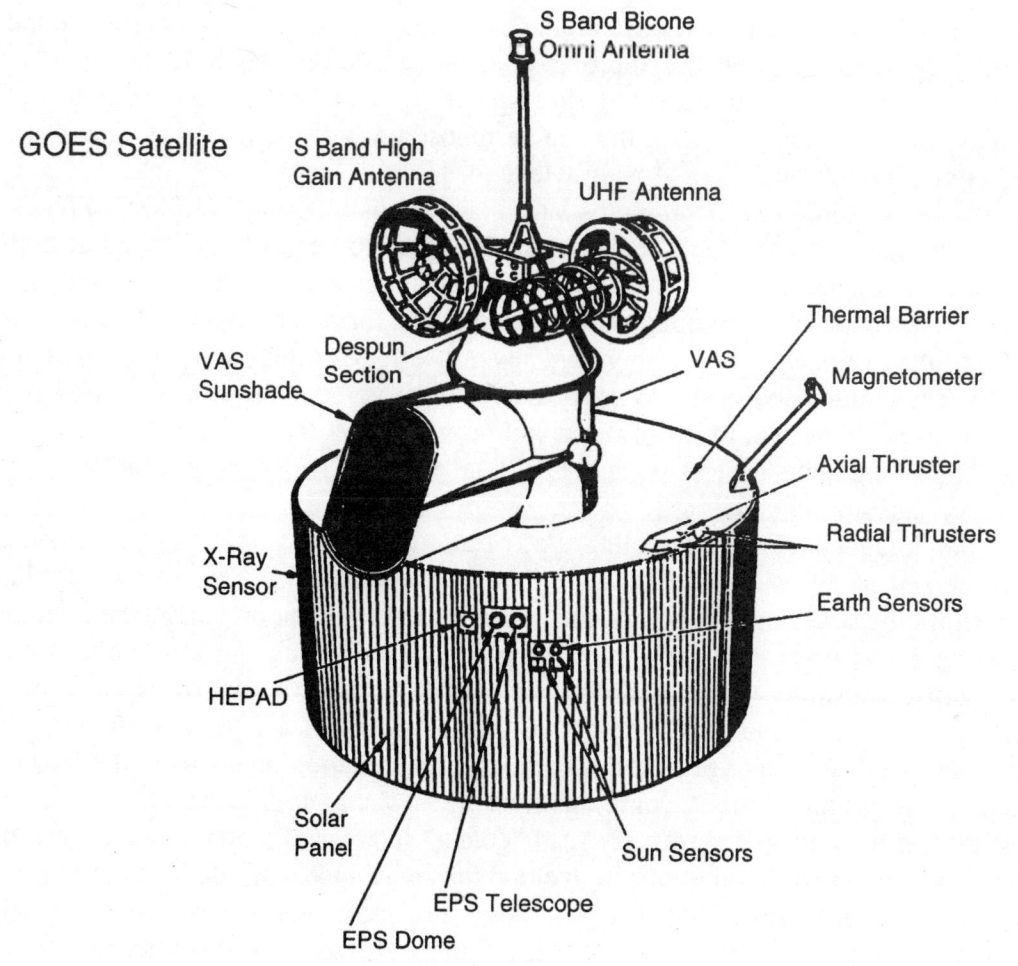

GOES Satellite

S Band Bicone Omni Antenna

S Band High Gain Antenna

UHF Antenna

VAS Sunshade

Despun Section

VAS

Thermal Barrier

Magnetometer

Axial Thruster

Radial Thrusters

Earth Sensors

X-Ray Sensor

HEPAD

Solar Panel

EPS Telescope

EPS Dome

Sun Sensors

Figure 10–5.

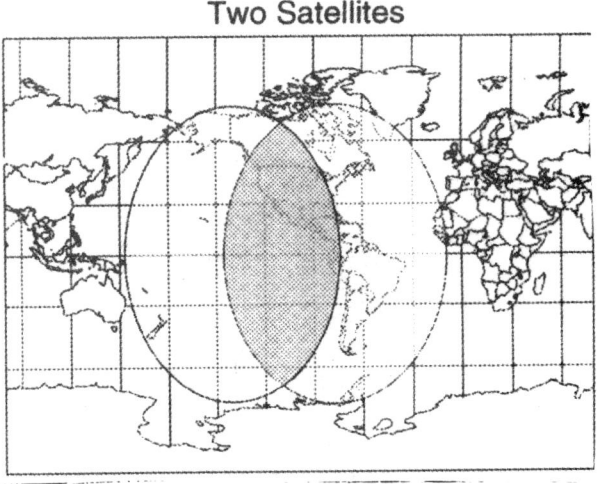

One Satellite   Two Satellites

Fiure 10–6. GOES geographic coverage of North America.

The visible and **thermal infrared images** present a powerful combination of cloud pattern and cloud height information. By comparing these data one to another and each with synoptic chart and other observation information, an extremely powerful analytic tool is revealed. This approach will be demonstrated.

Note in Figure 10–7 the areas that are totally devoid of clouds. It is likely these areas would coincide with high pressure cells or ridges. Some cloud masses, such as the two above the Gulf Coast and off New England, may be symptomatic of mesoscale thunderstorms as they are well defined and roughly circular. Evidence of the typical "comma shape" mid-latitude cyclonic storm is found east of Greenland and in the southernmost areas of the image. Other areas have much cloud cover, but the clouds could be low level or high.

In Figure 10–8, acquired only 15 minutes later than the visible image (1500 GMT vs. 1515 GMT), the cloud mass over the Gulf coast does not appear to be well developed at high altitudes, but the New England mass is still impressive at upper elevations. Two storm systems west of Chile at the bottom of the image show evidence of form and also height that indicates strong mid-latitude cyclones. Finally, there are several cells over the Pacific Ocean at approximately 10° north latitude that show symptoms of strong vertical development and possible hurricane formation. Note that much of the cloud cover on this image is revealed to be lower elevation and therefore indicates less vertical development and likely less of a factor in producing precipitation and stormy conditions.

Water vapor imagery represents another type of atmospheric tool that can be utilized along with the visible and the thermal IR. Figure 10–9 is an image that was created by recording the energy radiated in the 6.7 micron band. At this wavelength the radiation is absorbed and reradiated by water in the gaseous state. This makes it possible to record water in the atmosphere by visible, thermal IR, and water vapor windows whether it is in solid, liquid, or vapor form.

Interpreting water vapor imagery is complex, but the potential aid to weather analysis and forecasting is great. Water vapor radiation will modify the radiation levels from the Earth's surface and from clouds so that the most common effect when water vapor is present is to reduce the radiation level reading to make the imagery read "colder" in areas of water vapor concentration.

When reading a water vapor image the high cloud tops may appear similar (white) to how they would look on a normal thermal IR "window" image, except that the outlines are generally not as well defined. Conditions of snow or ice at the Earth's surface can also alter the gray level (brightness). Several generalizations of what the gray levels represent are potentially misleading, because the sensor indicates the temperature of the moisture that it sees. It is not always clear what the instrument is "seeing."

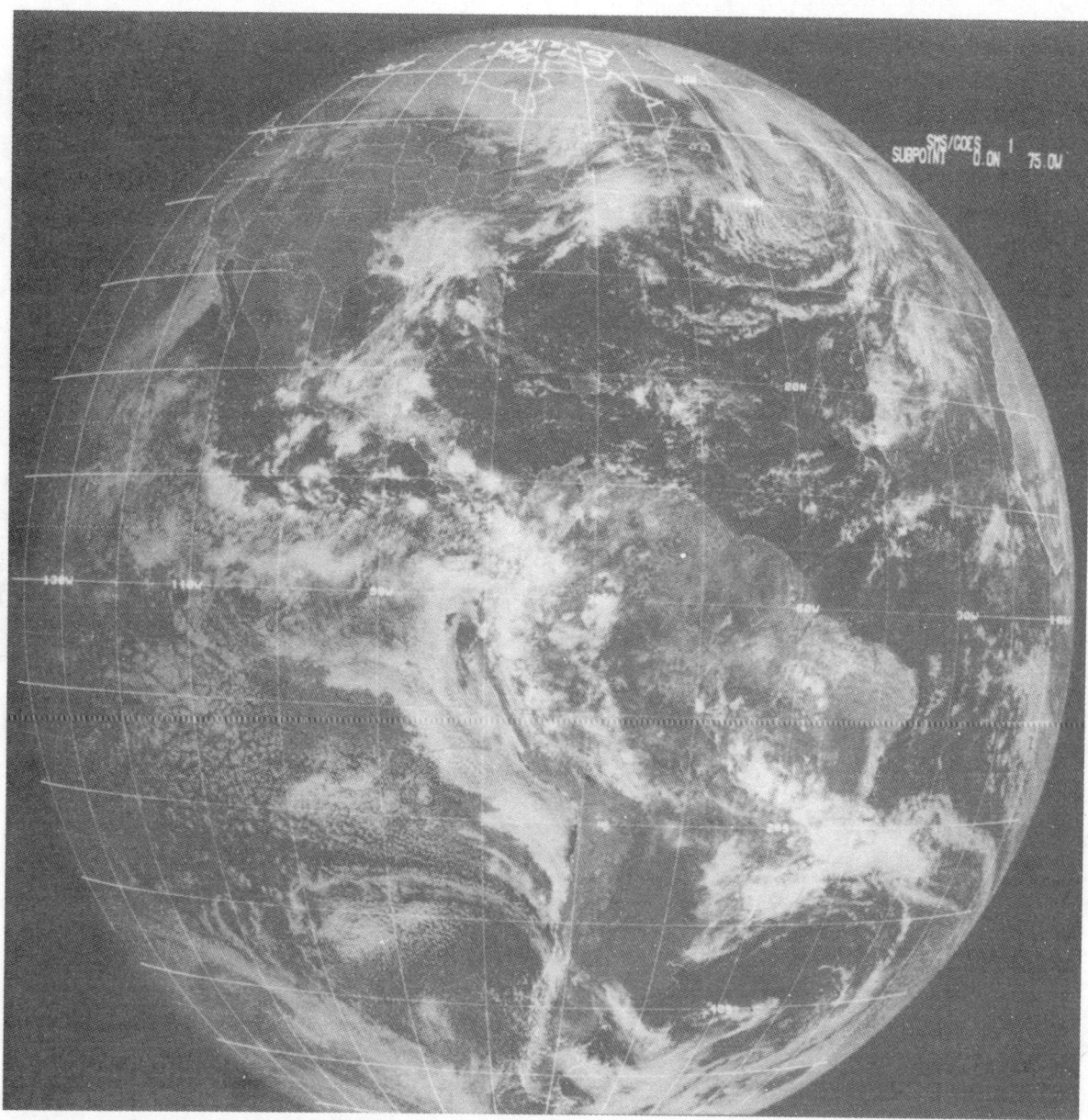

Figure 10–7. NOAA GOES full disc visible image. Centered on 75° west. Acquired 1500 GMT 10/06/75.

Some of the parameters involved in interpreting water vapor imagery are density of vapor, thickness of vapor layers, and surface condition of the Earth. Perhaps one generalization with some merit is that the gray levels in water vapor imagery indicate the topographic surface of the height of significant moisture in the atmosphere. Knowledge of temperature conditions in the atmosphere greatly increases interpretability of moisture present.

Compare Figure 10–9 with Figure 10–10, a **"standard IR"** image from GOES. Note that the area covered is the same and the time of acquisition is only 2 1/2 hours apart. On both images there appears to be a strong front extending from eastern Texas to the estuary of the St. Lawrence River. This front is more clearly defined on the water vapor image. Note the number of high cloud top concentrations across Mexico, Cuba, and the eastern Caribbean Sea on both images. Figure 10–10 has been modified by an enhancement curve that emphasizes convection and high cloud tops in particular. These high clouds are easily matched on both images.

The following material will illustrate the strength of the imagery from the GOES satellite when used in conjunction with the ***Daily Weather Maps*** published by **NOAA, NESDIS** (National Environmental Satellite, Data, and Information Service) as analytic and forecasting tools.

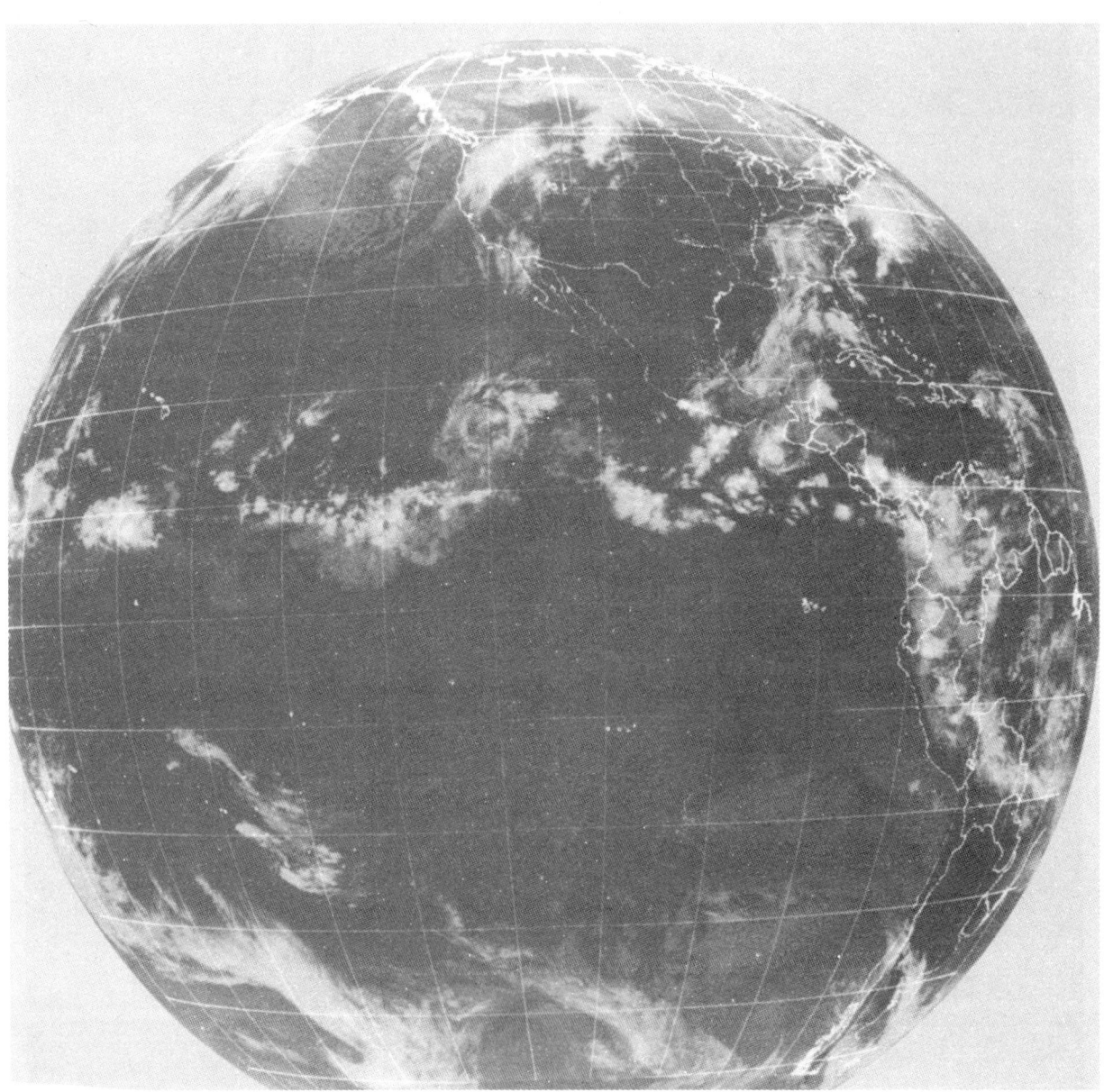

Figure 10–8. NOAA GOES full disc IR image, centered on 115° west. Acquired 1515 GMT 10/06/75.

Figures 10–11 through 10–13 are reproductions of the 7 AM EST surface weather maps for the period Thursday February 10 to Saturday February 12, 1983. Figures 10–14 through 10–16 are representations of the 500 millibar contour maps at 7 AM EST for the same period. Figure 10–17 is a GOES thermal IR image covering the eastern United States and acquired at 1600 GMT (1100 EST) on February 11, 1983. The enhancement curve (2CC) modified the image to emphasize areas of convection and lifting of air. Figure 10–18 is a visible GOES image obtained 30 minutes later for the same area. Figure 10–19 is a TIR, convection-enhanced image showing the area at 0400 GMT (11:00 PM EST same day) and Figure 10–20 is a visible image of the area on Saturday February 12, 1983 at 1700 GMT (12 noon EST).

The exercises at the end of the chapter will deal with comparison of these several data forms to enable relatively simple interpretation and forecasting for this short time period. See the station model and frontal symbols on the inside, back cover of the manual to assist you in answering the questions.

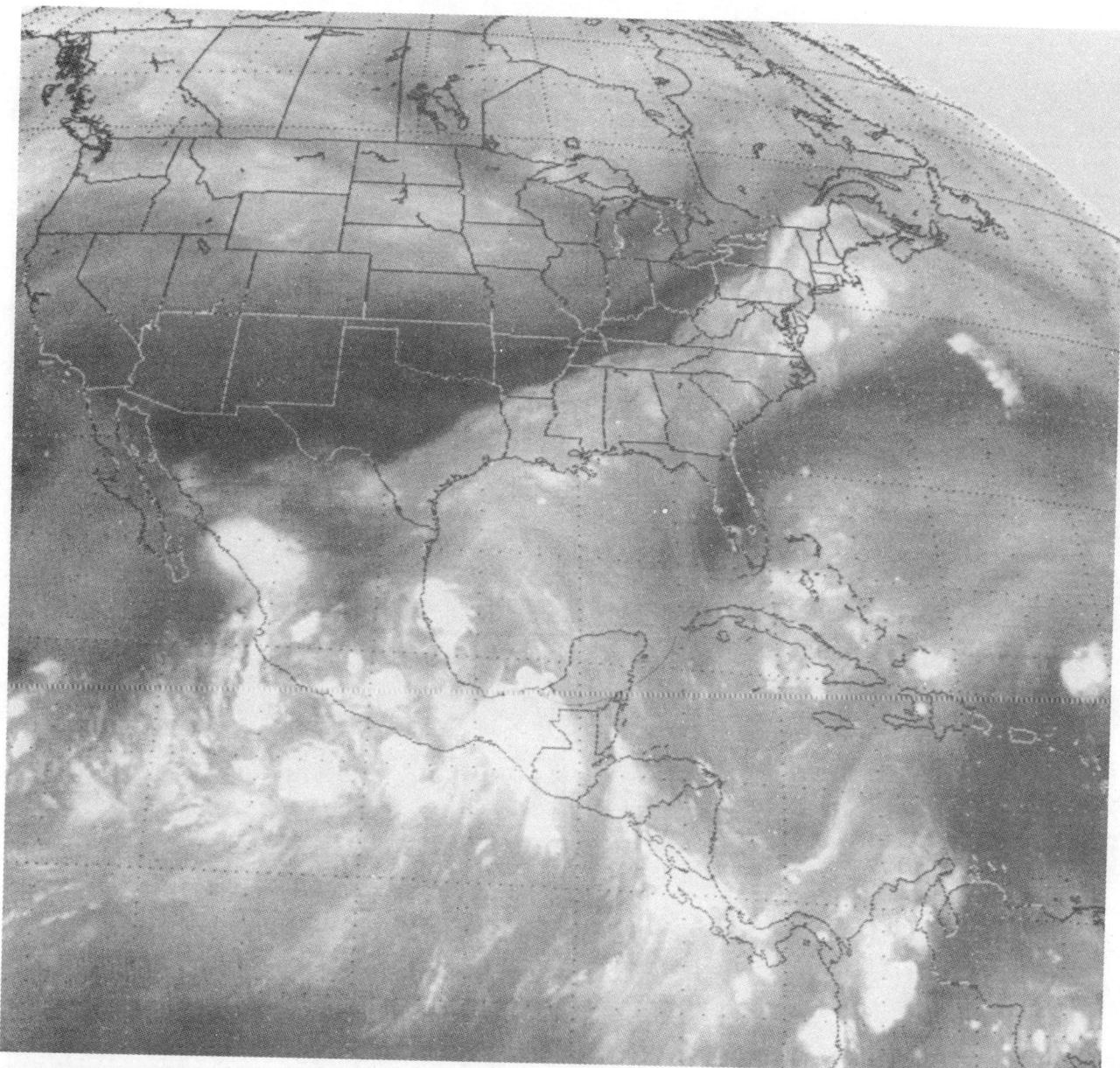

Figure 10–9. NOAA GOES water vapor image; coverage of U.S.A. and Central America. Acquired 0530 GMT 9/04/84.

POLAR ORBITER SATELLITES

NOAA Polar Orbiters. NOAA, NESDIS environmental data have been gathered by polar orbiting satellites since 1960. One of the principal vehicles is TIROS which is referred to as **NOAA-A, -B,** prior to successful launch and **NOAA-1, -2,** after insertion in orbit. Figure 10–21 is a diagram of **TIROS-N,** Advanced TIROS-N (ATN) which carries a number of instrument packages. The satellite follows a polar orbit (see Figure 10–4) at an average altitude of 530 miles (854 km) above the Earth. Normal operation has two such satellites in orbit at a time.

The satellites in the series collect many types of data that may be produced in the form of imagery or as statistical matrices such as atmospheric soundings. These data are relayed to Earth or to other satellites by several communications systems. The sensor systems on the satellite include the **TIROS** Operational Vertical Sounder **(TOVS),** Advanced Very High Resolution Radiometer **(AVHRR),** Space Environment Monitor **(SEM), ARGOS** Data Collection and Platform Location System **(DCS),** Search and Rescue System **(SARSAT).**

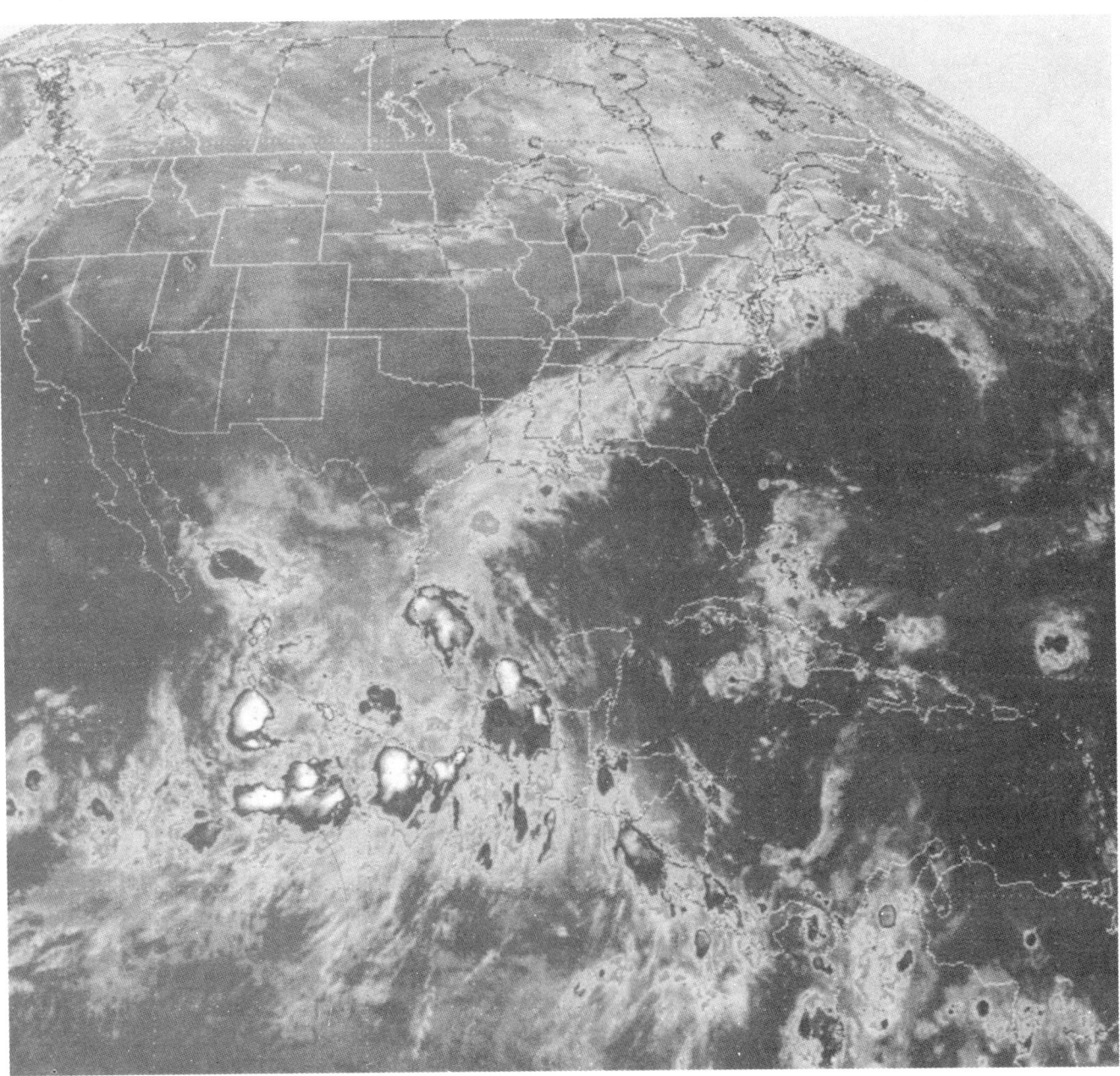

Figure 10–10. NOAA GOES standard IR image coverage of U.S.A. and Central America. Acquired 0800 GMT 9/04/84.

Some of the products yielded by these sensors are **temperature soundings** at 15 levels in the atmosphere, **sea surface temperature, ice analysis** in polar regions and in the Great Lakes of the United States, **vegetation index** of how "green" the surface land areas are, snow cover for the northern hemisphere, and the detection and tracking of **tropical cyclones.**

Figure 10–22 is an image acquired on October 9, 1994 by the satellite, NOAA-9. It displays the **Total Ozone concentrations** for the southern hemisphere and the data indicate that the level of ozone is slightly higher than in 1993. Changes in the amount of dust from the eruption of Mt. Pinatubo in the Philippines in 1991 and in the amount of sulphuric acid particles associated with it are believed to have contributed to the increase in ozone. Studies of the "ozone hole" have been conducted by NOAA's satellites for more than 15 years and provide the bulk of data available for scientists who are attempting to determine how seriously this problem may impact the Earth's environment.

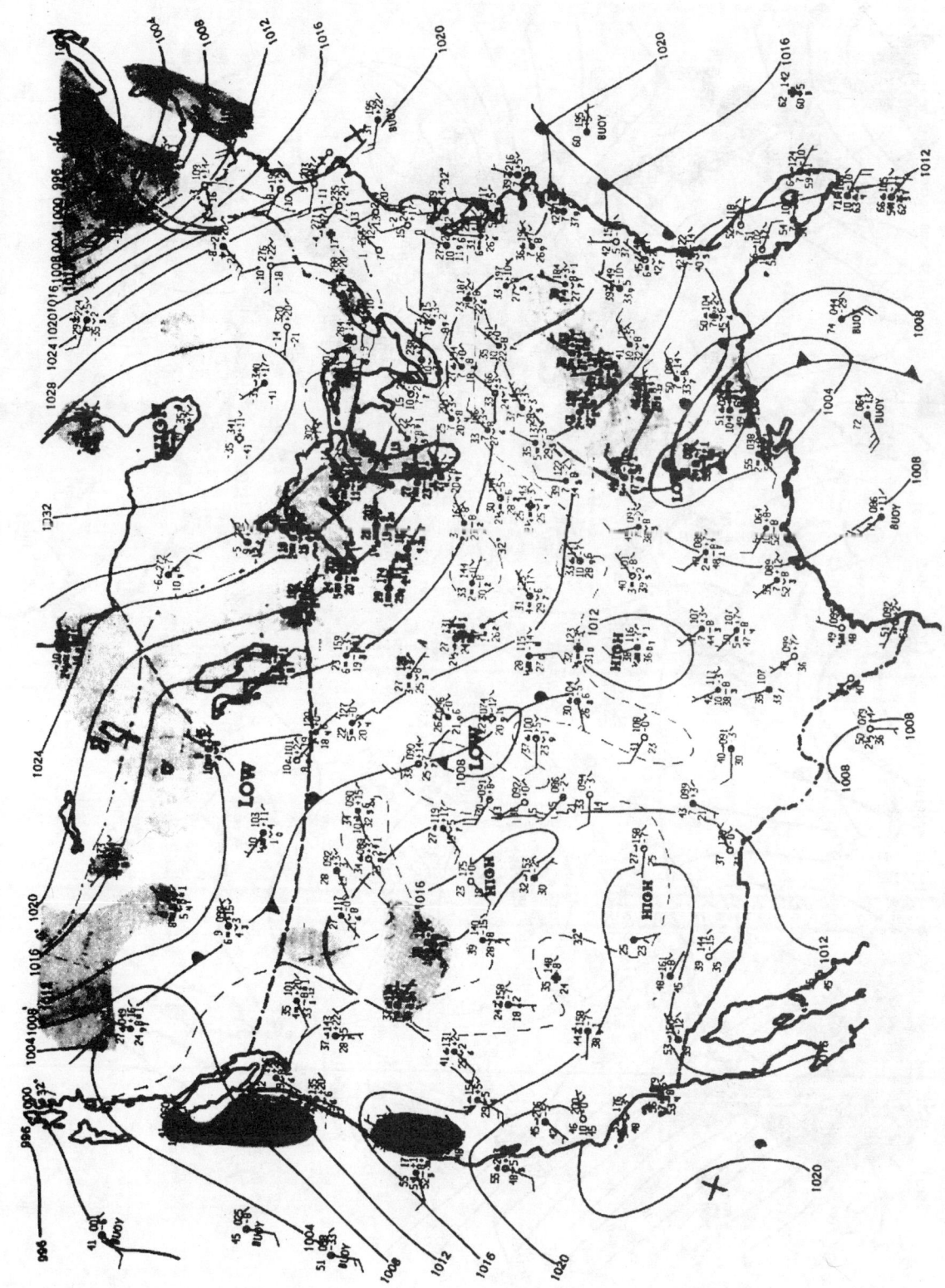

THURSDAY, FEBRUARY 10, 1983

Figure 10–11. U.S. Weather Bureau surface weather map. 7 AM EST 2 /10/83.

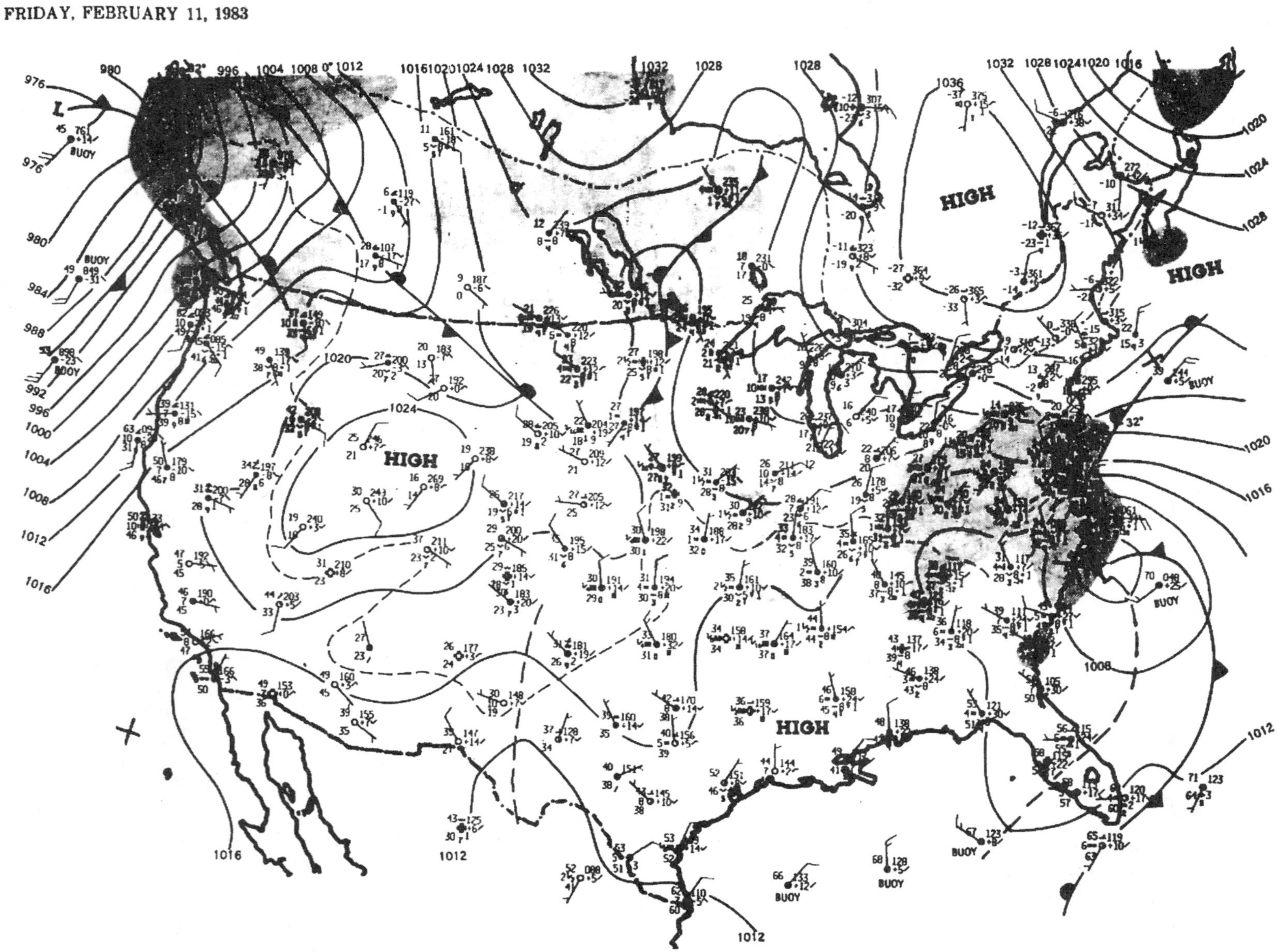

Figure 10–12. U.S. Weather Bureau surface weather map. 7 AM EST 2/11/83.

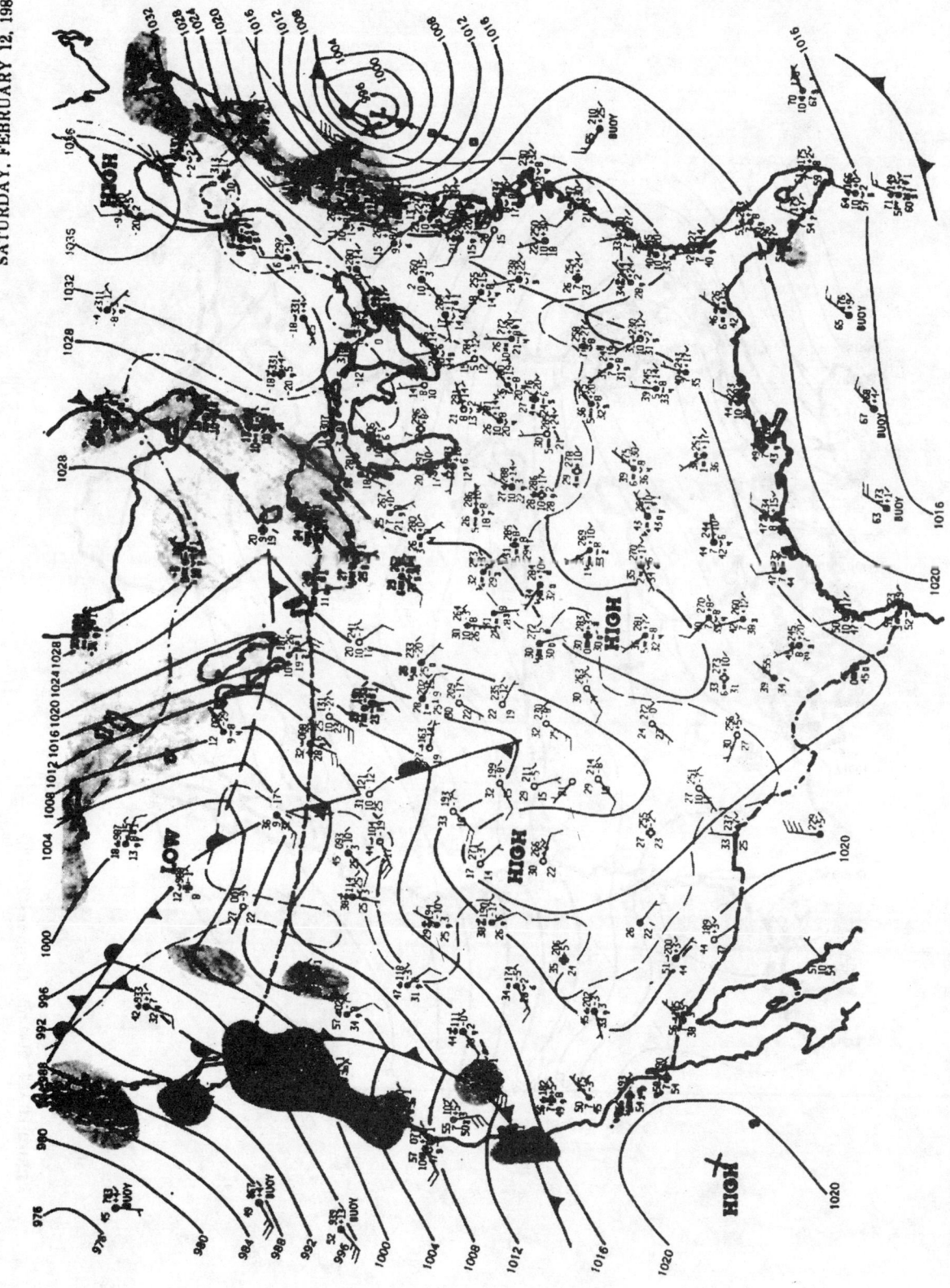

Figure 10–13. U.S. Weather Bureau surface weather map. 7 AM EST 2/12/83.

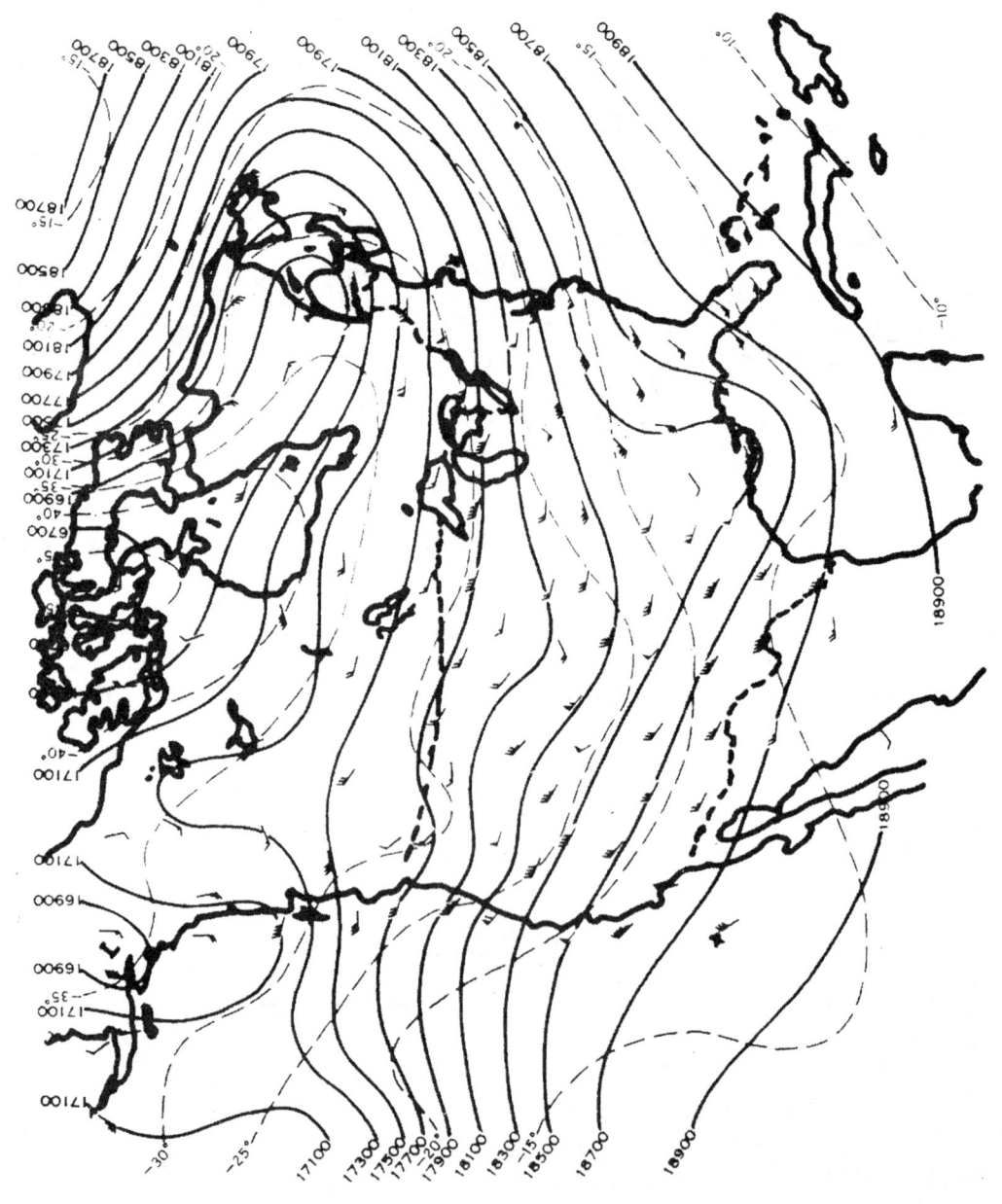

Figure 10–14. 500 mb. Contours 2/10/83

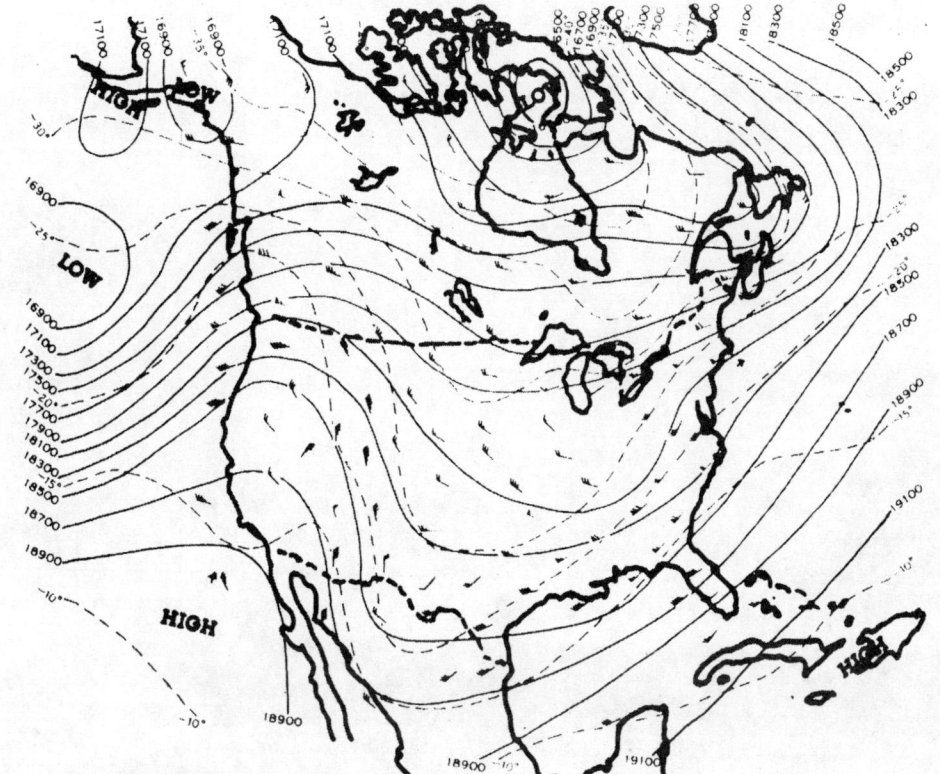

Figure 10–15. 500 mb. contours 2/11/83.

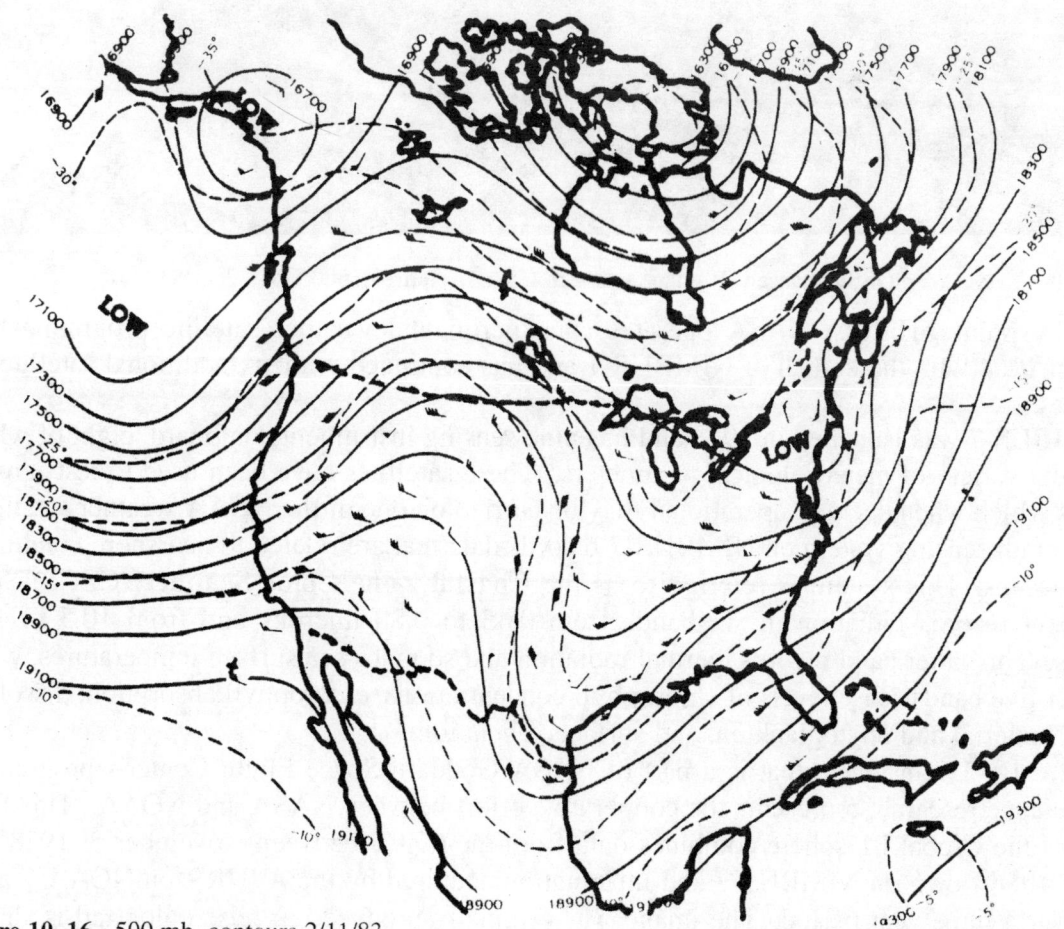

Figure 10–16. 500 mb. contours 2/11/83.

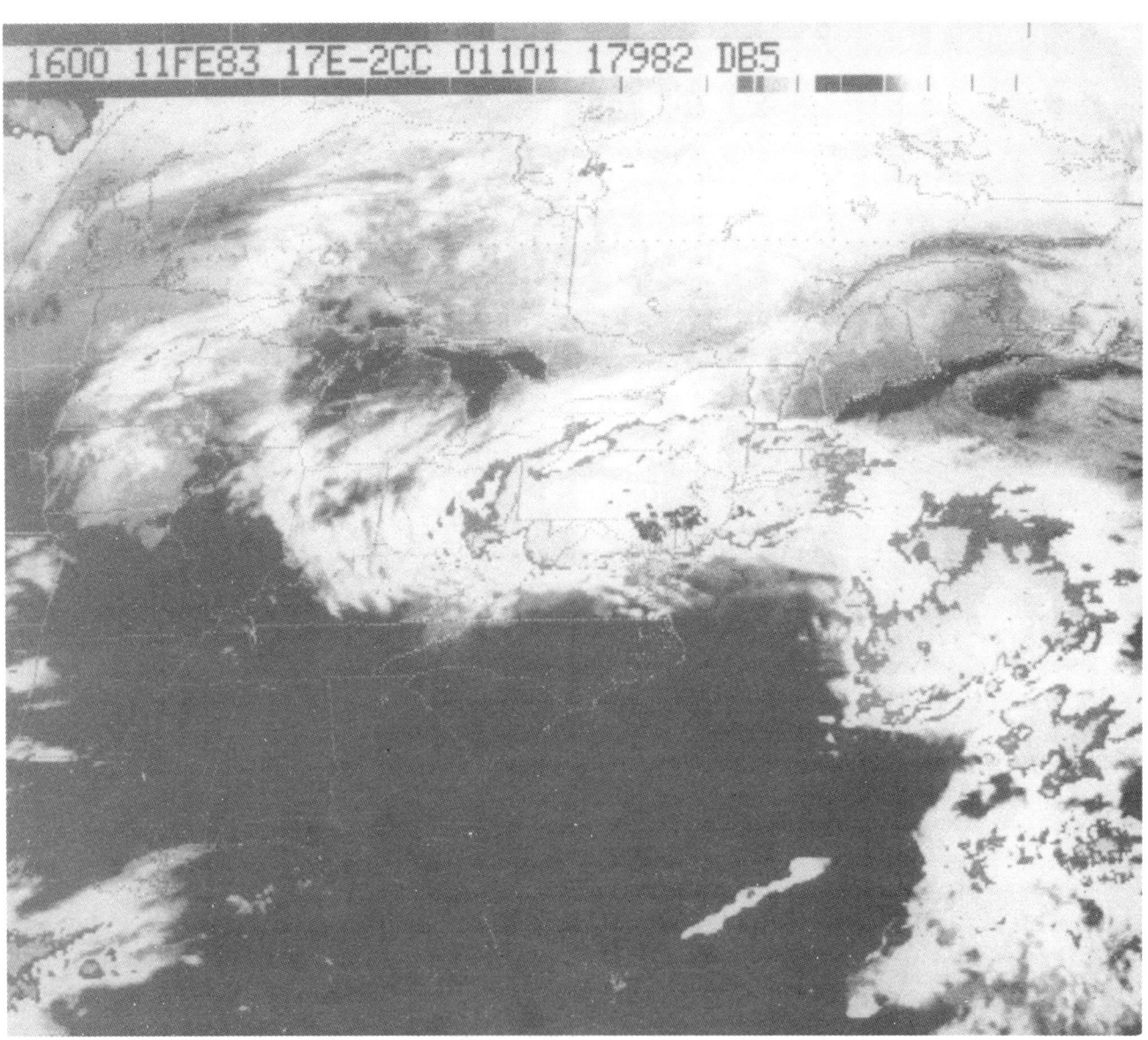

Figure 10–17. NOAA GOES enhanced IR image, eastern USA. Acquired 1600 GMT 2/11/83.

NASA polar orbiters. NASA operates an experimental weather satellite program which began in 1964 with the launch of NIMBUS-1 and has continued with six additional satellites in the series.

NIMBUS-7 was launched in 1978 and has nine sensing instruments on board, eight of which measure a variety of meteorological parameters. These satellites have been used to test sensing systems which when deemed operational may be carried on one of the NOAA weather satellites.

The ninth sensing system on NIMBUS-7 records data that are related to nearshore conditions of the oceans. This system is referred to as the **Coastal Zone Color Scanner** (CZCS). This instrument records radiation in six bands from 0.43 to 0.80 microns and from 10.5 to 12.5 microns. This latter band records thermal radiation and so maps sea surface temperatures, while the other five bands sense levels of chlorophyll concentrations, chlorophyll absorption, dissolved organic material and phytoplankton, and surface vegetation.

Figure 10–23, an image that is a part of NASA Goddard Space Flight Center's program of Earth science research, represents the cooperative effort between NASA and NOAA. This First Image of the Global Biosphere combines data from the CZCS between November of 1978 and June of 1986 flown on NIMBUS-7 and information acquired by the AVHRR on NOAA-7 after 1981 over a three year period. The image was originally produced in false color and is shown here in black and white.

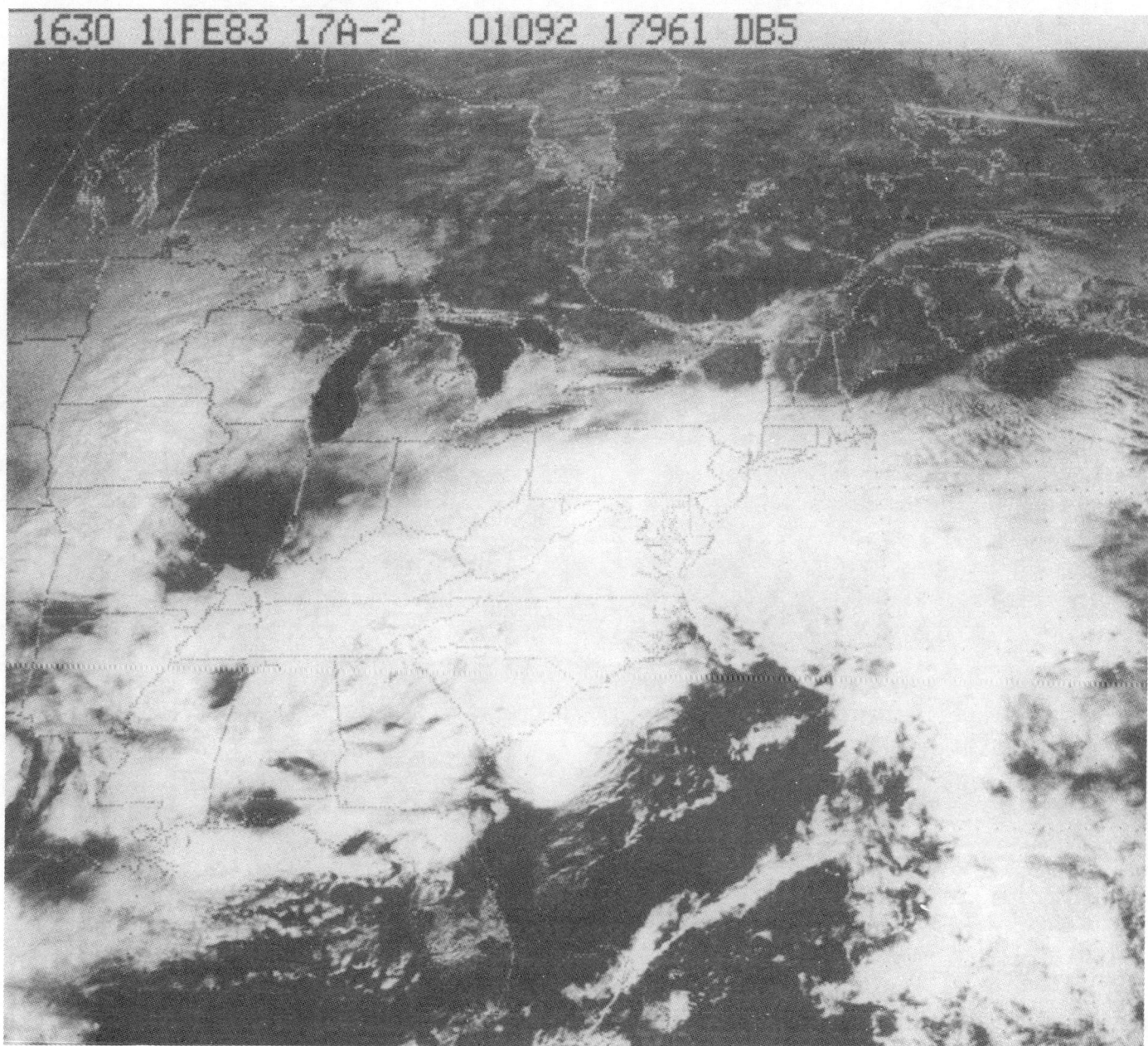

Figure 10–18. NOAA GOES visible image, eastern USA. Acquired 1630 GMT 2/11/83.

The different shades of gray on the ocean part of the image indicates the abundance of phytoplankton. These are microscopic organisms that are the bottom of the food chain for all marine life and so hold great importance. Since they contain a form of chlorophyll that changes the color of the ocean water, they can be detected by several narrow bands of the visible spectrum. The lightest shades of gray in high latitude water areas indicate the highest concentrations. Low latitude darker shades have the lowest amounts.

Land vegetation information was collected from 15,000 orbits of the AVHHR on NOAA-7 which measures land-surface radiation data which can be interpreted for potential vegetation coverage. The darkest gray tones on land areas represent the heaviest potential coverage of vegetation, and the lightest shades indicate drier, grassland areas. The shade closest to white coincides with the desert areas of the world.

Chapter 10   Satellite Mapping of Earth's Atmosphere and Water Resources   **121**

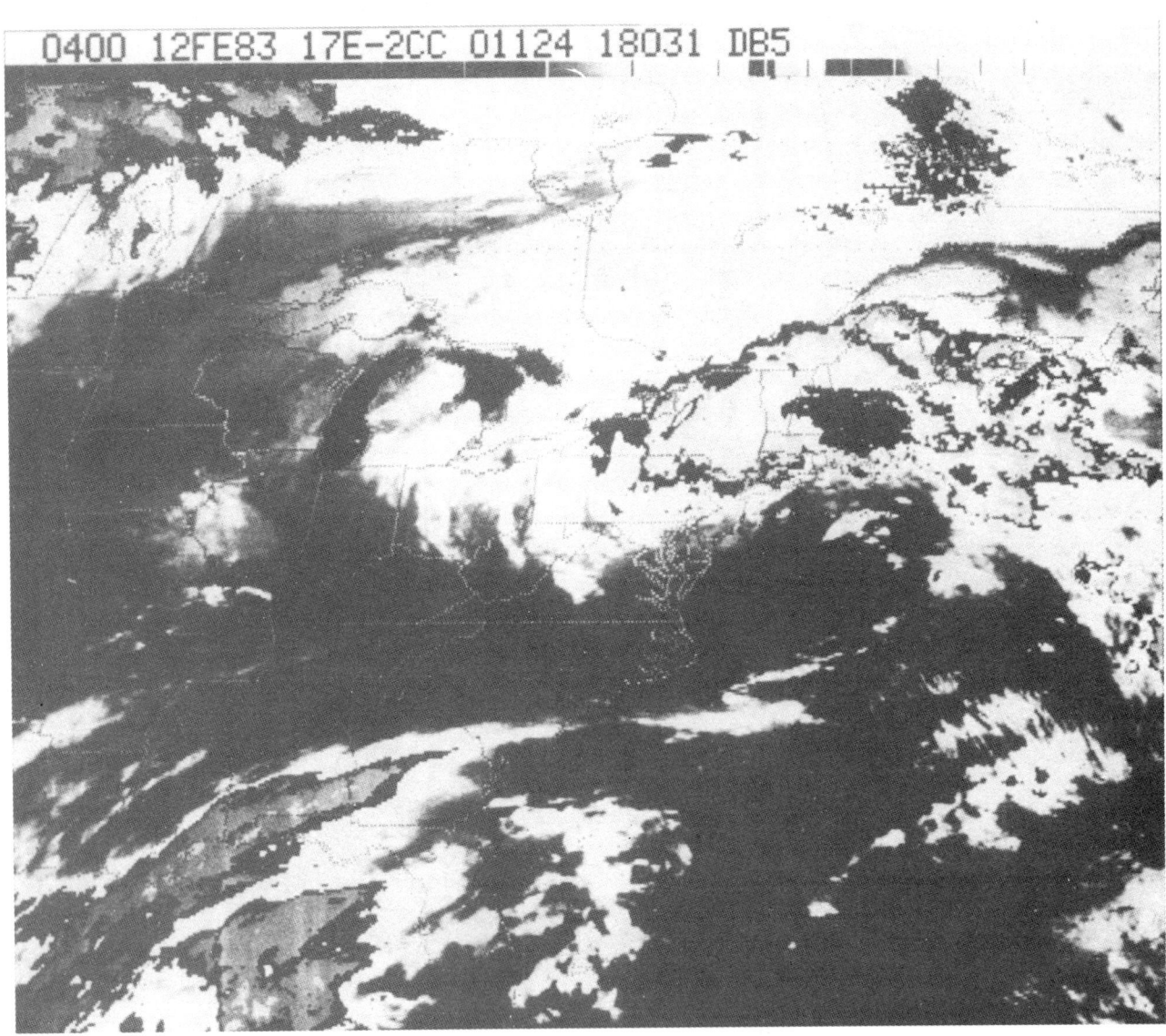
`0400 12FE83 17E-2CC 01124 18031 DB5`

Figure 10–19. NOAA GOES enhanced IR image, eastern USA. Acquired 0400 GMT 2/12/83.

This study is one of many being conducted by NASA and other cooperating agencies over a period of years as part of NASA's research program, **Mission to Planet Earth** (MTPE). That is a component of an even larger, international effort called **Global Change Research Program** (GCRP). A number of countries and many agencies of the United States are cooperating in an attempt to understand the Earth processes and their interaction with the activities of humans on the Earth in effecting global change.

Specific investigations are being focused upon the Upper Atmosphere Research Satellite Mission, Earth Probes such as the Tropical Rainfall Measuring Mission, the Total Ozone Mapping Spectrometer (TOMS), and the Earth Observing System (EOS). The archiving of land data will be carried out by the US Geological Survey (USGS), and NOAA will be the agency responsible for handling atmospheric and oceanic data.

U.S. Air Force Program. The U.S. Air Force conducts a **Defense Meteorological Satellite Program (DMSP)** which normally consists of two satellites in near-polar orbit at an altitude of approximately 515 miles (830 km). On board is an **Operational Linescan System (OLS)** which has the highest resolution of any meteorlogical satellite and is able to acquire daytime and night-time visible-near IR images (0.4 to 1.1 microns) with moonlight or any other ambient light. The OLS also has a thermal IR band operating at 10.5 to 12.5 microns.

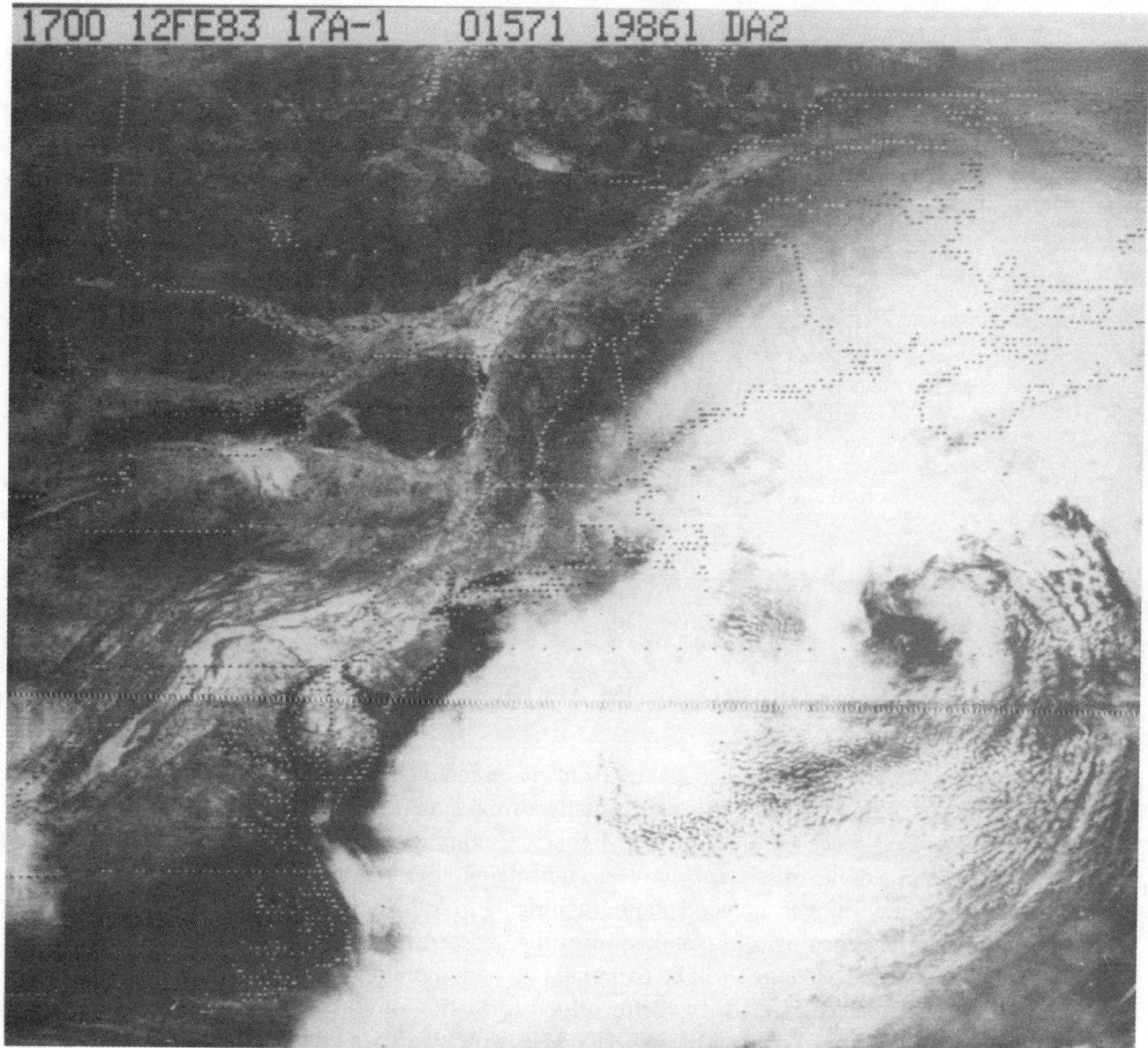

1700 12FE83 17A-1   01571 19861 DA2

Figure 10–20. NOAA GOES visible image, eastern USA. Acquired 1700 GMT 2/12/83.

Figure 10–24 is a DMSP image mosaic acquired on cloudless nights over the United States. They were acquired at local midnight during three, separate passes in the winter months of January through March to obtain a clean **visible-infrared** image that is an excellent display of urban lights from southernmost Canada to Mexico City. The same characteristics of the sensor that make it possible to record city lights are also ideal to acquire the pattern of lightning and fires at night.

Combining satellite image resources. When a naturally occurring event holding the potential for human disaster can be monitored from space, it is imperative that all available resources be made available to this end. Figures 10–25 and 10–26 are images of Hurricane Andrew which struck Florida and the Gulf coast late in August of 1992.

Figure 10–25 is an image acquired by NOAA-10 at 8:34 AM EDT as the eye of the storm was exiting the western coast of Florida. Note the cloud pattern indicating circulation around the eye. In Figure 10–26 a night (10 PM EDT) image composite of visual and infrared from the Air Force DMSP satellite shows the hurricane is over the Gulf of Mexico on its path to the Louisiana coast. Note the absence of lights south of Miami's bright signature and the pattern of lights across the southeastern United States.

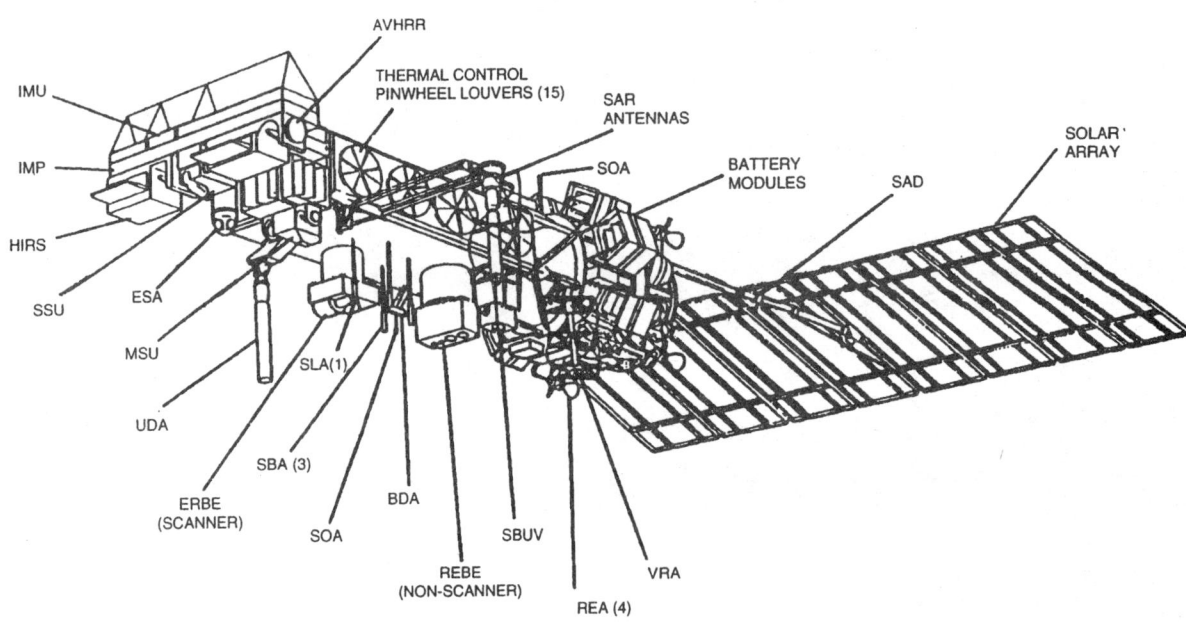

Figure 10–21. Advanced TIROS-N (ATN) vehicle.

HAND HELD PHOTOGRAPHY FROM SPACE

A few thousand photographs have been taken from space and most of these have been **"hand held photography"** acquired by the astronauts, usually using a modified 70mm Hasselblad camera. Many of the manned space missions of the Mercury, Gemini, Apollo, Skylab, and Space Shuttle programs had as one of the mission objectives acquiring photographs related to a theme. Some of the various mission objectives included photographing hydrologic features, cloud formations, fires on the Earth's surface, geologic formations, oil spills, storms, lightning, and evidence of human activities. Some of these features will be examined on hand-held photography in Chapter 12.

In Figure 10–27 one can see the top of the cloud formation of Hurricane Elena in an oblique photograph acquired from Shuttle Mission STS 51I. Note that the eye is easily discernible and the motion and direction of the wind is evidenced by the cloud pattern.

Exercise 10   Satellite Mapping of Earth's Atmosphere and Water Resources

1. Compare and contrast the characteristics of the Polar Orbiter satellites of the United States with the Geostationary satellites. What are the advantages and disadvantages of the two types?

2. Review Figure 10–6. Why is it important for the United States to have both GOES East and GOES West functioning at the same time? If only one satellite is operating, how might that impact people's lives?

3. Examine Figure 10–7. Use a piece of drafting film or vellum to trace some of the prominent cloud features. Locate as many mid-latitude cyclonic storms as you can in both Northern and Southern hemispheres. Can you locate any cloud forms that owe their origin in part to orographic control? Do you see any evidence of tropical

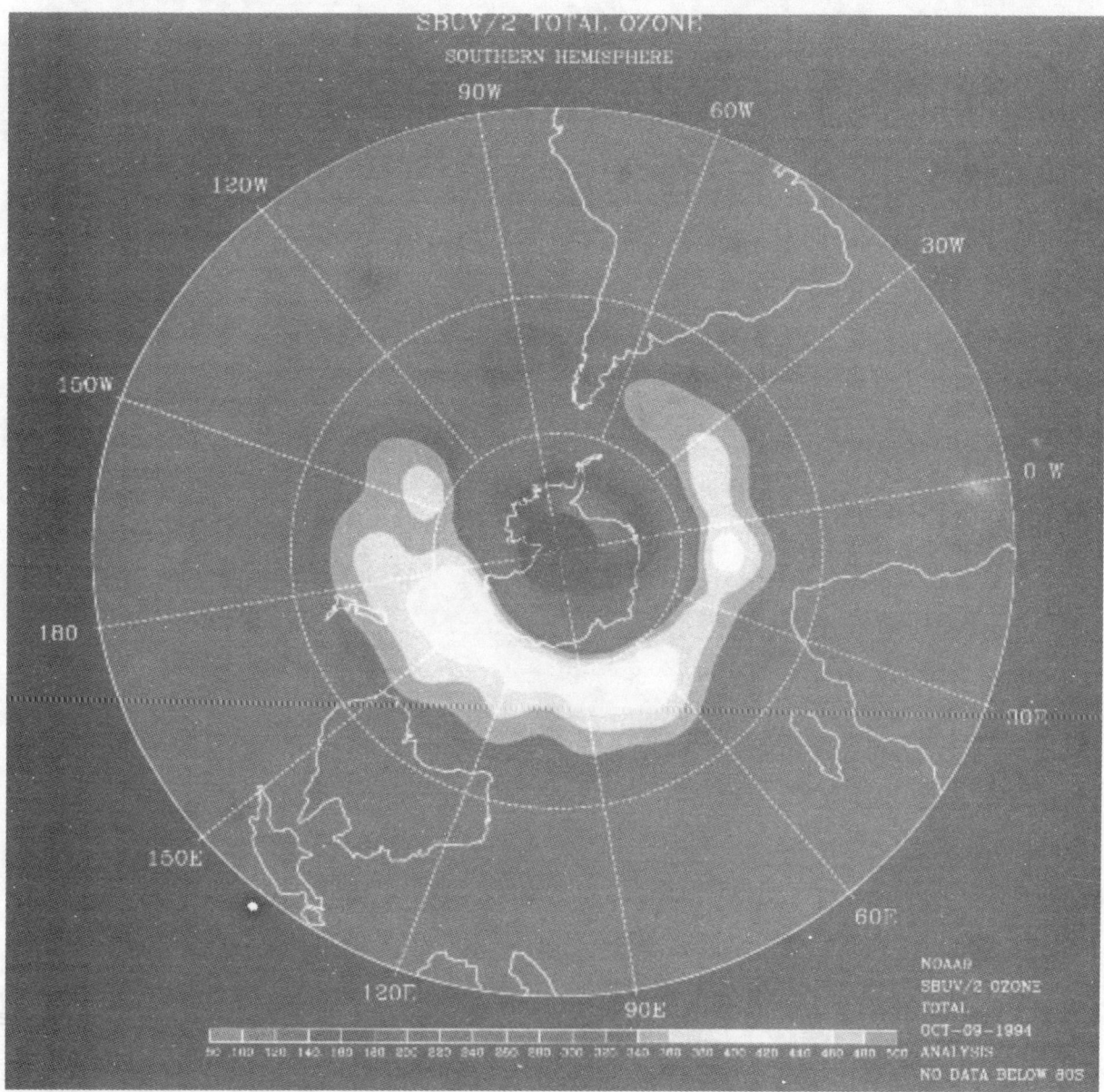

Figure 10–22. NOAA9 Total Ozone Concentration. Southern Hemisphere. Acquired 10/9/94.

cyclones? Are there any cloud masses that might be explained as meso-thunderstorms? Plot your results on the overlay. Also locate any areas that are likely the center of a high pressure cell or ridge.

4. Figure 10–8 is a full disc IR GOES image. Review the in formation supplied in the text of Chapter 10, if necessary, then answer these questions. Are there areas in this image that might have had considerable cloud cover and yet clouds are not conspicuous? Explain. Can you find a good example of such an area? Locate that area on an overlay sheet as well as any areas of strong frontal development. Can you find any cloud masses that have experienced strong orographic lifting to bring them to high altitudes? Plot them on the overlay

What is the evidence for fairly strong development of tropical depressions over the Pacific Ocean at about 10° north latitude? Explain how the thermal IR image adds information that is not shown on the visible GOES image.

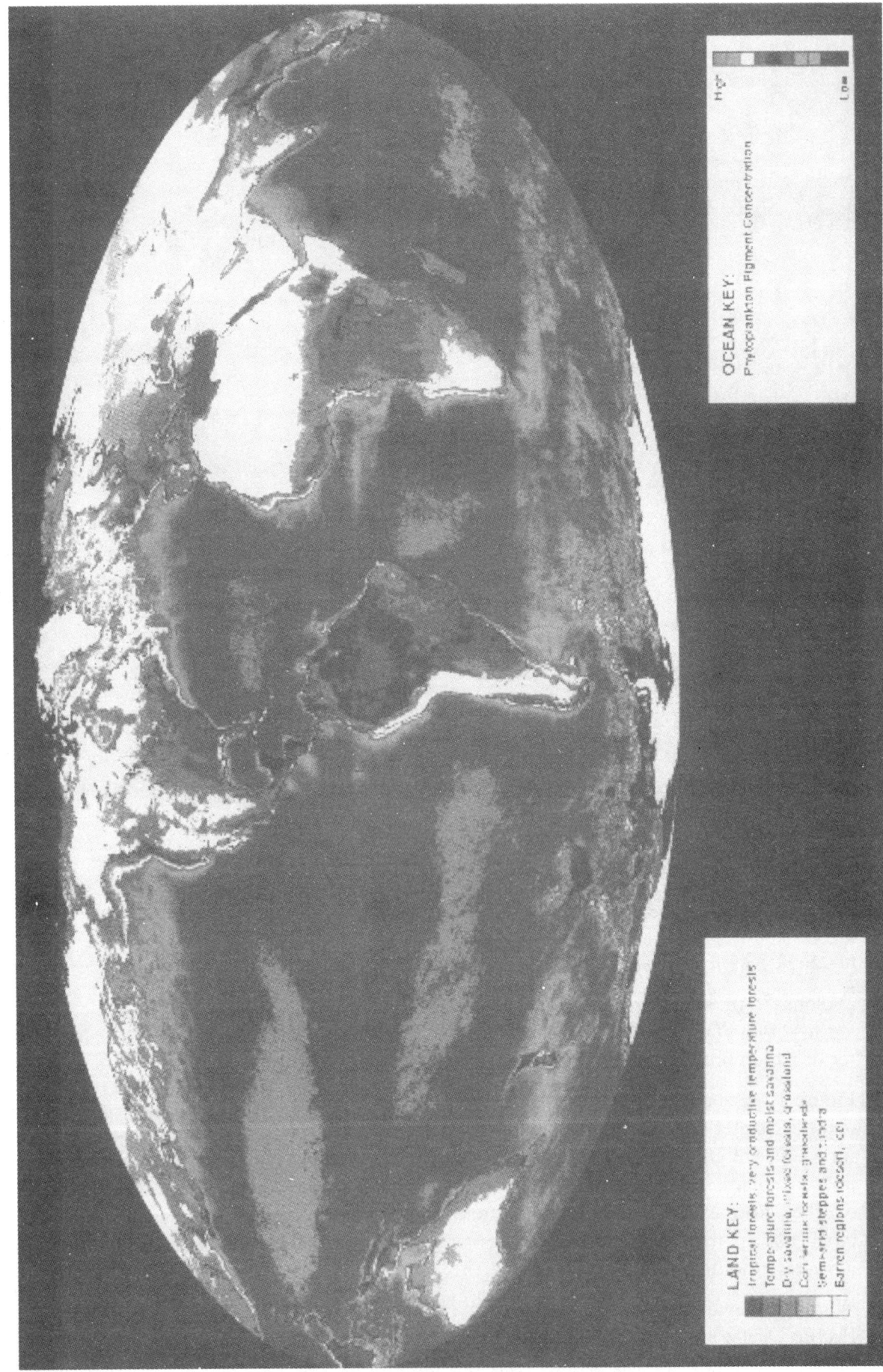

OCEAN KEY:

Phytoplankton Pigment Concentration

High

Low

LAND KEY:

Tropical forests, very productive temperature forests

Temperature forests, mixed forests, moist savanna

Dry savanna, mixed forests, grassland

Coniferous trunk forests, grasslands

Semi-arid steppes and tundra

Barren regions (desert, ice)

Figure 10–23. NASA NOAA First Image of the Global Biosphere.

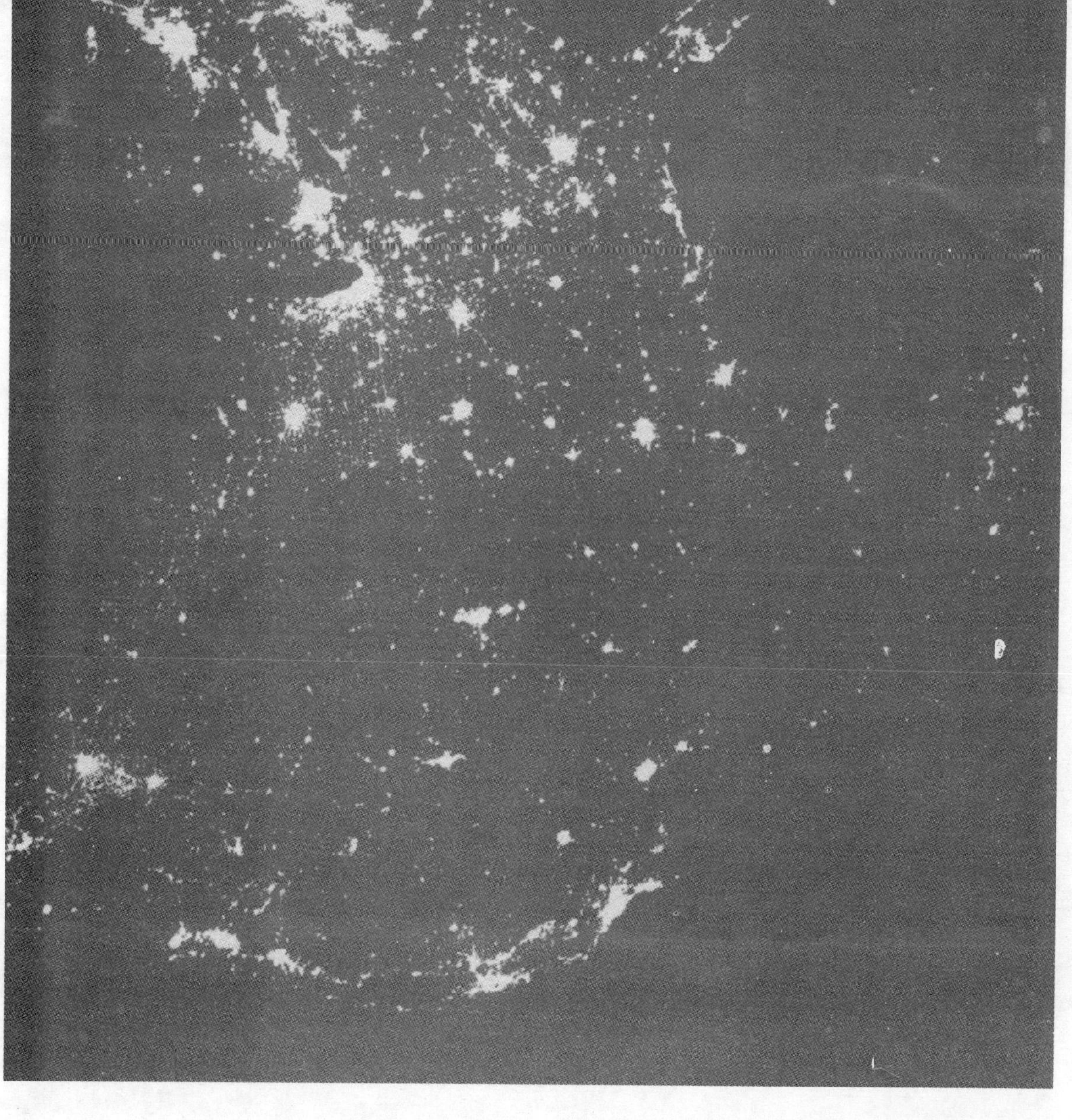

Figure 10–24. U.S. Air Force DMSP OLS image nighttime visible-infrared. Acquired January–March Courtesy Hank Brandli, Satellite Meteorologist.

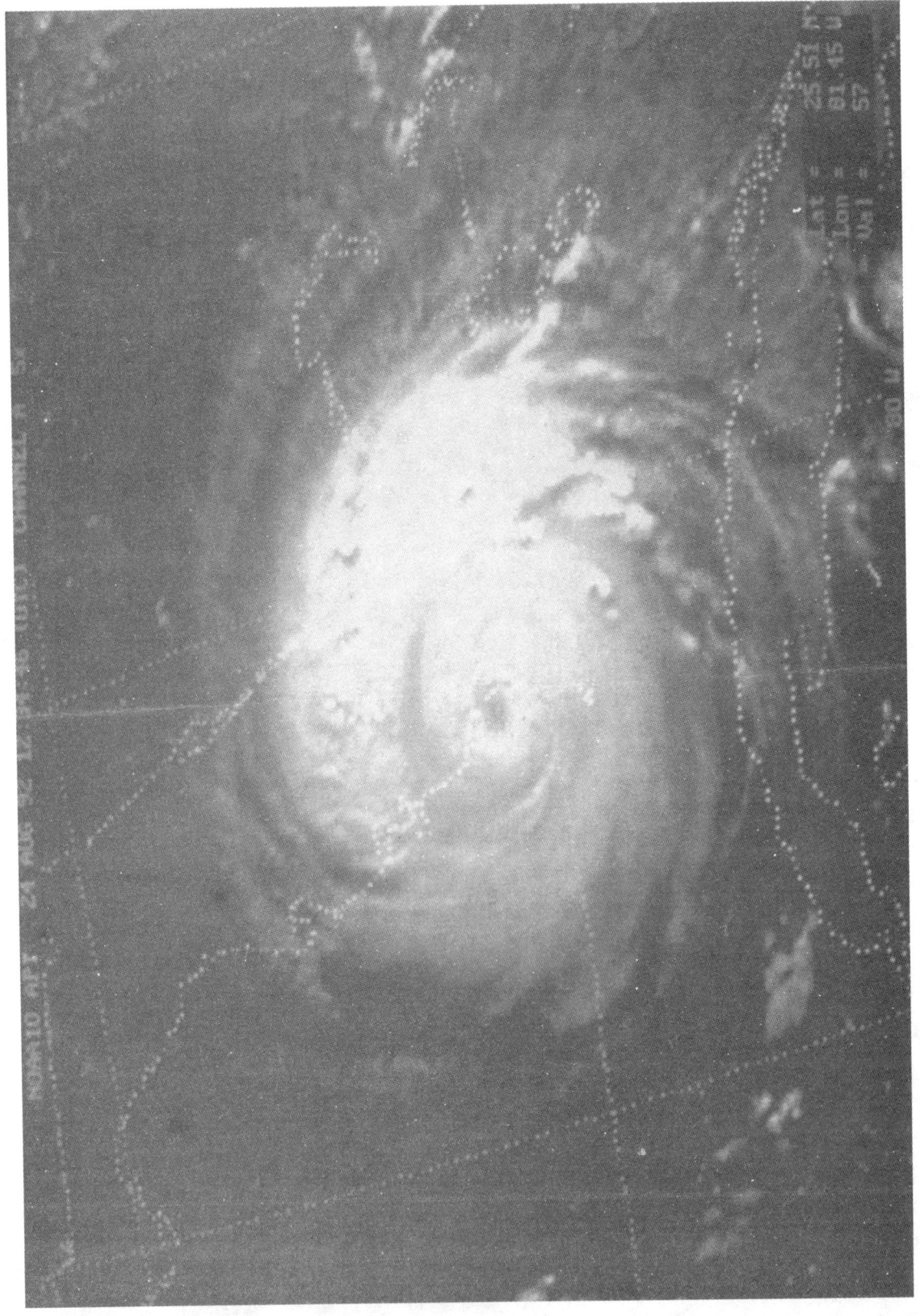

Figure 10–25. NOAA10 image of Hurricane Andrew eye off west coast of Florida. Acquired 8:34 AM EDT 8/24/92 Courtesy Hank Brandli, Satellite Meteorologist.

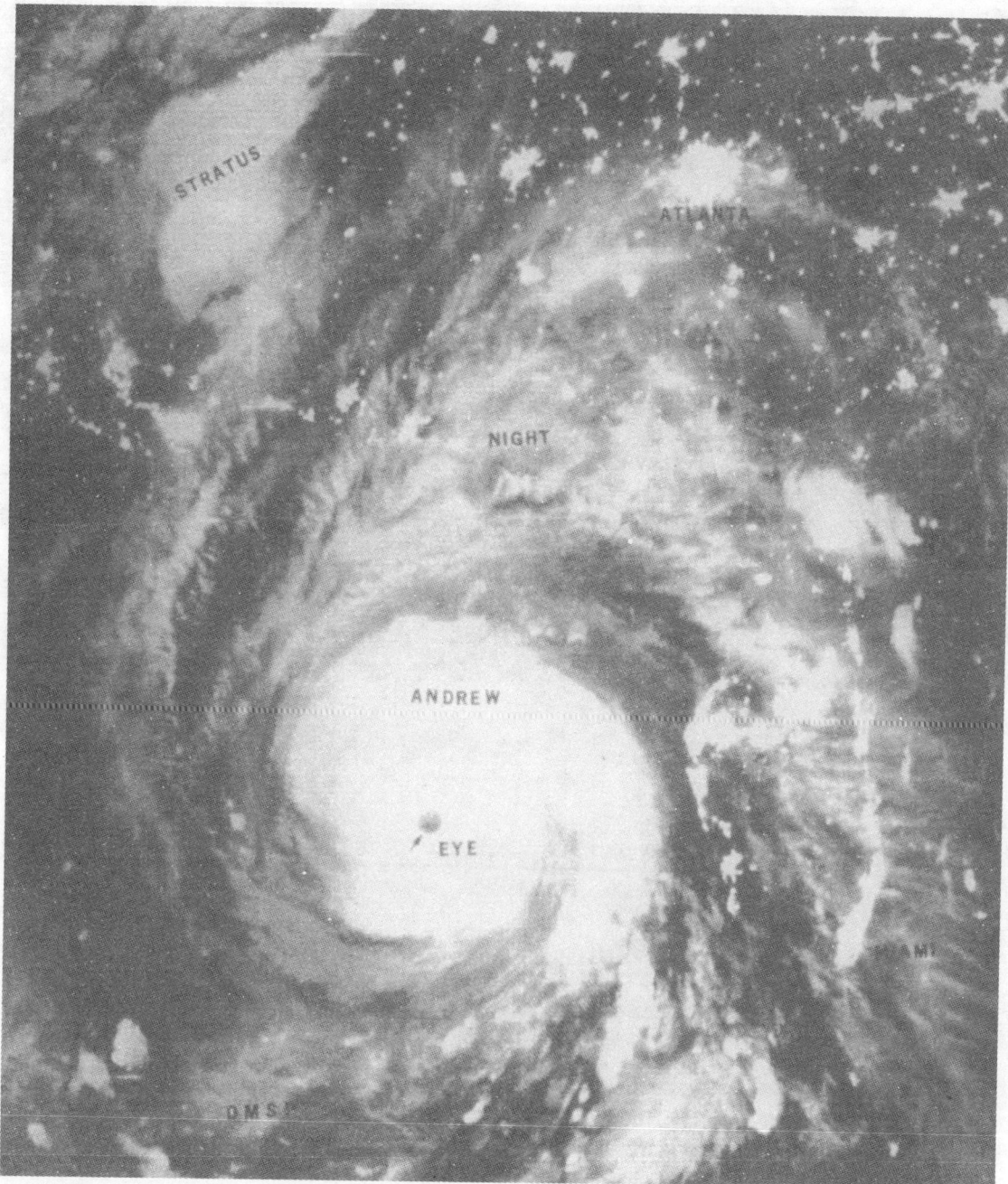

Figure 10–26. U.S. Air Force DMSP OLS image nighttime visible-infrared. Acquired 10 PM 8/24/92. Courtesy of Hank Brandli, Satellite Meteorologist.

5. Compare Figures 10–9 and 10–10. The former portrays water vapor and the latter is a thermal infrared version of the same area that was recorded 2 1/2 hours later—both are nighttime images from the GOES satellite. Note the position of the front extending from eastern Texas to the estuary of the St. Lawrence River. If lighter gray tones indicate more water vapor and darker tones represent less water vapor, how does that match up with the front on Figure 10–9? If this is a cold front, where would you expect to find the highest concentrations of water vapor? Compare the area west of the front on each image. Is the information provided on these images compatible? Explain.

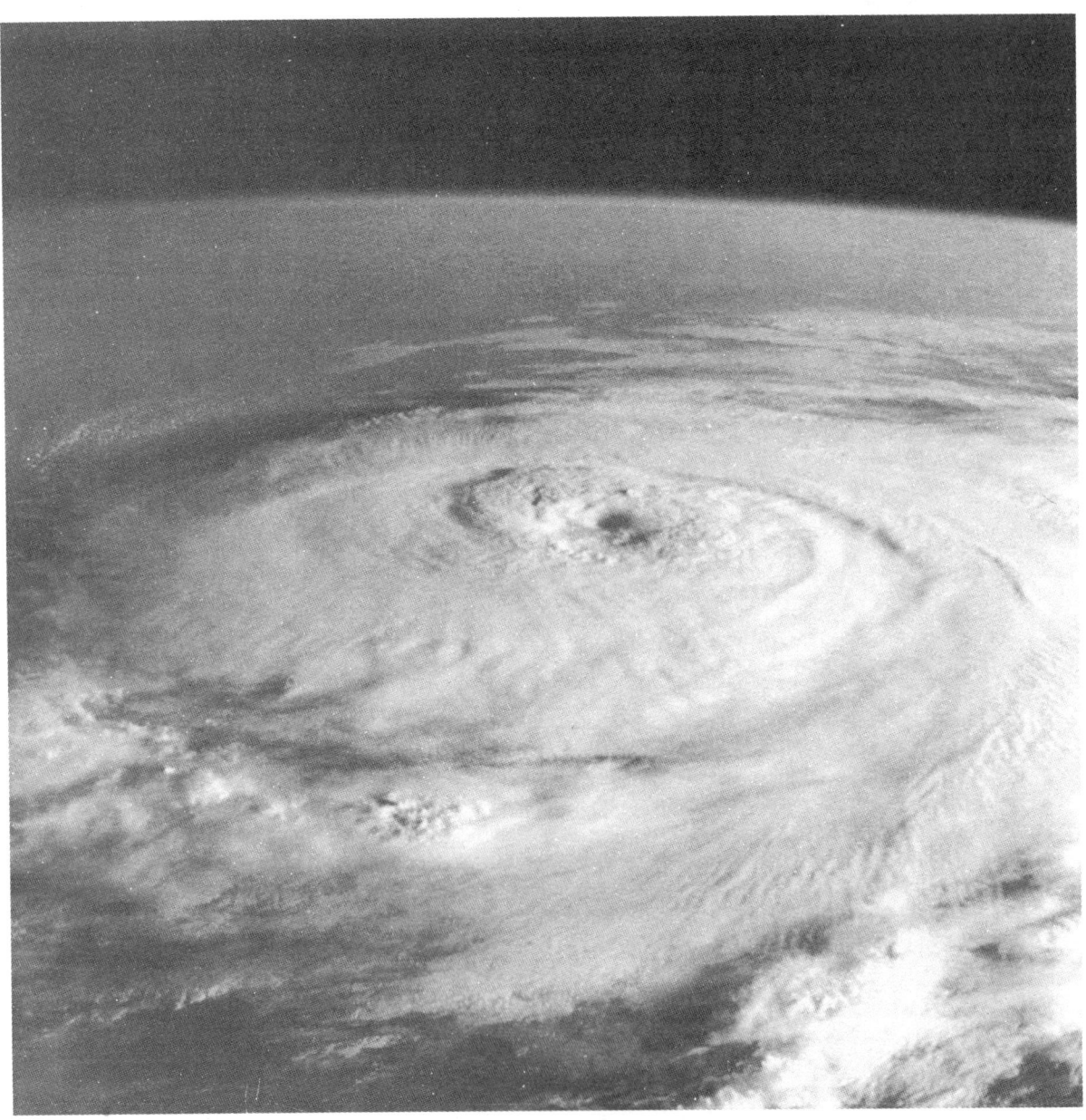

Figure 10–27. NASA STS 51I-44-0052 Hurricane Elena over Gulf of Mexico, September, 1985. Courtesy Media Services, Johnson Space Center

6. Figures 10–11 through 10–16 are reproductions of the 7 AM EST Surface Weather Map and the 500 millibar height contours for the period Thursday February 10 through Saturday February 13, 1983. Note the location of the low pressure over the Gulf coast of Florida–Alabama at the surface, its movement, and the frontal development during the 48 hour period. Note also the direction of wind flow at the 500 mb level. Compare the weather maps mentioned to the four GOES images, Figures 10–17 to 10–20 taking care to observe the times of image acquisition (GMT is 5 hours later than EST). Also, be aware that images annotated 17E-2CC are thermal IR and those marked 17A-2 are visible. Using drafting film or vellum for overlay material, plot the location of the fronts (see in side cover for symbol information), centers of high and low pressure, and likely areas of precipitation for each of the images. Use the weather maps to assist you in interpolation of data.

7. See Figure 10–22, Total Ozone for the southern hemisphere on 10/09/94. Considering the amount of data collected, the area and volume of the atmosphere involved, and the number of variables involved, is it surprising that the existence and severity of the "ozone hole" is controversial? Why is there likely to be political debate over this subject? Is potential cost of programs to ameliorate this situation a likely factor in the debate? Explain.

8. Examine Figure 10–23, *First Image of the Global Biosphere*, in regard to oceanic distribution of phytoplankton. Note the highest concentrations are at middle and higher latitudes or along the coasts of land masses. What might be factors that would account for this distribution? How does the pattern of vegetation on land correspond with climate regions of the Earth? Explain. What does this tell us about scientific efforts to classify Earth characteristics, such as climate, before such views from space were available?

9. In Figure 10–24 the nighttime pattern of lights is displayed in this DMSP mosaic. Compare this image with a map of North America or the United States. Are all of the major cities represented by the lights?

Note the area from just north of Boston, MA (A-3 area) to Washington, DC. This has been referred to as "Megalopolis," one continuous metropolitan area. What is the evidence in the image for this designation? There are several other areas in the image where other such megalopolitan urban entities seem to be growing. Can you identify them?

Note the pattern of distribution of the largest cities (areas of most light). There are many theories related to urban location and growth. Can you explain why some of these cities have located where they did? How might such an image as this lend itself to understanding urban location patterns?

10. Figures 10-25 and 10-27 are images of two hurricanes in the visible part of the spectrum. The former is a NOAA-10 daylight scan of Hurricane Andrew and the latter is a handheld photograph from the Space Shuttle of Hurricane Elena.

Note the pattern of circulation evident in both storms. If it is true that hurricanes become stronger as the diameter of the "eye" grows smaller and as the cloud mass becomes more compact, what would you estimate the danger of these storms to be? Note the eye of Andrew is just exiting the west coast of Florida. If the storm was moving from east to west and the wind was moving in a counter-clockwise inspiral near the Earth's surface, what part of the west coast of Florida relative to the eye would likely experience the highest speed winds?

How does tracking these storms from space assist in the prediction of areas of greatest danger and potential damage? Explain. If these were thermal infrared images that were enhanced to show areas of greatest convection, how would that assist in evaluating potential danger and damage?

11. In Figure 10–26 another image of Hurricane Andrew, this one acquired by the Air Force DMSP satellite, is shown as it approaches the coast of Louisiana at 10 PM EDT on 8/24/92. This is a visual/infrared composite that shows city lights, lightning, and cloud patterns. Compare this image with a map of the southeastern United States. What might explain the lack of city lights south of Miami and in the Florida Keys?

The highest sustained wind in the storm was 140 mph at this point. What part of the coastal area would likely be hit hardest relative to the eye left of eye, along the path of the eye, or right of eye? Explain.

Based upon the diameter of the eye and the compactness of the cloud pattern, would you say the storm maintained most of its force or did it appear to be weakening?

If such storms draw most of their energy from the release of latent heat of condensation (when water vapor changes to liquid), what is likely to happen to the storm when it moves over land and away from the Gulf waters, a major source of water vapor?

chapter *11*

INTERPRETATION OF RADAR IMAGERY

TERMINOLOGY

Radar
active system
passive system
plan position indicator (PPI)
Side Looking Airborne Radar
   SLAR
Ka, X, L band radars
Rayleigh criterion
depression angle
grazing angle
real aperture radar
synthetic aperture (SAR)
look direction
azimuth direction
ground resolution cell
range resolution
azimuth resolution

beam width
Doppler principle
slant range distance
radar return
specular reflection
diffuse reflection
site factors
situation factors
horizontal polarization (HH)
vertical polarization (VV)
cross polarization (HV, VH)
banding
Seasat
layover
Space Shuttle
Shuttle Imaging Radar (SIR-A)
complex dielectric constant

HISTORICAL PERSPECTIVE

In the late 1800s and early 1900s, it was discovered that transmitted radio waves would reflect off solid objects. During the 1920s Great Britain and the United States began study of this phenomenon to use in detecting ships and aircraft. This led to the creation of **RADAR (Radio Detection and Ranging)**.

Radar is unusual compared to most other remote sensing systems in that it is an **active system.** The greatest portion of sensors that have been used to collect data without being in contact with the target have recorded reflected or emitted energy that was present at the target. That is, they are **passive systems.** In contrast, radar sends out an electronic signal at various wavelengths that bounces off the target and the energy that returns to the radar device is recorded. It may be used when clouds are present or when there is no visible light as it "paints" the target with its own beam of energy.

During World War II a form of radar was developed to use for navigation and target location. This was called **Plan Position Indicator (PPI)** and featured a rotating antenna and images produced on a circular viewing screen (cathode ray tube). The image was displayed with a contin-

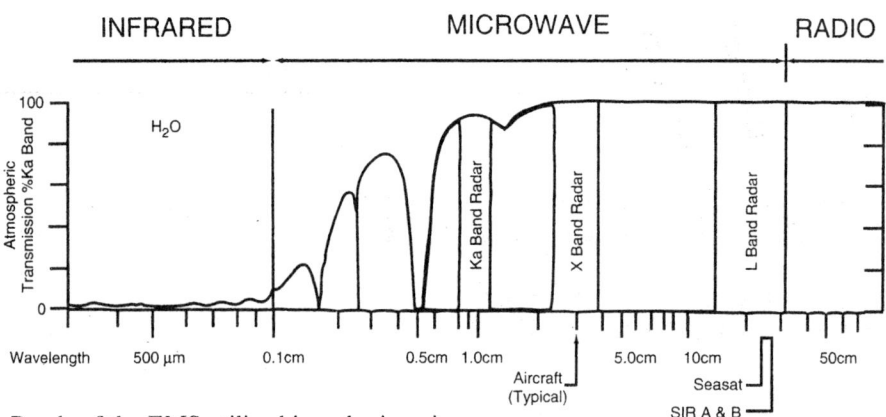

Figure 11–1 Bands of the EMS utilized in radar imaging.

uous 360° sweep designed with timed picture decay. This type system is still utilized for weather radar as well as on ships and aircraft.

In the 1950s **SLAR (Side Looking Airborne Radar)** was developed with a continuous strip data-collection capability. This made it possible to carry out reconnaissance of an area without actually flying over it. This opened up the possibility of mapping any area at any time since radar has day/night capability.

SYSTEM CHARACTERISTICS

Figure 11–1 shows the portion of the electromagnetic spectrum that is involved with radar imaging. There are three wavelengths that are employed in radar imaging: **Ka** (0.8–1.1 cm), **X** (2.4–3.8 cm), and **L** (15.0–30.0 cm). Each of these wavelengths is likely to produce a different image of the same target area, as the apparent roughness of the target surface texture will vary according to the length of the wave painting it. The shorter Ka and X bands are more often used for imaging as they provide better resolution while the L band is used in cases where greater penetration of the target surface is desired.

The **Rayleigh criterion** shown below and in Figure 11–2 is an expression of target surface irregularities, radar wavelength, and **depression angle** (the angle between a horizontal at aircraft and target).

$$h < \frac{\lambda}{8 \sin \gamma}$$

h = height of surface irregularities or roughness
λ = (lambda) wavelength
γ = (gamma) depression angle

Figure 11–2 illustrates the effect of radar wavelength in producing different responses to the same ground surface. The result of this relationship is that a target surface that appears smooth on an L band radar could appear somewhat rough on an X band version, and rougher still on a Ka band image. Figure 11–3 illustrates the concept of depression angle and **grazing angle**. The angle at which the radar beam strikes the ground is also a factor in the type of image produced. This is analogous to changing the angle at which a flashlight illuminates a basketball resting on the ground, thus creating shorter or longer shadows.

There are two basic types of SLAR systems, **brute force radar** or **real aperture** and **synthetic aperture radar (SAR)**. Brute force radar is considerably less expensive, but is not as desirable as the SAR type because of changing resolution values with increasing distance from the plane.

Resolution on radar images has two directional components, one in the **look direction** or **range** and another in the **azimuth direction** (or flight path). The radar transmitter sends out bursts

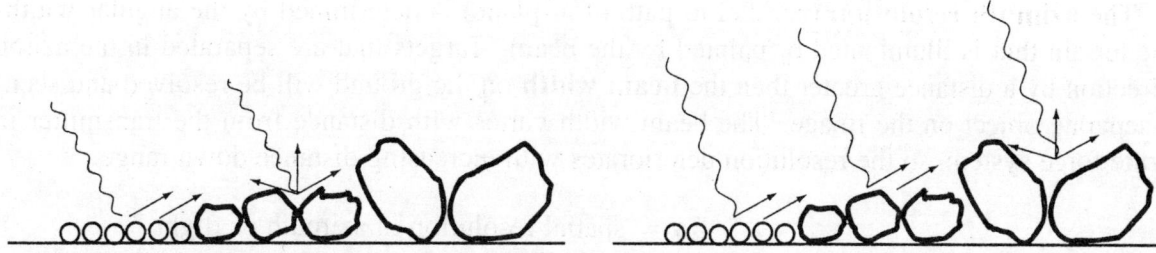

Figure 11–2 Radar wavelength response to surface roughness.

of electromagnetic energy that illuminate a narrow strip of terrain to the right or to the left of the aircraft. This energy strikes objects on the surface that send strong or weak returns to the aircraft based upon several factors.

The **ground resolution cell** is determined by the range resolution plus the azimuth resolution. **Range resolution** is a function of the pulse length in microseconds, the speed of electromagnetic radiation, and the depression angle of the transmitted beam. The range resolution rule of thumb is that the resolution is approximately equal to one-half the duration or length of the transmitted pulse. So if two objects are farther apart in the look direction than half the length of the pulse, the two objects will be seen as two. If they are closer together than half of the pulse length, they will be seen as one object on the radar image.

$$Rr = \frac{\tau c}{2 \cos \gamma}$$

Rr = spatial resolution in range direction
τ = (tau) pulse length in microseconds
c = speed of electromagnetic radiation
γ = (gamma) depression angle from a horizontal at aircraft to target

Example: A depression angle of 50° and a pulse length of 0.1 microseconds would yield a range resolution of 23.4 meters.

$$Rr = \frac{(0.1 \times 10^{-6} \text{ sec})(3 \times 10^8 \text{ m} \times \text{sec}^{-1})}{2 \cos 50°} = \frac{30 \text{ m}}{2 \times 0.64} = 23.4 \text{ m}$$

Therefore objects closer than 23 meters to each other would be resolved as one object, whereas those farther apart than 23 meters would appear separately on the radar image.

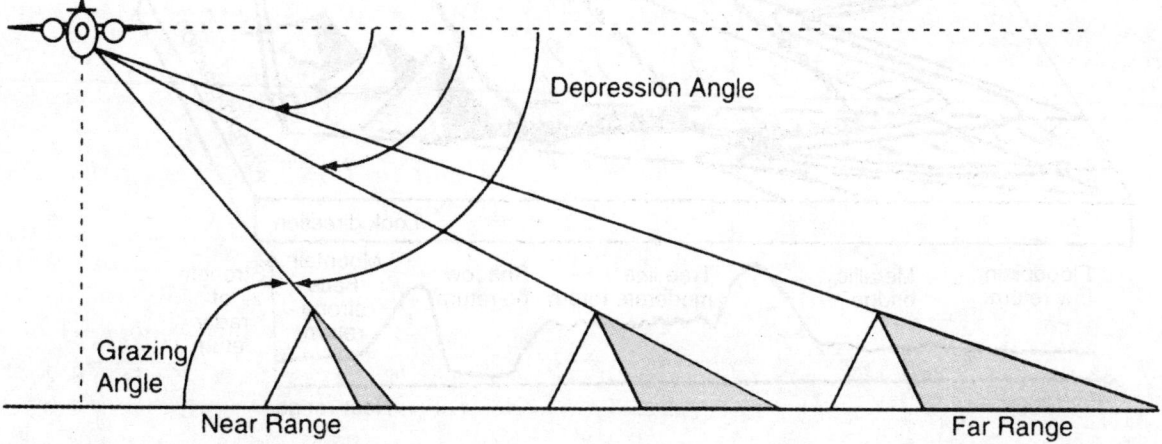

Figure 11–3 Depression angle, grazing angle, radar shadow.

The **azimuth resolution** (parallel to path of airplane) is determined by the angular width of the terrain that is illuminated or painted by the beam. Targets that are separated in the azimuth direction by a distance greater than the **beam width** on the ground will be resolved and seen as a separate object on the image. The beam width varies with distance from the transmitter in a brute force system, so the resolution deteriorates with increasing distance down range.

$$Ra = \frac{0.7 \ SR \ \lambda}{D}$$

Ra = spatial resolution in azimuth or flight path direction

SR = slant range distance (direct distance) from transmitter to target

λ = (lambda) wavelength of transmitted energy

D = antenna length

Example: Using a typical wavelength of 0.86 cm for a Ka band radar system and an antenna length of 500 cm, a near slant range distance of 5 km would yield an azimuth resolution of 6 m while a far slant range distance of 20 km would produce an azimuth resolution of 24 m.

$$Ra = \frac{0.7 \ (5 \ km \times 0.86 \ cm)}{500 \ cm} = 6m \qquad Ra = \frac{0.7 \ (20 \ km \times 0.86 \ cm)}{500 \ cm} = 24m$$

In a synthetic aperture system (SAR) the antenna length is electronically produced as being quite long which results in narrow beam width and high resolution on the ground. This is accomplished by using the **Doppler principle** to produce a beam width that is not only narrow, but remains the same at varying distances from the aircraft. Obviously a SAR system is to be preferred over a brute force, real aperture system. This is especially true for radar systems on satellites as the **slant range distance** will very likely be quite long. Figure 11–4 illustrates the basic features of a SLAR system as it would operate from an aircraft.

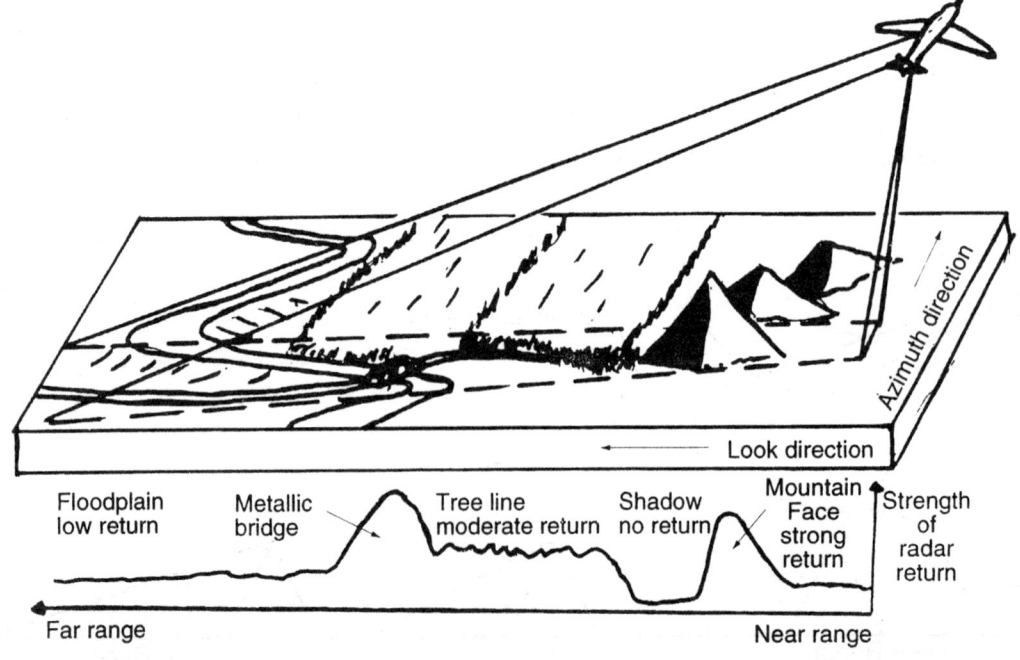

Figure 11–4 Impact of terrain of radar return.

RADAR RETURN AND IMAGE SIGNATURES

Radar images often appear to be photographs as there are varying textures, brightness levels, and shadows. This is not the case, although many radar images possess an amazing level of detail. It is most important to remember that the radar image is acquired by bouncing an electronic impulse off the features in a ground scene. Bright areas on the image are produced by a strong **radar return** and the darker areas by weak returns. **Radar shadows** are produced in areas that are not illuminated by the radar beam.

Characteristics that can produce a strong or weak return are related to the orientation and surface slope of features, their surface texture, and the material of which they are made. Moisture levels are also a factor. Generally the resolution parameters previously mentioned will determine whether an object will be resolved and appear on the image. However, sometimes a metallic feature such as a bridge or chain link fence will show up on the image as a strong return despite its size.

Smooth objects that present a surface at right angles to the radar beam will yield a very strong return and appear bright on the image. This can occur when topographic features are oriented in just the right compass view and slope aspect as seen from the aircraft. Urban areas typically provide many strong returns that are referred to as **corner reflectors.** When the radar beam strikes the side of a building it will bounce off at the same angle as the incoming energy, hit the street, and then rebound from the street directly back to the radar receiver. (The radar receiver and transmitter are part of the same device that alternately sends and receives energy.)

Figure 11–5 illustrates the strong returns mentioned above and also displays other surface return characteristics. When an area on the imaged surface is smooth, the radar signal tends to bounce off the ground at the same angle as it approached and so little or no return gets back to the receiver. This is referred to as **specular reflection** and is similar to the way in which visible

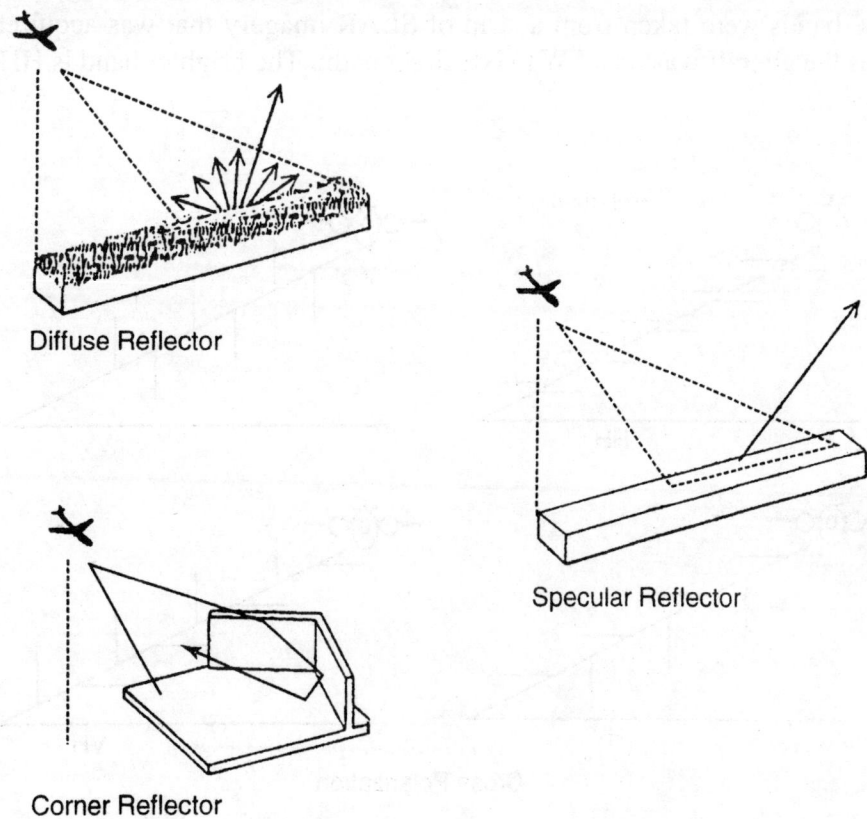

Diffuse Reflector

Specular Reflector

Corner Reflector

Figure 11–5 Diffuse, specular, and corner radius.

light reflects off a mirror. If the ground area is rough or broken up, the radar beam will be deflected in many different directions, some of the signal returning to the receiver. This circumstance is called **diffuse reflection.**

Note that the strength of the radar return depends upon several characteristics of the target and this must be considered in carrying out interpretation of radar images. The surface of water bodies or ice-covered areas may be smooth or rough and so the appearance on the image can vary although the material has not. One may perceive a collection of very strong returns that is typical of an urban area, but similar signatures can be provided by other "culture" phenomena such as industrial areas. It is also possible that physical landscapes of many surface angles and aspects could yield a comparable signature. Other characteristics of **site, situation,** and **associated features** must be considered.

Another factor that affects the data recorded by radar systems is the **polarization** of the radar signal. The radar energy is transmitted with electronic vibration in a **horizontal** or **vertical** plane. Upon striking the target surface most of the energy will remain in the form of polarization in which it was transmitted, returning to the receiver in the same orientation. A common designation for energy sent and received in horizontal vibration is **HH.** Likewise energy sent and received in vertical mode is termed **VV.**

Some of the energy transmitted will change its orientation upon striking the ground and will return to the receiver in the opposite orientation, e.g., **HV** or **VH.** The terminology for this phenomenon is **cross-polarization.** See Figure 11–6.

This occurrence is quite significant as radar returns that are in HH mode will provide a different image of the target area than returns in HV form, likewise for VV vs. VH. Typical SLAR imagery acquired from aircraft will be produced in the form of a continuous roll with the polarized image, either HH or VV, presented next to the cross-polarized HV or VH version. A comparison of the two portrayals with maps or other information related to the target area allows for more detailed interpretation. See Figure 11–7 for an example of SLAR imagery in HH and HV format. These bands were taken from a strip of SLAR imagery that was acquired just west of Boston, MA as the aircraft was on a SW to NE flight path. The brighter band is HH and the other is HV.

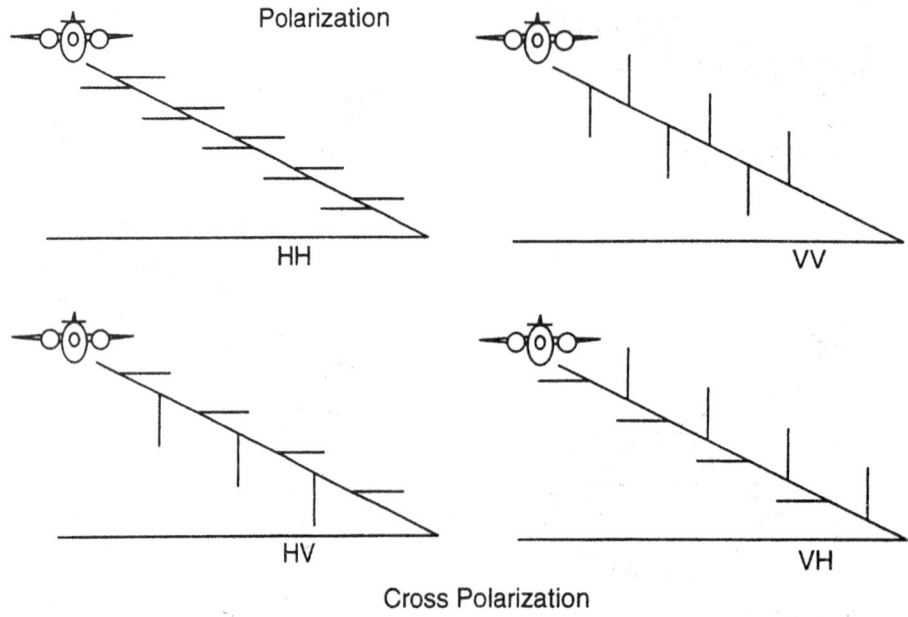

Figure 11-6. SLAR signal and return.

One other image characteristic of such SLAR products should be noted here as it may affect interpretation. Often the HV band will present an interference or "static" phenomenon referred to as **banding.** In Figure 11–7 the banding occurs on the lighter part of the darker band and appears as parallel "waves" or ridges. The effect is usually readily apparent, but one should be aware of it especially when viewing rolling or hilly topography with parallel ridge lines.

Figure 11–8 is a copy of a portion of the Boston USGS topographic map published at a scale of RF 1:250,000 and enlarged here to aid comparison with the SLAR images of the same area. The features that are most easily recognized initially are lakes and reservoirs and large rivers due to shoreline shapes and contrast in tones. Take care in making this comparison, as there is often some distortion or "rubber sheet stretching" of shapes.

Another area acquired on this same mission of July 22, 1966 is displayed in Figure 11–9. The lighter band at the top of the page is HH and the lower band is HV. The image displays the path followed by the Hudson River and also by a more recent glacial advance across the Hudson Highlands north of New York City. These mountains have a number of local names, but are part of the Taconic Mountains and represent some of the oldest rock material in North America. This is an extremely interesting area in regard to geomorphology and history and will be further examined in the questions at the end of this chapter.

Note that both bands in Figure 11–9 have an area of image loss near the center of the scene shown. This was caused by the aircraft turning and the result was no signal return could be recorded. This turn was minor, but on some images the blackout area is much longer corresponding to a major alteration of course.

Figure 11–10 is a portion of the USGS topographic maps at a scale of 1:250,000 that cover the image area. The map has been enlarged for better comparison here. Note the high density of major roads and how prominently they appear on the map and their lower visibility on the images.

SEASAT RADAR IMAGERY

NASA launched the **SEASAT** satellite on June 26, 1978 in order to investigate the capabilities of a space-borne radar system. A major focus of this mission was to acquire microwave imagery of oceans and marine phenomena. One of the five sensors aboard the satellite was an L-band (22 cm wavelength) SAR radar with 25 meter resolution. The depression angle of the SAR system varied from 67° to 73° from the horizontal which resulted in less shadow, more **layover,** and pronounced image displacement. Layover is the displacement of ridge and mountain summits towards the radar system because those areas receive and reflect the radar signal before slope areas at lower elevations do. See Figure 11–11.

The orbit was near polar and averaged 500 miles (800 km) altitude above the Earth which allowed for considerable coverage of the United States, Alaska, central America, and western Europe.

Each scene covered 62 × 62 miles (100 × 100 km) acquired at a 20° angle off nadir. Unfortunately the satellite experienced a major electrical failure on October 10, 1978 and could not be reactivated. Nevertheless, considerable SAR imagery and other information had been recorded and transmitted to Earth during the life of SEASAT. These data are still being analyzed.

Figure 11–11 is a SEASAT image of the Denver, CO area that displays all of the characteristics of spaceborne SAR over land areas. Note the distinct boundary between the high plains and the Rocky Mountains along the Rocky Mountain "front." The layover effect is quite noticeable in the presentation of the "flatirons" just west of Denver. Note also the differences in patterns shown within the urban setting as opposed to the hydrologic information east and south of Denver.

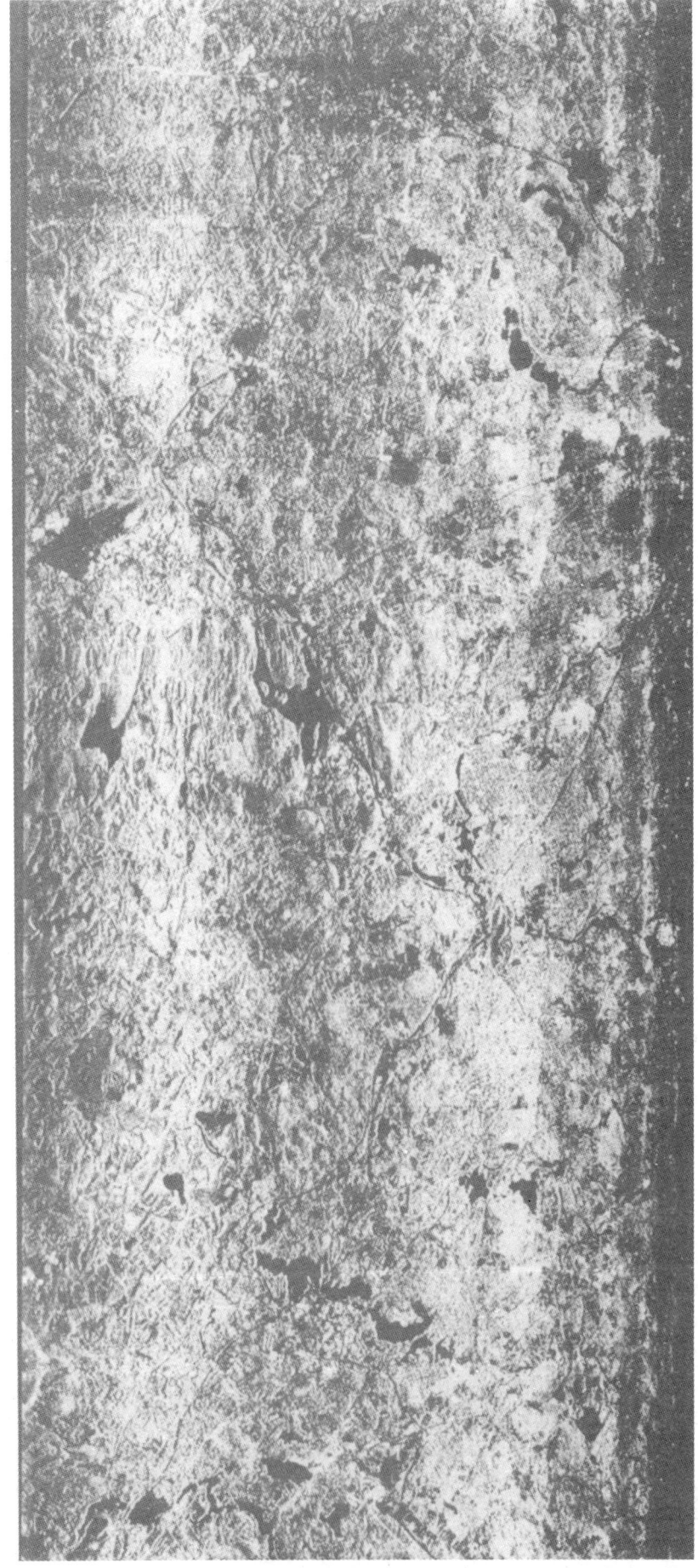

Figure 11–7. (Top of page looks southeast) SLAR image west of Boston, MA. Acquired 7/22/66.

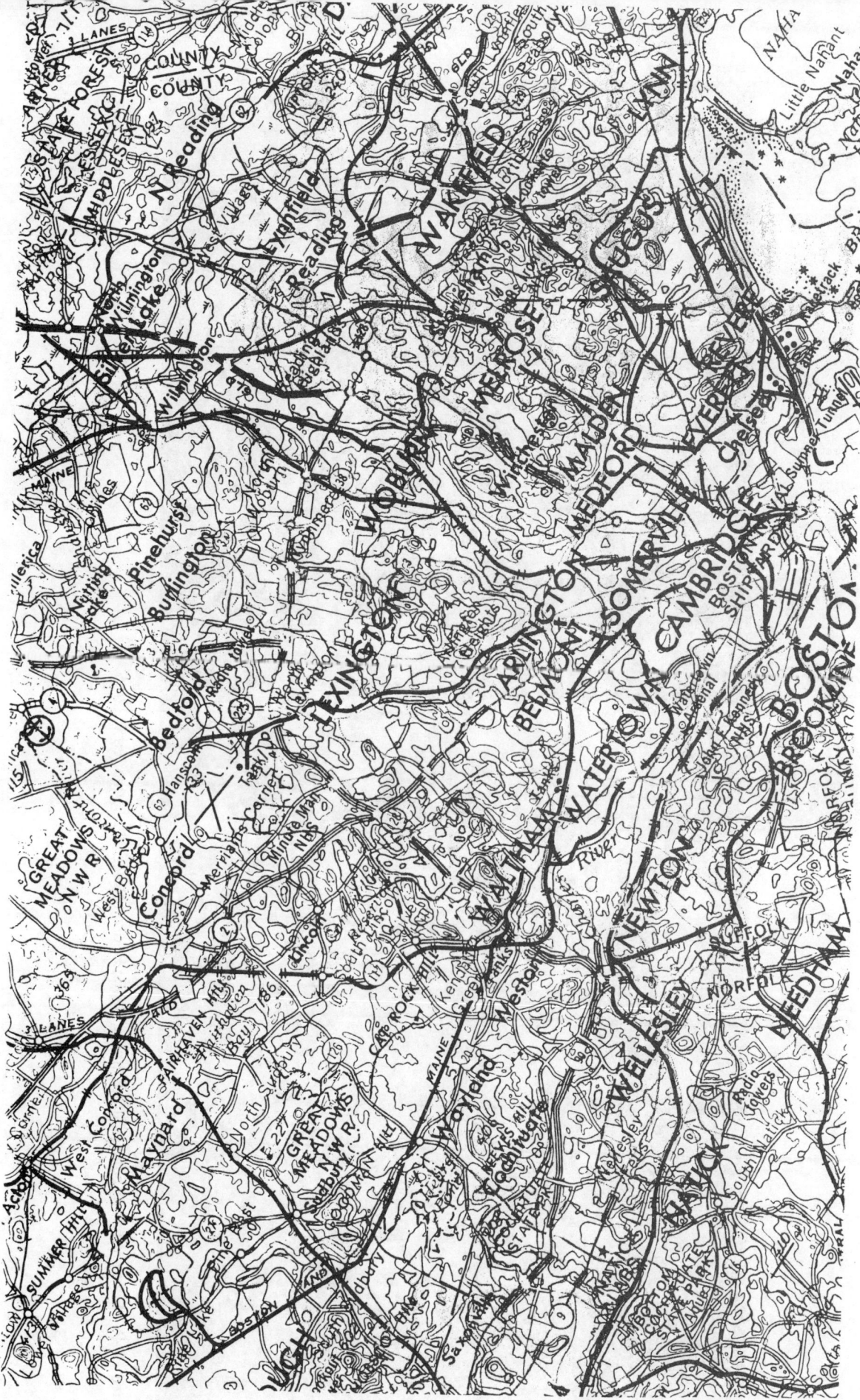

Figure 11–8. USGS topographic map Boston, MA (part). RF 1:250,000 enlarged to approximately 1:160,000.

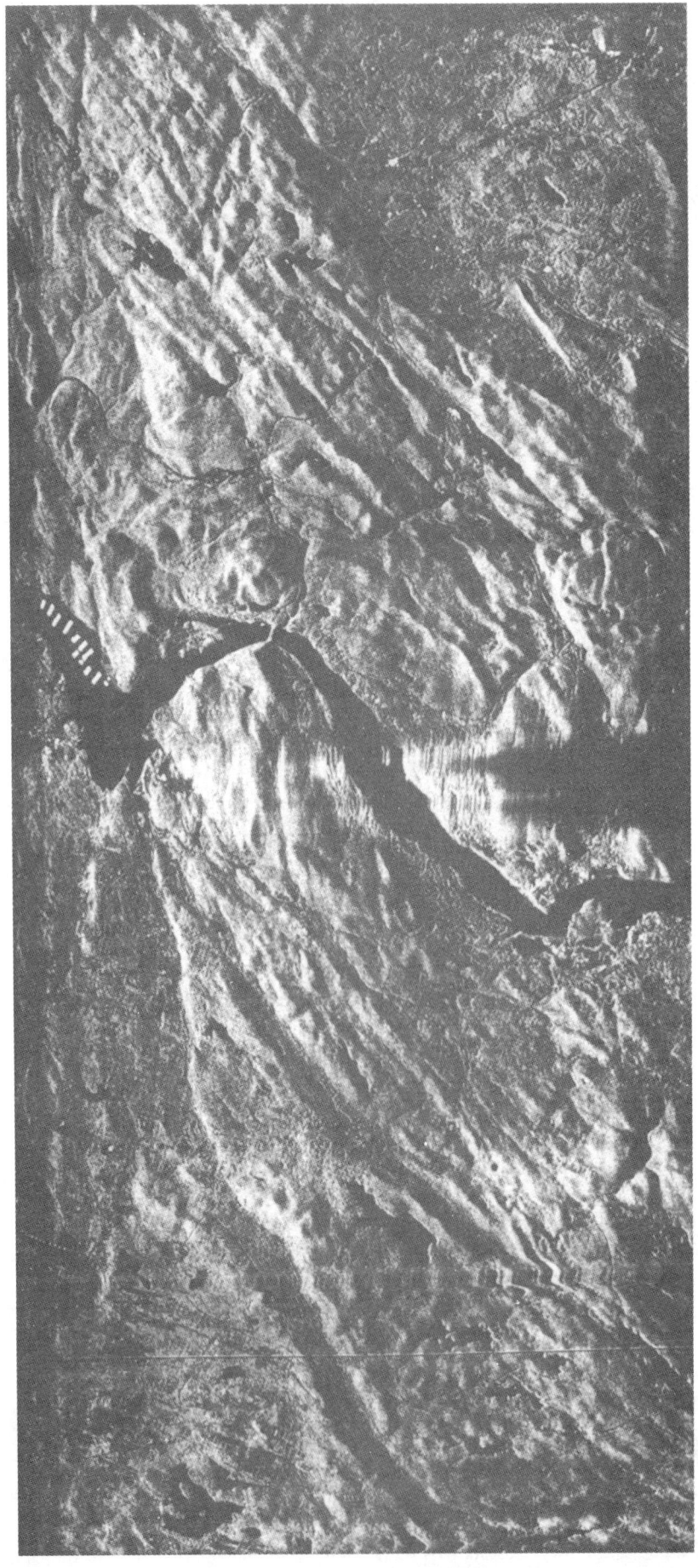

Figure 11–9. (South at left of page.) SLAR image lower Hudson River, NY. Acquired 7/22/66.

A very different appearance is produced by a SEASAT image of the New Orleans, LA area in Figure 11–12. Here the topography is almost flat and the major differentiation of surface features is between land and water areas. The Mississippi River flows through New Orleans to the left of image center, Lake Pontchartrain is at A-2 in the image, and the access to the Gulf of Mexico is at C-3. The wetland areas that abound in this scene are best exemplified in A-1, B-1, and C-1,2. Water depths and/or sediment in suspension are illustrated by variation in gray tones.

Many cultural features are also clearly evident. The impact of the French arpent system of land layout (long lot) is easily visible at right angles to the banks of the Mississippi River in and adjacent to New Orleans (see Chapter 12 discussion of the St. Lawrence River-Island of Orleans for more information on this system of land division). The 24 mile long Lake Pontchartrain Causeway stands out strongly in its path across that water body. It is obvious from comparison of this image with that of Denver that SEASAT SAR imagery is a powerful analytical tool whether or not there is considerable terrain relief.

Shuttle Imaging Radar

On November 12, 1981 NASA launched the second **Space Shuttle** mission, (**STS 2**), which carried a SAR radar system as part of its experimental payload. This SAR is referred to as **Shuttle Imaging Radar A (SIR-A)** and is an L-band (23.5 cm wavelength) system with a resolution of 125 feet (38 m) flying at a nominal altitude of 163 miles (262 km). It had a depression angle of 37° to 43° and acquired HH imagery at a scale of 1:500,000.

Data take 24C was acquired over South America, Central America, and southwestern United States on November 11–12, 1981. Two images are presented here as examples of this system's products. Figure 11–13 is an SIR-A image of the U.S./Mexico border in the vicinity of Imperial Valley and Salton Sea, CA and Mexicali, Mexico. This is a desert area that supplies wintertime vegetables to all of the U.S. due to the availability of Colorado River water for irrigation. The Imperial Valley is framed by the San Jacinto Mountains on the west and the Chocolate Mountains on the east, while the Mexican portion has the Sierra Madre Occidental on the west and the Sierra del Hueso to the east.

The image displays fine detail of the commercial agriculture on the U.S. side of the border and the small, fragmented land use in Mexico. Field shapes and textures are clear and the several, small urban areas to the North of the border stand out sharply as does the larger city of Mexicali to the South.

Figure 11–14 is a SIR-A image of a large part of the Los Angeles basin and much of the city that was obtained on the same data take as the previous image. This area is bounded in part by the Santa Monica and San Gabriel Mountains on the north and the Santa Ana Mountains to the southeast. The hilly and mountainous parts of the image are presented in strong contrast to the urban texture of Los Angeles itself. Note the areas within the basin are highly developed with some notable, contrasting areas as examples to the contrary. A high concentration of corner reflectors in A-2 marks the location of "downtown" LA.

Exercise 11   Interpretation of Radar Imagery

Refer to Appendix A to cross reference other images in this manual that treat the areas dealt with in this chapter. The goal here is to emphasize some of the types of image information to be found in SLAR imagery, especially those data that are unique to radar. Comparison with other kinds of imagery and maps will enhance this effort.

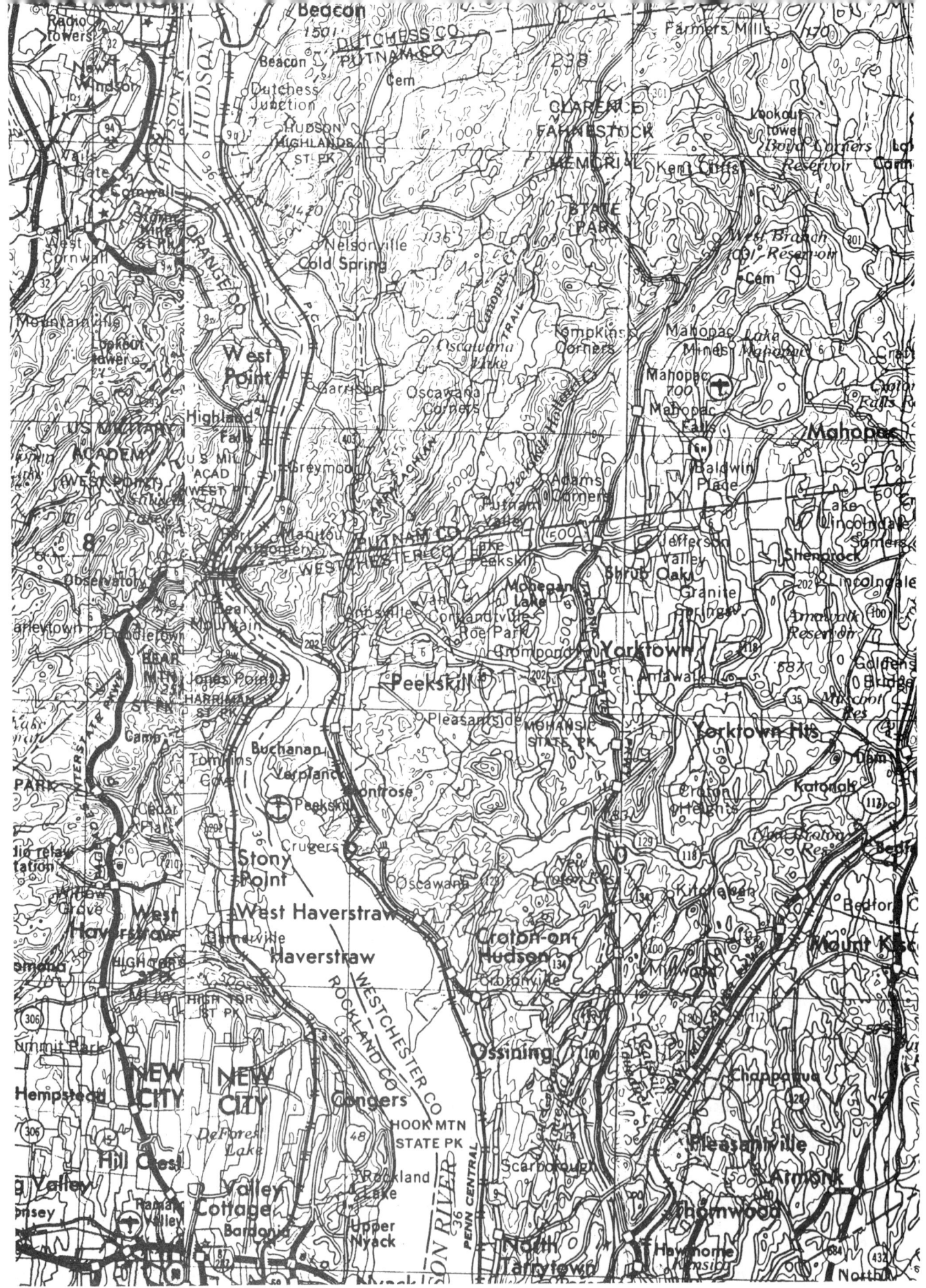

Figure 11–10. USGS topographic map lower Hudson River. RF 1:250,000 enlarged to approximately 1:180,000.

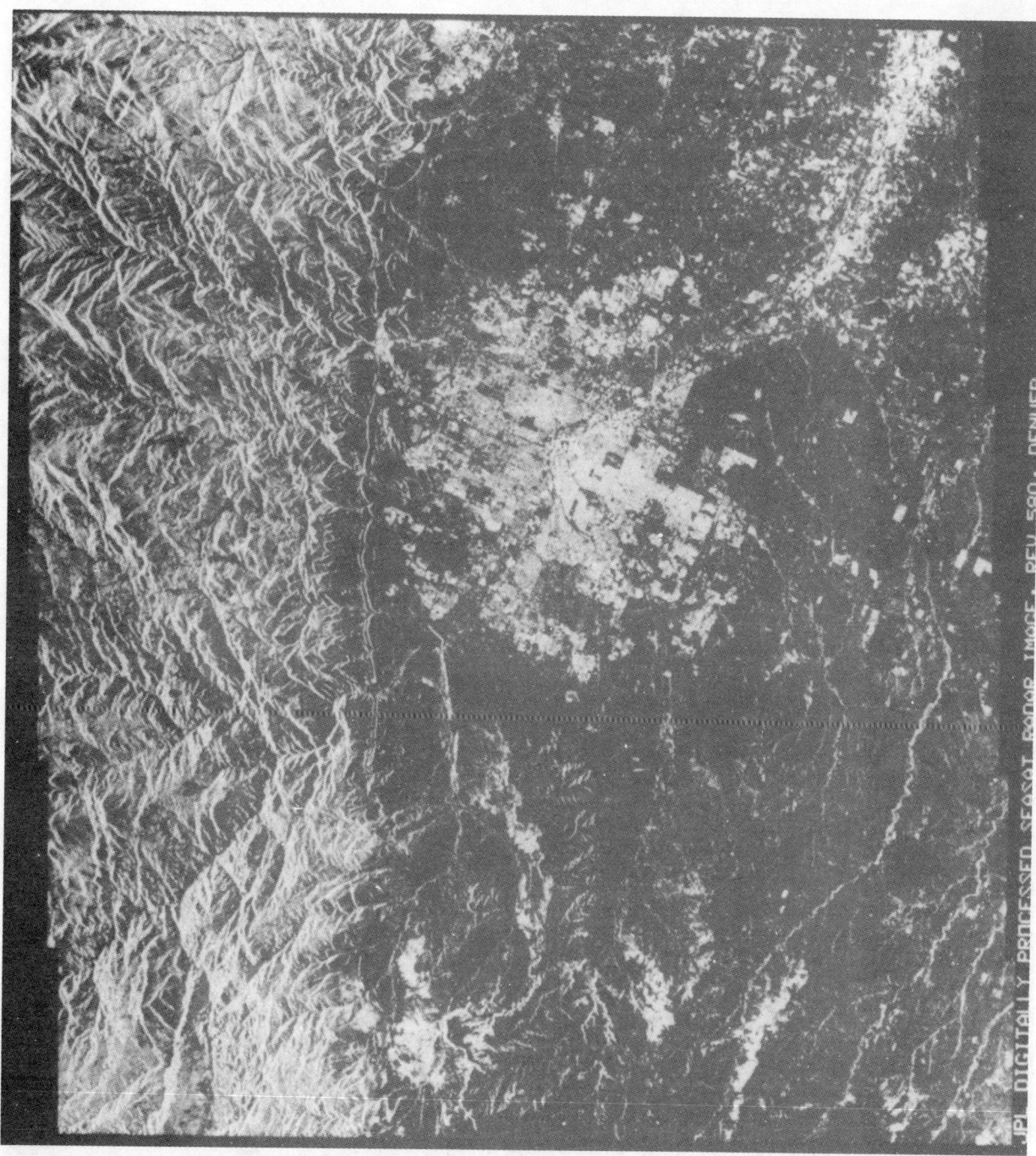

Figure 11–11. SEASAT SAR image of Denver, CO. Acquired August, 1978.
Courtesy NOAA/NESDIS/NCDC/SDSD.

A. Airborne SLAR imagery.

1. Figure 11–7 on page 140 is a SLAR image of an area just west of Boston, MA. The HH band is at the top of the page and the HV at the bottom. Examine the map in Figure 11–8 and note the location of some of the major roads such as I-90, 128, I-93, and Route 2. How many of these roads are visible on the HH image? Do they show up any better on the HV image in Figure 11–8?

2. What other types of linear data might be visible on a SLAR image? What are the surface characteristics that would make it possible for them to appear on radar? What specific linear features can you see in each of the bands in Figure 11–7? List by A,B,C location method and by type of feature.

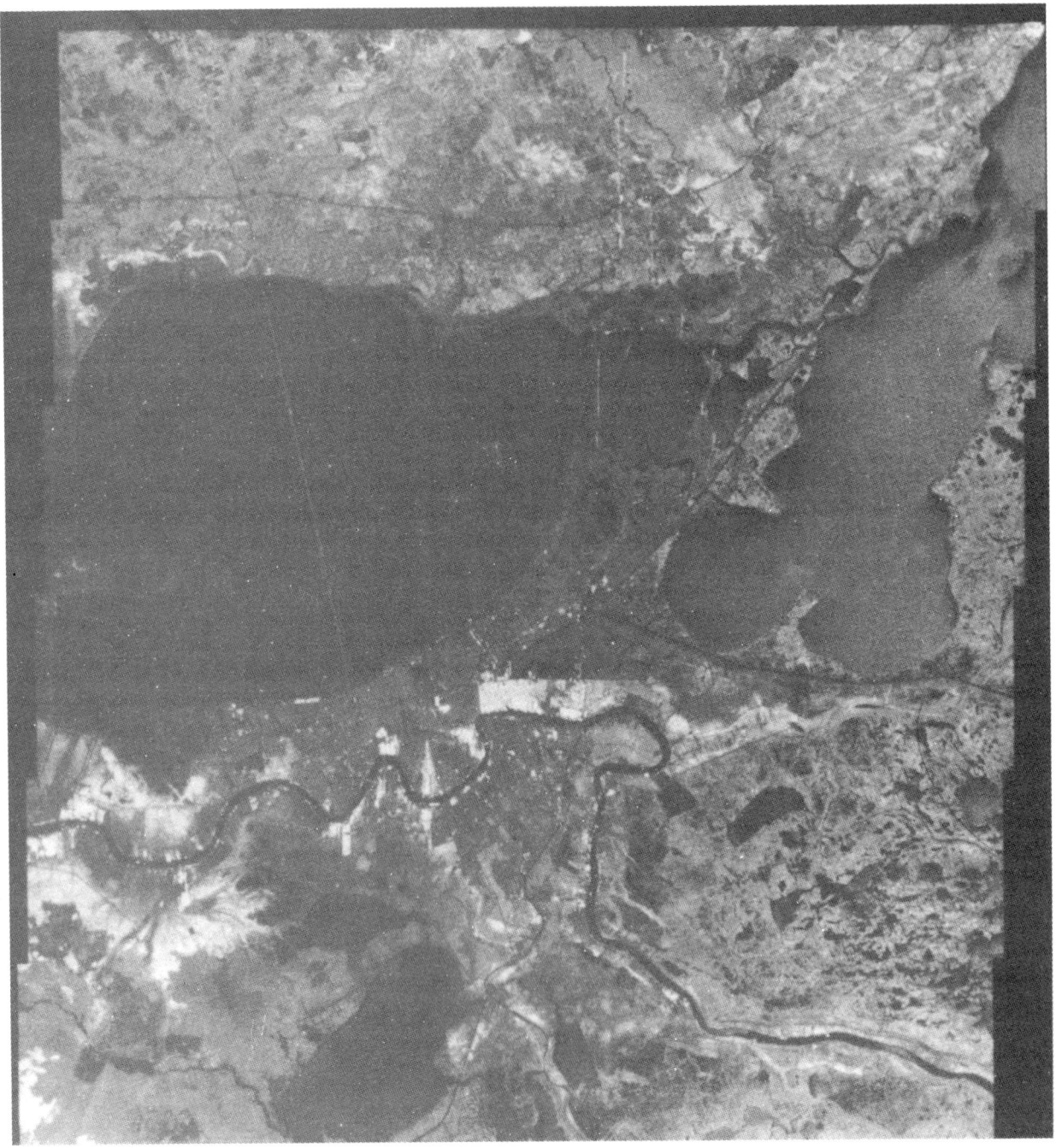

Figure 11–12. SEASAT SAR image of New Orleans, LA. Acquired 8/21/78.
Courtesy NOAA/NESDIS/NCDC/SDSD.

3. In the B-1 section of both images at the left side of the page there is a dark area extending down and away from a curved section of Route 128. Knowing the general characteristics of radar returns, what might the surface of this area be? If you were informed that it was a sparsely wooded wetland, would that fit in with the nature of radar signatures?

4. Hanscom Air Base is in the C-2 area of each band near the edge. Can you distinguish between runways and taxiways vs. other surfaces? Based upon the number of corner reflectors, do you think there are many buildings on this facility?

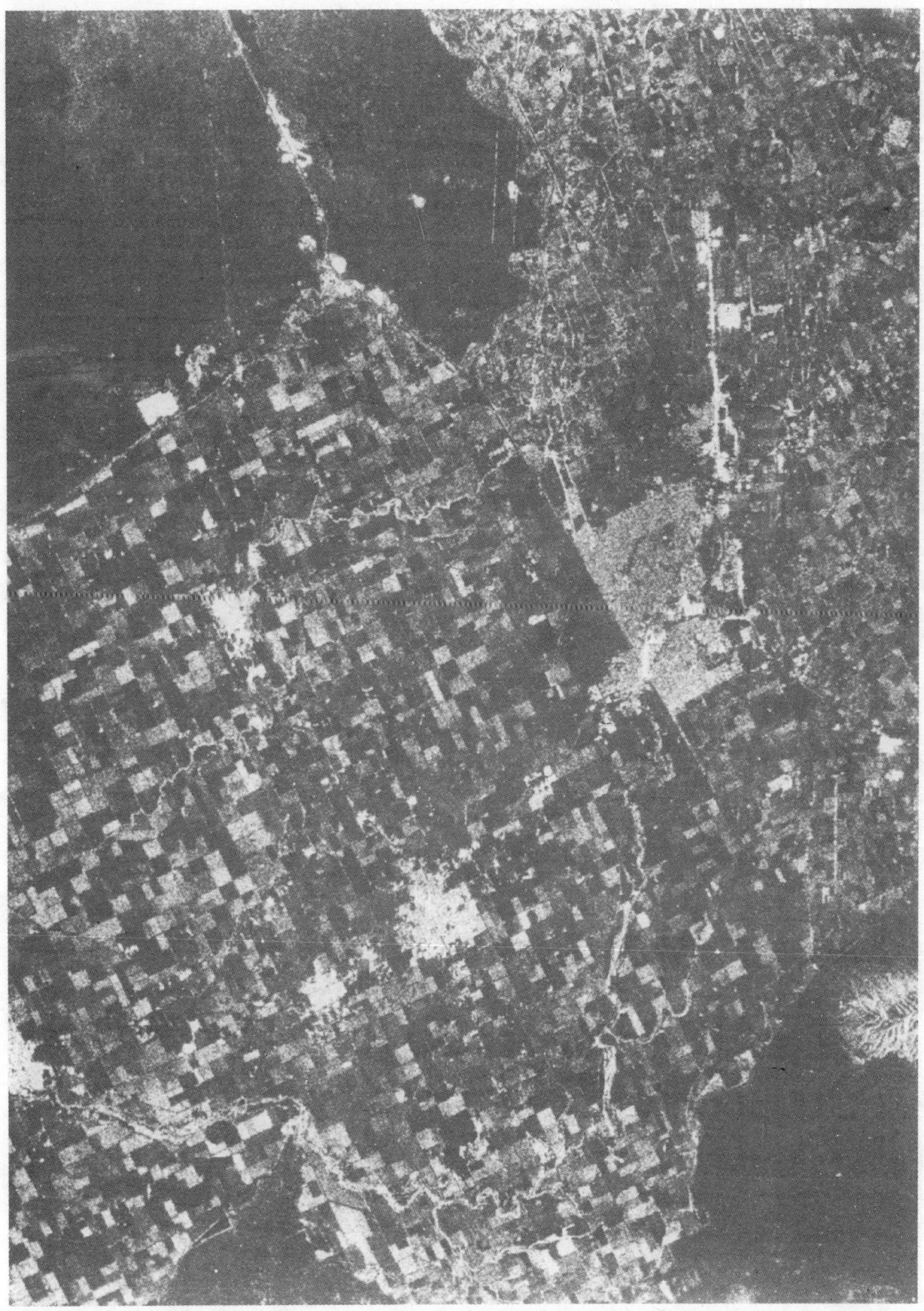

Figure 11–13. SIR-A SAR L-band image of CA–Mexico border. Acquired 11/11–12/81.
Courtesy Charles Elachi and NASA/NSSDC/GSFC.

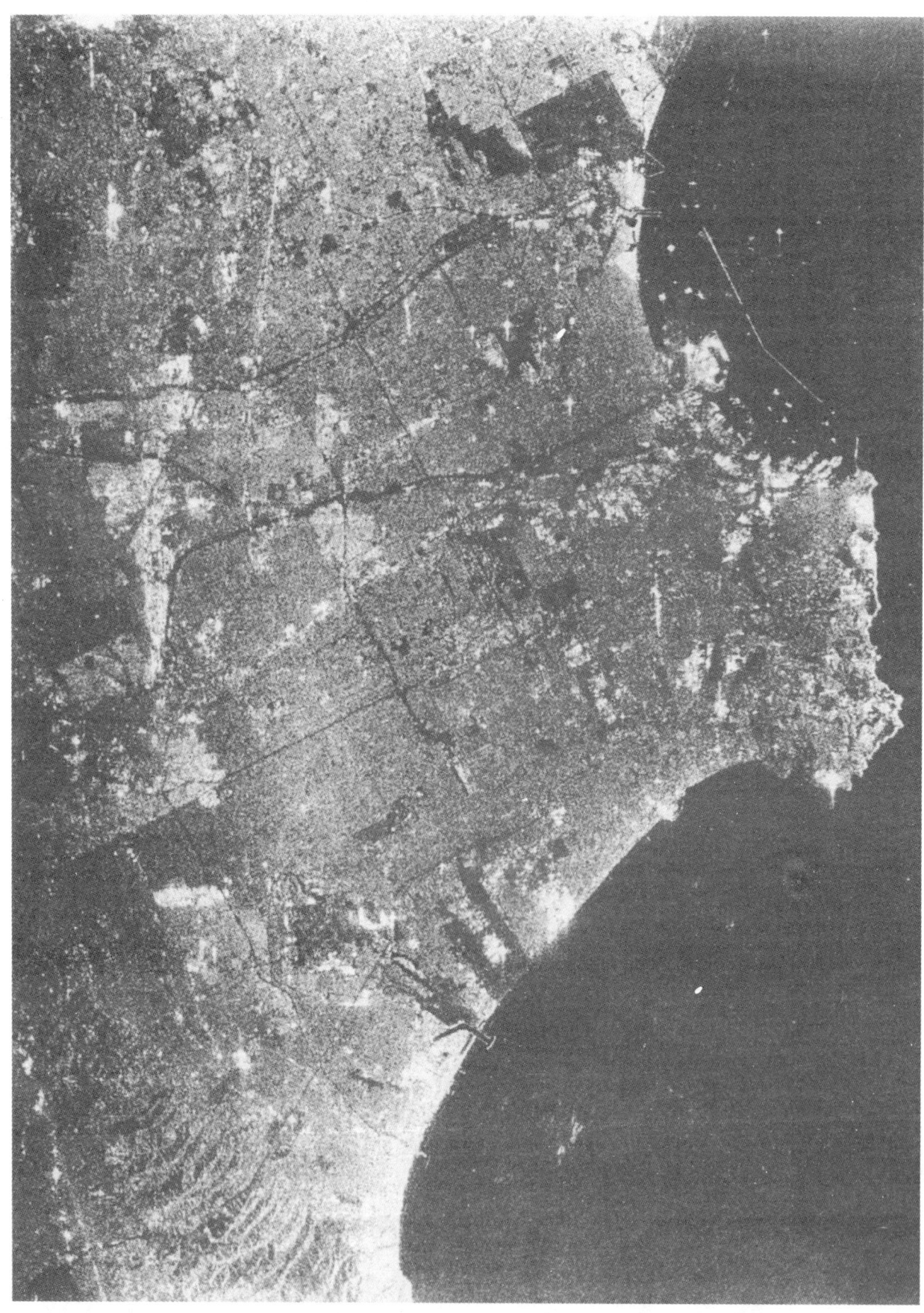

Figure 11–14. SIR-A SAR L-band image Los Angeles basin. Acquired 11/11–12/81. Courtesy Charles Elachi and NASA/NSSDC/GSFC.

5. In the A-3 area of each image there appears to be a junction of several linear features. From all the evidence that is available to you, what would you surmise this feature to be? Explain your reasoning.

Figure 11–9 displays HH (top) and HV bands of a SLAR image of the Hudson River in the vicinity of West Point Military Academy. Figure 11–10 is a portion of the USGS map of the area at approximately the same scale as the images. Compare both figures and then answer the following questions.

6. There are three major physiographic regions of the United States represented on these images. They are distinctly different in their appearance due to variations in topography, structure, and surface material. Can you draw boundaries on overlay material where these regions meet? What evidence is there in the SLAR image to indicate variations in land use related to regional differences? Explain.

7. The Hudson River traverses the Hudson Highlands in this area and a tongue of the most recent continental glacier followed the same path over 10,000 years ago. If you knew that the deepest part of the Hudson River was at West Point in the upper left corner of C-2 and that the ice made two right angle turns at this part of the river, what would that tell you about the hardness and resistance of the bedrock material? What other evidence in the image supports your conclusion?

8. Note the objects in the river at A-2. They are parallel, at a protected arm of the river north of New York City, have strong return characteristics, and the image date of acquisition was 1966. They are made of metal. What might they be? (This is not something normal radar signature information is likely to reveal—try your luck!)

B. SEASAT Radar Imagery.

1. Figure 11–11 is a SEASAT image of the Denver, CO area. In the image two very different physiographic regions are represented, the high plains area of the Great Plains, and the Rocky Mountains. Locate and map on an overlay the boundary between these two regions.

2. Immediately west of Denver along the Rocky Mountain front are features known as "flatirons," former horizontal, sedimentary rock layers that now are tilted upward (the amount of tilt is exaggerated by layover effect). If you knew that the mountains to the west were igneous, hard rock, what would these facts reveal about the mountain-building process in this area? Radar is obviously very helpful in revealing variations in rock type and rock structure.

3. Examine the surface hydrology in B-1, C-1, and C-2. The river patterns are revealed rather well. If you knew that it is an area of semi-arid climate and that it is likely that most of the rivers in the image were dry, why do the rivers show up so well? What is there about the surface and about radar characteristics that allow these river channels to be revealed? Might this same phenomenon occur in other dry regions of the world?

4. Study the Denver urbanized area. Can you distinguish between different parts of the city in regard to density of development or variation in land use? What image signature characteristics allow you to make these observations?

5. Most cities in the central part of the United States are characterized by a peripheral zone where urban land uses change gradually or abruptly into agricultural or rural

uses. This zone has been referred to as the "rural-urban fringe." Can you detect such a zone here? What radar signature characteristics assist you? Explain.

Figure 11–12 is a SEASAT image of the New Orleans, LA area. The appearance of this area is decidedly different from that of Denver. Here the topography is almost flat and there are large amounts of water visible. In fact, the land/water interface is the most important aspect of the area, affecting daily livelihood, urban development, and safety.

New Orleans is bisected by the Mississippi River, lies immediately north and west of the Gulf of Mexico, and has the 640 square mile Lake Pontchartrain on its doorstep to the north. Much of New Orleans has elevations that are below sea level and pumps water on a regular basis to avoid flooding. The average elevation of the city is only one foot above the mean sea level of the Gulf of Mexico and is below the high water levels of both the Mississippi River and Lake Pontchartrain. The Mississippi River surface elevation in the New Orleans area is higher than many second floor elevations just beyond the levees.

6. If Hurricane Andrew had hit the Louisiana coast several miles west of the city center coinciding with high tide on August 25, 1992 what prediction might one make for flooding, destruction, and possible loss of life?

7. New Orleans is considered to be one of the highest risk areas in the United States for hurricane threat. How might the presence of the extensive wetlands as shown in Figure 11–12 help to ameliorate this threat? Explain.

C. Shuttle Imaging Radar (SIR-A).

Examine Figure 11–13, a SIR-A image of the U.S./Mexican border in the Imperial Valley—Mexicali, Mexico area. This was acquired on November 11 or 12, 1981 by an L-band radar system. The **complex dielectric constant** is a value controlled by the electrical properties of a surface material and is an indication of an objects' likelihood of reflecting or allowing penetration of radar energy. In natural substances the primary characteristic affecting this relationship is moisture, wet areas reflecting more strongly than dry. In materials produced by people, metallic objects produce a stronger reflection than their size would normally call for. Consider these points and answer the following questions.

1. If this desert area had essentially the same soil distribution in the agricultural areas, can you determine whether or not the radar beam has penetrated the vegetative canopy? Take note of the variation in field tones to aid you in your answer. Explain your reasoning.

2. How many cities can you locate based upon corner reflectors and possible metallic objects present? Give their locations using the A,B,C method and give relative city sizes. If the city of Mexicali is approximately 200,000 population, how do the cities on the U.S. side of the border compare?

3. Note the variation in size and shape of agricultural fields in the United States and in Mexico. What might account for this?

4. Is there any evidence in this image that might help to explain why the irrigated croplands stop where they do? That is, why don't they extend farther outwards?

5. What kinds of linear features can you locate on the image and then identify? What is your evidence?

Figure 11–14 is a SIR-A image of much of the Los Angeles basin and Figure 9–28 is part of a USGS topographic map that covers most of the same area. The SLAR image extends from the Santa Monica Mts. west of Santa Monica to Newport Beach on the coast and almost to the foothills of the San Gabriel Mts. in the interior.

6. If military facilities are usually distinct from nearby areas, can you locate 2 such areas on or near the coast SE of Long Beach? What image/signature characteristics set them apart? What is the nature of such bases that produces this identification clue?

7. What kinds of linear features are present on the map in Figure 9–28? How many of these can you locate on the image? How do they vary?

8. L.A. International Airport, LAX, is on the coast near Marina del Rey. What are the radar signature characteristics of LAX on the image? What do they indicate about amount of buildings, smooth areas, overall size, and relation to the surrounding area? Do they fit with your concept of a large international airport? Explain.

9. Based upon the number and distribution of corner reflectors, how would you describe the pattern of development of the Los Angeles basin? How does this compare with any other large cities of the U.S.A. with which you are familiar? What impact might this type of urban pattern have on the daily functioning of Los Angeles and its suburbs?

chapter *12*

INTERPRETATION OF SELECTED EARTH SURFACE PHENOMENA

TERMINOLOGY

mental library
patterns and processes
human/environment interface
Industrial Revolution
old age desert mountains
inselbergs
erosional remnants
bajada
alluvial fans
intermittent streams
intermontane region
ridge and valley
great valley
physiographic regions
anticlinal fold
synclinal fold
superposed stream
plunging anticline

karst topography
sinkholes
barrier islands
old age river
central pivot agriculture
Taconic Mountains
Hudson Highlands
Triassic lowland
Palisades
compatible land uses
change through time
sequent occupance
soil marks
French long lot
rotures
ridge and furrow
   agriculture
plunging syncline

RECOGNITION/IDENTIFICATION FACTORS

In Chapter 3 the concept of image signatures was discussed as a tool to use in breaking down and identifying image features. The general approach to image identification alluded to the use of an interpreter's past experiences and travel. Any information derived through on-site observation is valuable in image identification. Also important in this regard are data gained from research with any of the imaging media that deal with the representation of all or part of the Earth's surface.

The **mental library of experiences and signatures** may be further enhanced by the creation of interpretation logs as explained in Chapter 6. This chapter will make use of all of these methods of recording, remembering, and using signature information. Examples of a variety of Earth surface features will be examined with a focus upon **patterns** and **processes** that aid the identification effort. Another collateral frame of reference is the **relationship between the natural environment and human occupance.** That is, how does each impact the other at the **human/environment interface?**

The length of time that mankind has been on this planet is very small compared to the geologic time scale during which natural forces have been at work. Even during their period of habitation the impact upon the Earth by humans has been relatively minor until 300 years ago. Since the **Industrial Revolution** the ability of people to modify the Earth's environment has increased dramatically and this power seems to be growing exponentially.

EXAMPLES OF STRONG ENVIRONMENTAL IMPACT

Phoenix, AZ. Figure 12–2 is a Landsat Thematic Mapper image acquired 7/23/85 and presented here as a black and white version of a composite image of bands 2, 3, and 4. Figure 12–3 is a part of the USGS topographic coverage of this area.

The Phoenix metropolitan area is one of the fastest growing urban areas in the United States. The attractiveness of the climate and the beauty of the desert have lured many retirees to the region. Future development is tied in large measure to the availability of potable water for the region overall. Locally, the added growth of agriculture and, to some extent, residential building is impacted by the terrain.

The TM image reveals both the lack of surface water as well as the features of advanced **old age desert mountains.** Fields devoted to agriculture are confined to valleys at elevations where pumping irrigation water is feasible. They show up as dark and light rectangles on the image. Drainage patterns are clearly evident on the map and the image.

Figure 12–1 is a diagram of a similar area in a late stage of erosion. Compare the features in the diagram with the map and the image of the Phoenix area.

Exercise 12.1   Interpretation of Selected Earth Surface Phenomena: Areas of Strong Environmental Impact—Phoenix, AZ

1. **Inselbergs** are **erosional remnants** in mountainous desert areas that are gradually buried in their own erosional debris. By comparison of Figures 12–1, 12–2, and 12–3 can you identify several good examples of such features? Give their location using the A, B, C method where they are found on the TM image.

2. **Alluvial fans** are features that are constructed below the mouth of a canyon or stream valley emanating from highlands. Such features are built by surface streams depositing material back and forth across the fan. Radiating stream courses are an indication of such features. A good example of such a feature may be found northwest of the Sacaton Mountains at C-3 on the map and B-3 on the image. Can you find several other good examples? Give their locations on the image.

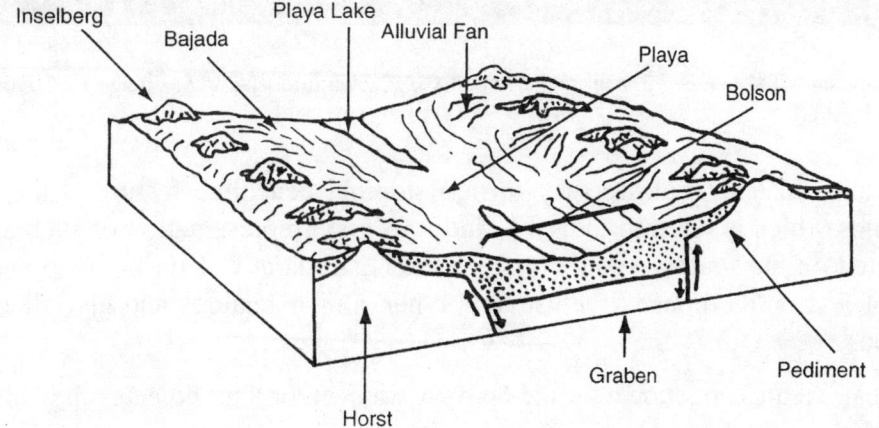

Figure 12–1. Arid region-block fault mountains—late stage erosion.

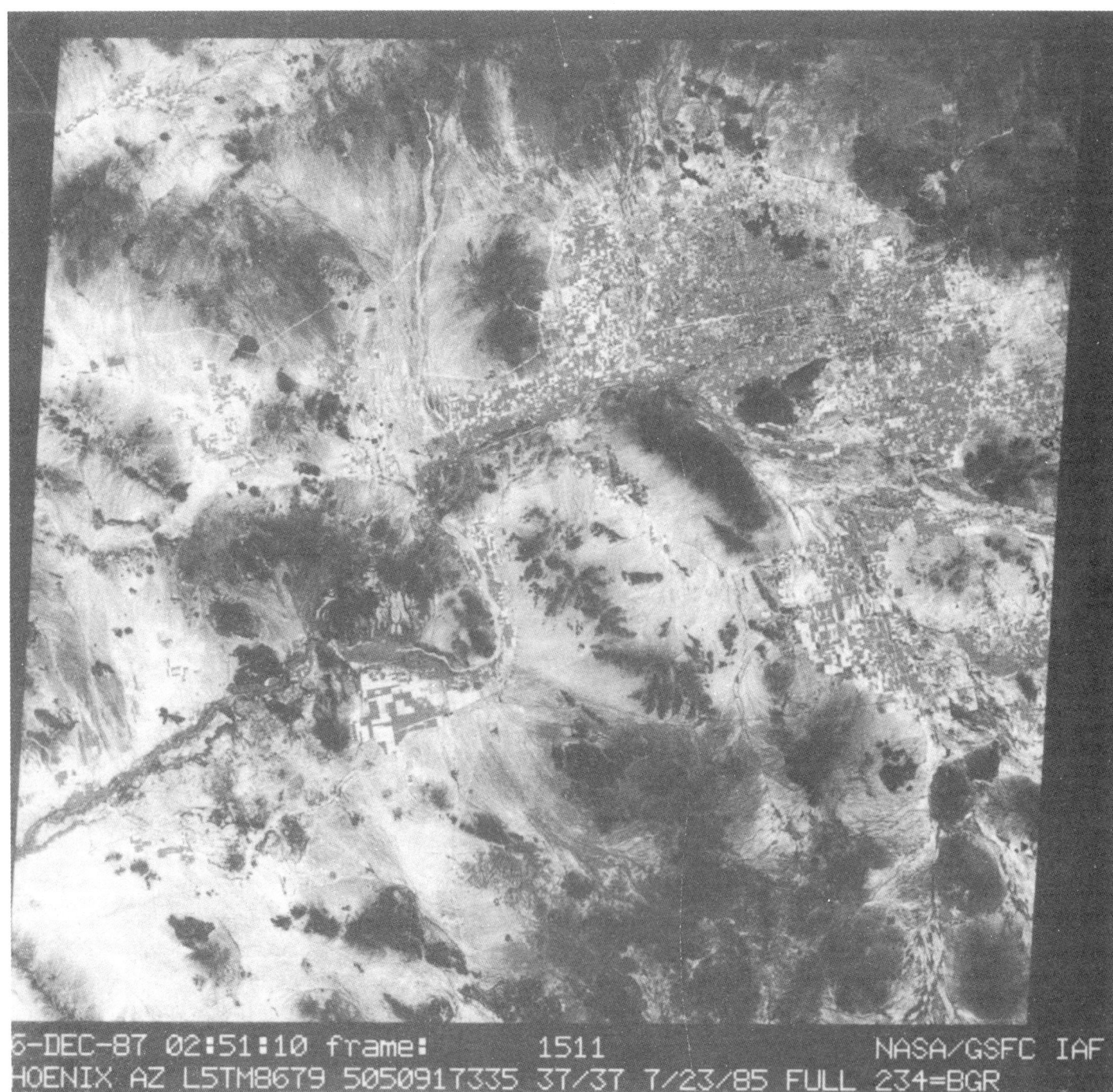

6-DEC-87 02:51:10 frame: 1511 NASA/GSFC IAF
HOENIX AZ L5TM8679 5050917335 37/37 7/23/85 FULL 234=BGR

Figure 12–2. Landsat 5 TM bands 2,3,4 subscene Phoenix, AZ. Acquired 7/23/85. Image ID 5050917335. Courtesy NASA/GSFC.

3. When alluvial fans coalesce they form a sloping, benchlike feature at the base of highlands which is referred to as a **bajada.** A good representation of such a feature is located on the southwest side of the Sierra Estrella at C-1 on the map and upper right of B-2 on the image. Find several other similar features and give their image locations.

4. Note that streams are shown on the map with a symbol that indicates they are **intermittent** (line interrupted by three dots and line continued). Such streams flow for

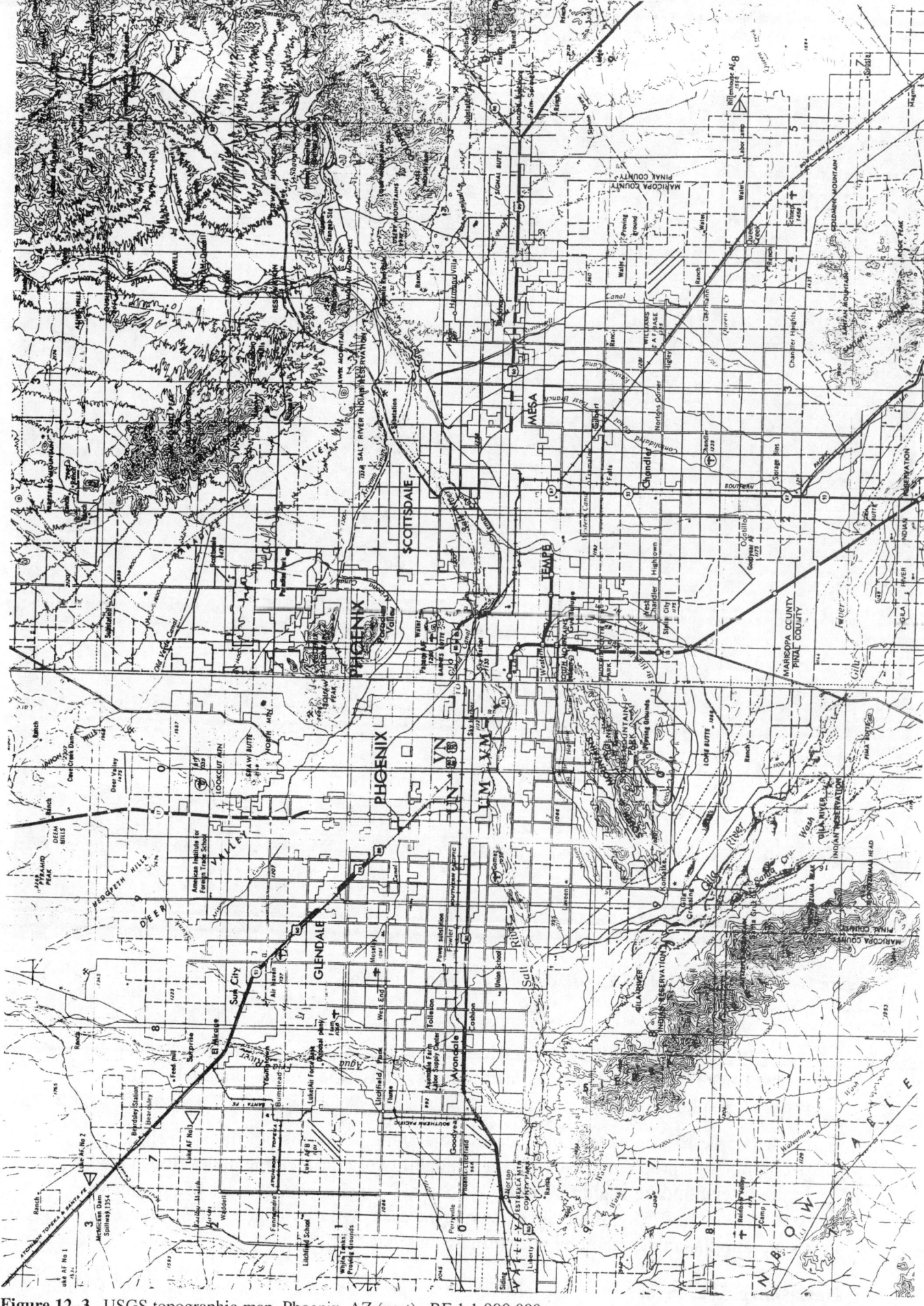

Figure 12–3. USGS topographic map, Phoenix, AZ (part). RF 1:1,000,000.

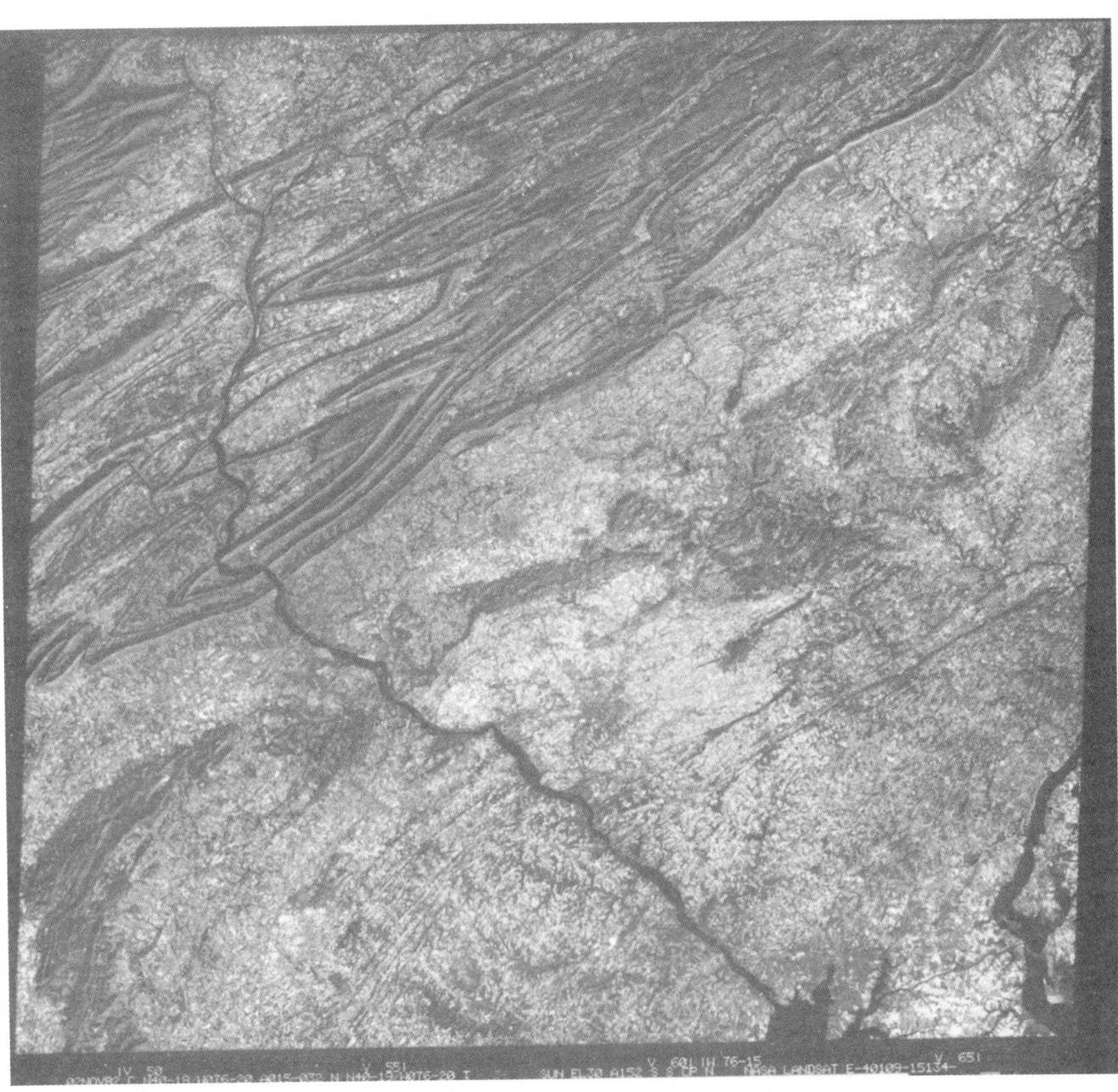

Figure 12–4. Landsat 4 TM bands 2,3,4 Harrisburg, PA. Acquired 11/2/82. Image ID 4010915134. Courtesy NASA/GSFC.

only part of the time. If that is so, why is it so easy to pick out the stream channels on the image? What happens in arid areas to cause the stream channels to be so visible? Explain.

5. Some people have advanced the idea that the United States is not likely to suffer an overpopulation problem because of the availability of considerable undeveloped land in the **intermontane region** between the Sierra Nevada and Cascades Mountains to the west and the Rocky Mountain system to the east. What evidence is available from examining this image and map that refutes that idea? Explain.

Harrisburg, PA. Figure 12–4 is a TM image acquired on 11/2/82 and is a black and white version of the false color composite of bands 2, 3, and 4. Figure 12–5 is a reduction of USGS

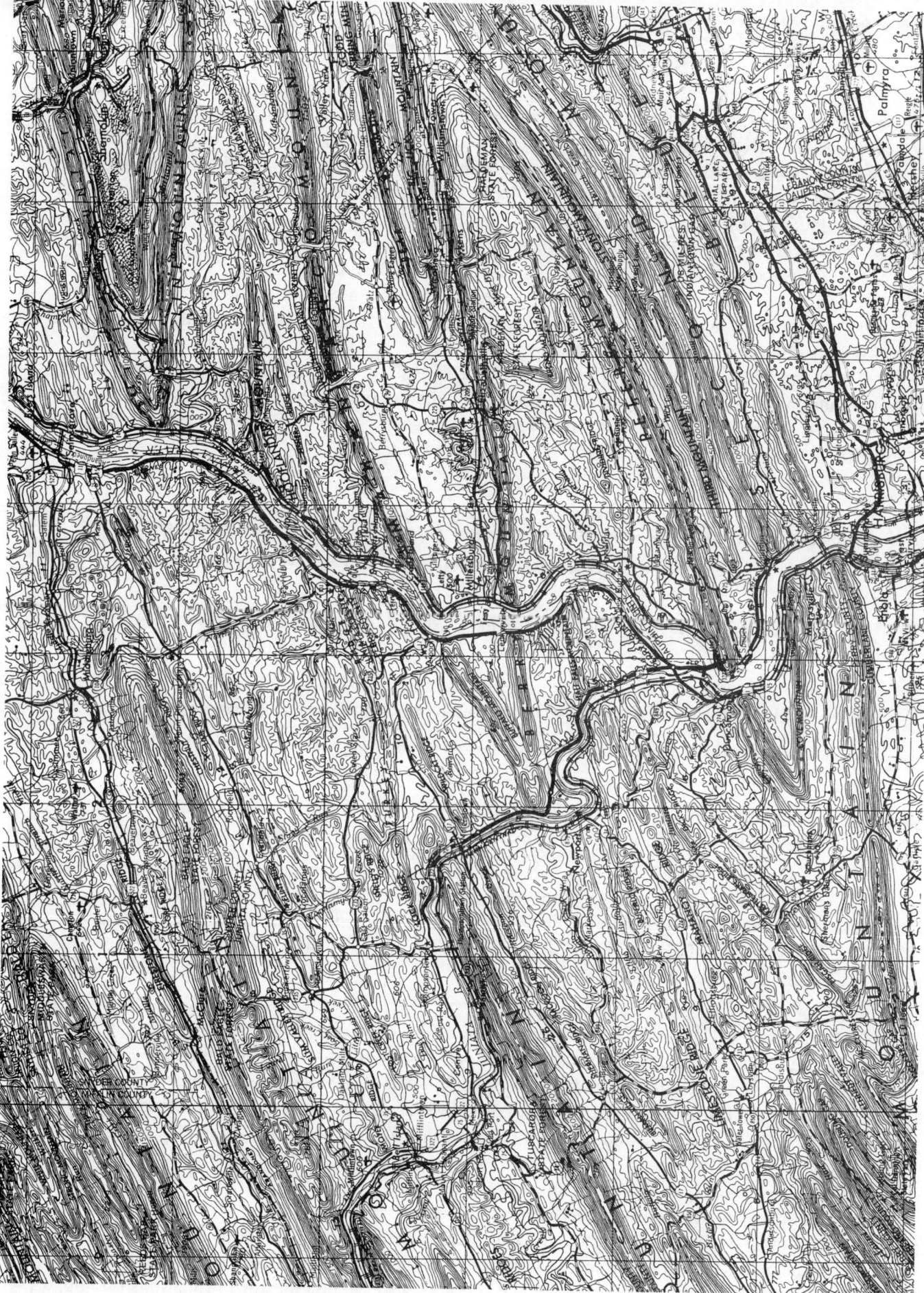

Figure 12–5. USGS topographic map, Harrisburg, PA (part).

topographic coverage of part of the area shown on the image. This is a portion of the **ridge and valley** region of the Appalachian system that includes the Appalachian Plateau, the ridge and valley, and the **Great Valley** that is bounded to the east by the Blue Ridge Mountains. The Great Valley is represented by the area below and to the right of the last ridge. These **physiographic regions** extend from the town of New Paltz, NY on the Hudson River to the vicinity of Birmingham, AL.

The river flowing from A-1 to C-3 in Figure 12–4 is the Susquehanna which empties into the upper Chesapeake Bay. The river on the east side of C-3 is the Delaware River which runs to Delaware Bay. The Susquehanna predates the landform features of the present surface and was let down onto these ridges and valleys as they were being formed into **anticlinal and synclinal folds,** making it a **superposed** stream.

The city of Harrisburg, PA is located on the east bank of the Susquehanna River and lies on the margin of the great valley and the ridge and valley regions. The darker tones of the nearby ridges in Figure 12–4 indicate forest cover and the lighter shades of the lowlands are indicative of fields that have been harvested in this November scene.

Exercise 12.2 Interpretation of Selected Earth Surface Phenomena: Areas of Strong Environmental Impact—Harrisburg, PA

1. Compare Figures 12–4 and 12–5. What evidence is presented there to explain why farms and settlements are located almost exclusively in the lowlands? Explain.

2. How would you describe the typical ridge feature common to both figures? What are its characteristics? What image or map evidence assists you in that regard?

3. Harrisburg is the site for fairly large railroad switching yards and facilities. What is there about its location that makes this a logical location for such activity?

4. How important might the Susquehanna River have been as a means of transportation and communication in the early history of this region of the United States? Why?

5. Folded rock strata that tilt or plunge downward my be detected by the shapes of the end of the fold. If **plunging anticlines** are characterized by a **cigar-shaped** outside and **plunging synclines** are denoted by a **canoe-shaped** end, can you identify examples of each of these features on the image with the aid of the map? Give image location using the A,B,C method.

Orlando, FL. This TM image of east central Florida was acquired on 5/14/84 and is shown in Figure 12–6 as a black and white presentation of a false color composite of bands 2, 3, and 4. Figure 12–7 is a USGS topographic map covering much of the area.

The cultural features shown include Cape Canaveral and the Kennedy Space Center in B-3, C-3, the city of Orlando at the top center of C-2, and the Walt Disney Resort Complex to the immediate southwest. The land use/land cover of the inland area is devoted primarily to citrus orchards and extensive fields for cattle grazing, and wetlands. The physical makeup of the area has had and continues to exert considerable influence upon human activity.

The north–south spine of central Florida is characterized by features of **karst topography.** The name of this type of landscape was derived from the Dalmatian coast area in the former country of Yugoslavia. It is characterized by the action of water upon limestone formations leading to sinkholes and a variety of other solution features.

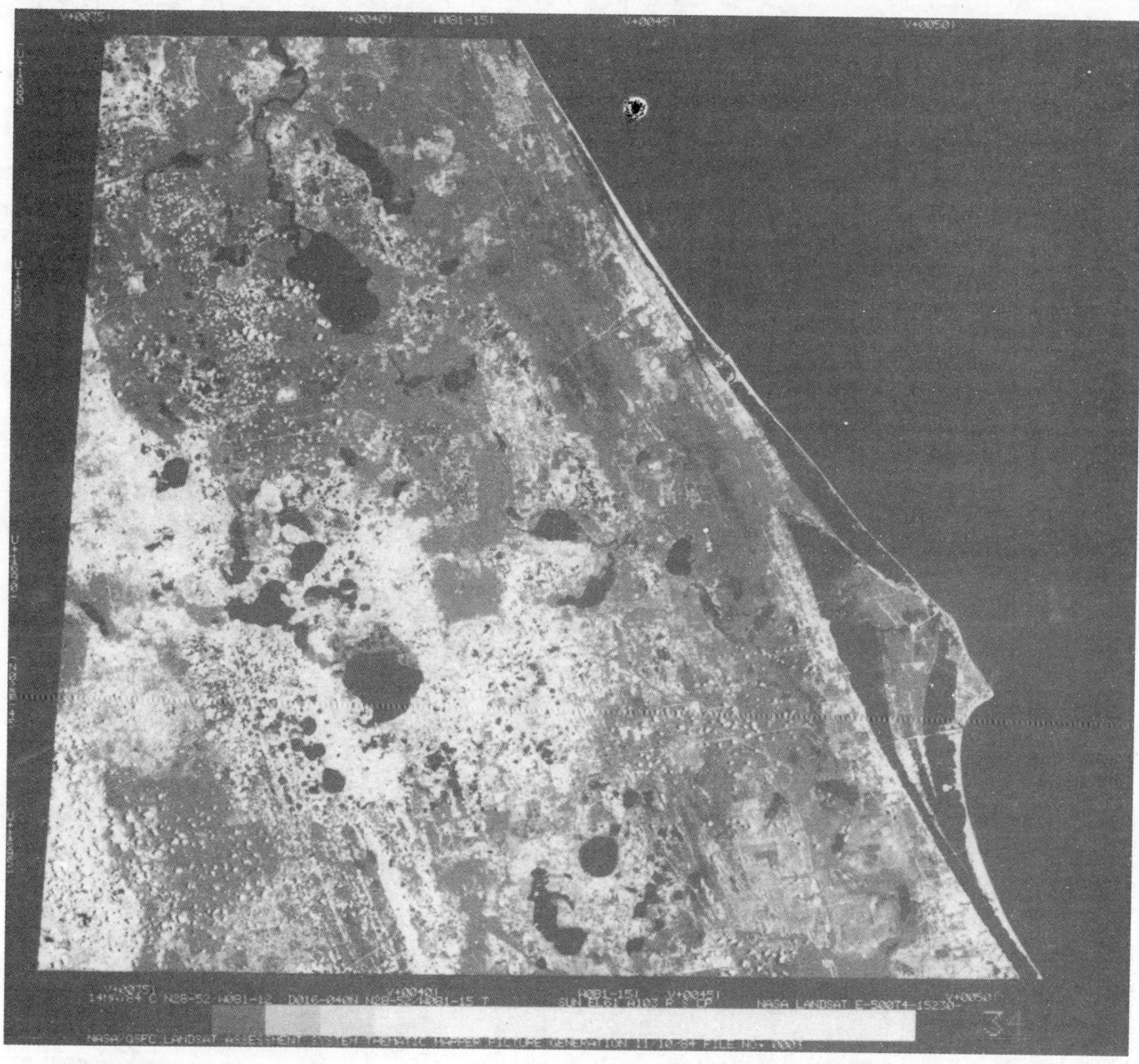

Figure 12–6. Landsat 5 TM bands 2,3,4, Orlando, FL. Acquired 5/14/84. Image ID 5007415230. Courtesy NASA/GSFC.

There are many places in the world where such landforms have been created. Other notable examples are found on the island of Puerto Rico and on the Yucatan Peninsula. The type of karst forms created depends upon hardness and thickness of limestone beds, climate, and level of the ground water table. In Florida, the most pervasive features are **sinkholes** that are dry or contain water.

Figure 12–8 is a black and white print of an oblique, color photo of Florida taken by a hand-held Hasselblad camera from Space Shuttle STS 51-C in January of 1985. The altitude was 219 miles (352 km.). In this view one can see the many sinkhole lakes that extend northward from Lake Okeechobee at the upper right corner of C-2.

Lake County, FL; Lake Nellie Quadrangle. Figure 12–9 is a black and white airphoto of the Lake Nellie area of Lake County, FL and Figure 12–10 is part of the Lake Nellie USGS Quadrangle for the same location. The nominal scale of the photo is 1:20,000 and the map is 1:24,000. This large scale view of karst features is located north of the Orlando image and map.

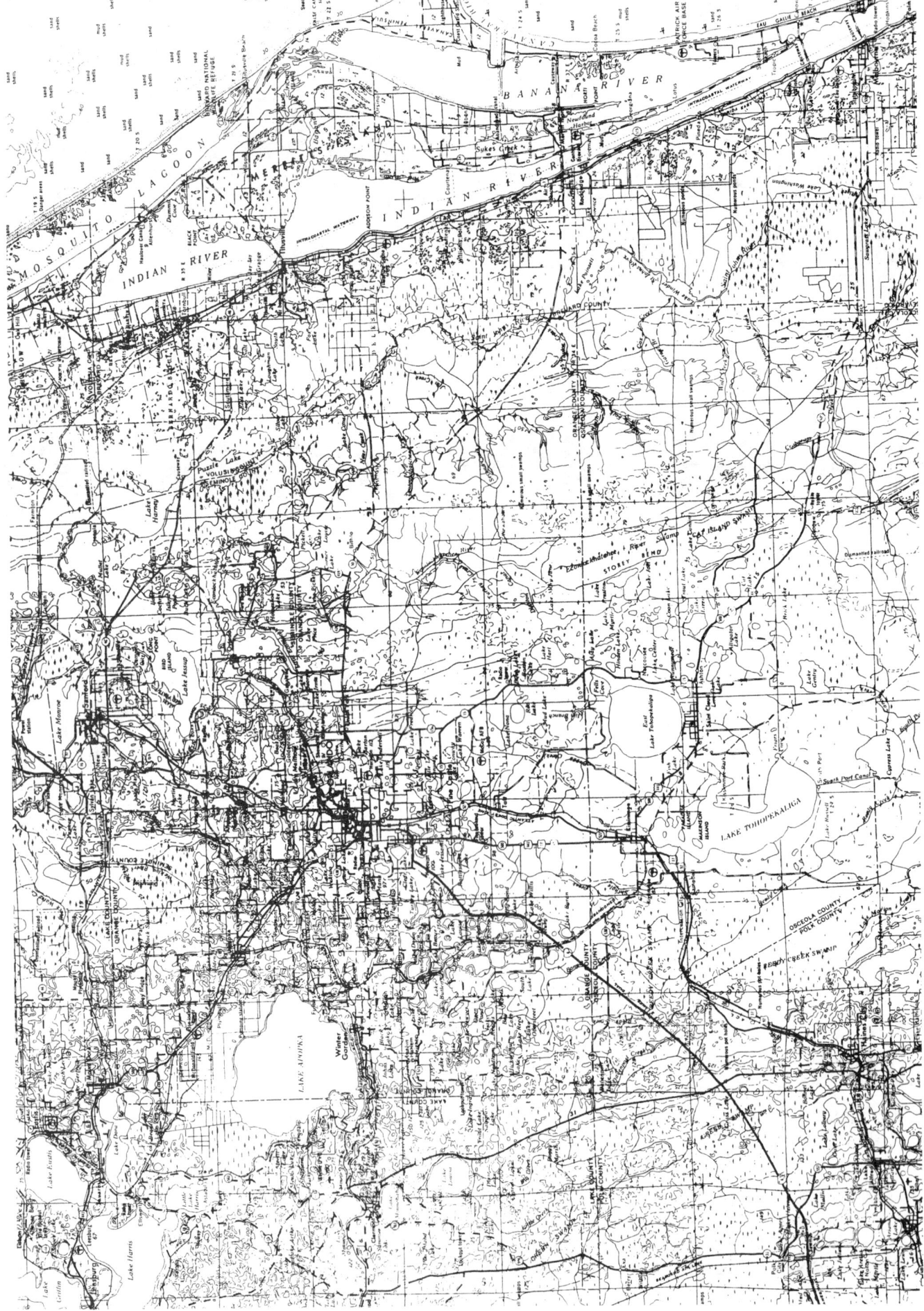

Figure 12–7. USGS topographic map, Orlando, FL. RF 1:250,000 (reduced here) 1962.

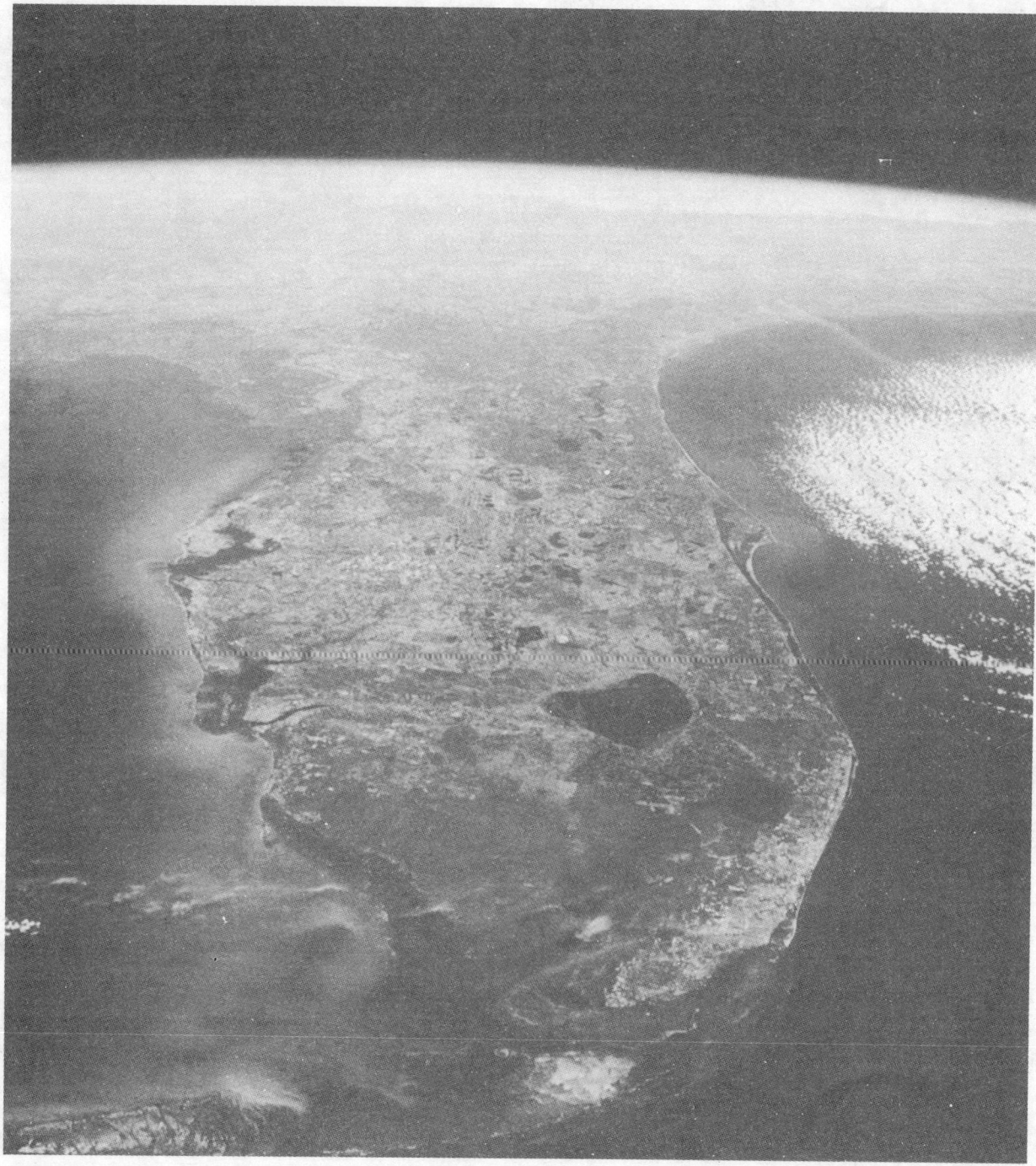

Figure 12–8. NASA STS 51-C scene of Florida. Oblique view. Original color photo acquired 1/24–27/85 altitude of 190 nautical miles.

Exercise 12.3   Interpretation of Selected Earth Surface Phenomena: Areas of Strong Environmental Impact—Orlando and Lake County, FL

1. In examining the three images and the accompanying maps, is there evidence that the karst topography is concentrated more in some areas than in others. What is the evidence and what might account for these variations? Explain.

2. What do these images and maps reveal about the overall drainage pattern in the Florida karst area? Why aren't there more surface streams?

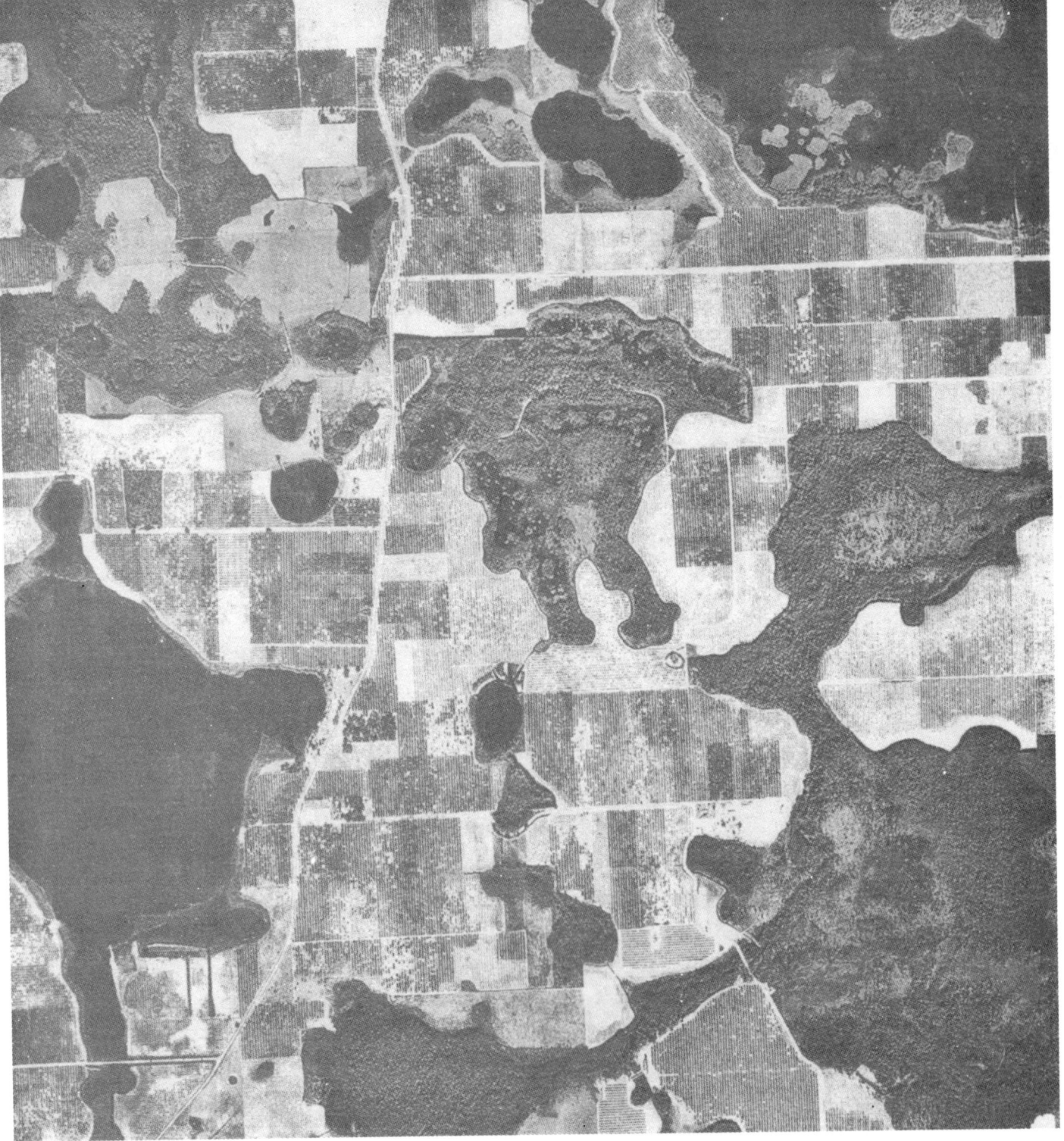

Figure 12–9. Black and white airphoto. Lake Nellie, FL. Acquired 1/17/66. ID CTS-1GG-256. Courtesy USDA, ASCS.

3. On all three images the size and shape of sinkhole lakes seems to vary considerably. What might account for this?

4. Figure 12–9 reveals that virtually all of the land area that is not water or wetland is devoted to citrus orchards. This photo is thirty years old. If you visited this area now would you expect to see more or less farmland? Explain.

5. Examine the water and wetland areas displayed in Figure 12–9. What evidence is present on the image and accompanying map that would allow you to make predictions about the size and acreage of such areas in the future?

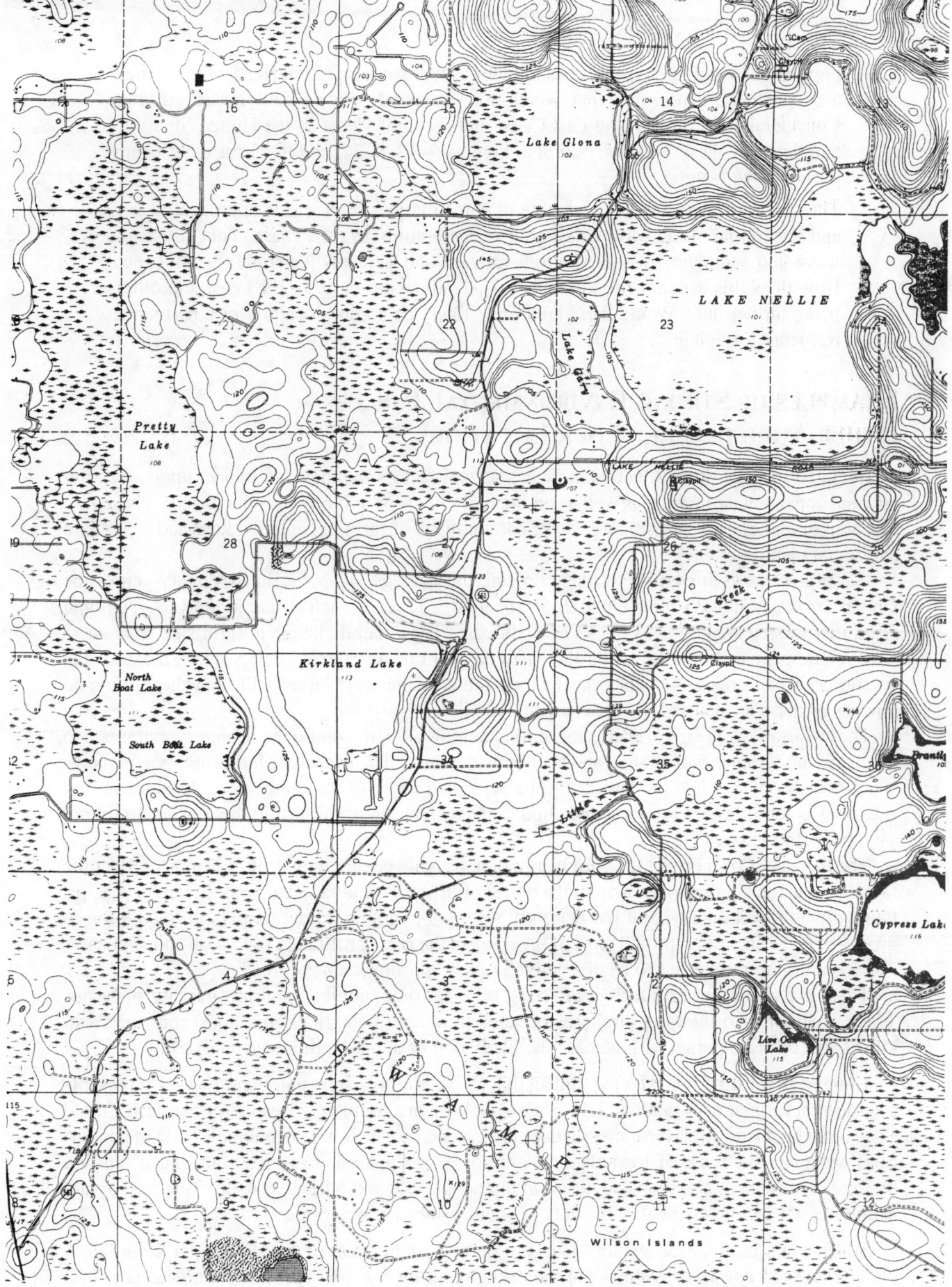

Figure 12–10. USGS topographic map, Lake Nellie, FL (part). RF 1:24,000. 1959. Photorevised 1980.

6. Figures 12–6 and 12–8 reveal coastal features that are termed **barrier islands.** These are ephemeral landforms that were built by wind, waves, and ocean currents. Considerable sections of the East Coast and the Gulf Coast of the United States are paralleled by these features. See also Figures 9–10, 9–22, and 9–23 for other examples of these features.

The origin of these landforms is due in part to large ocean waves approaching land and the bottom of the wave striking the ocean bottom near the coast. This thrusts the wave and sediment from the bottom forward and deposits it parallel to the beach. How does this account for some barrier islands being close to the coast and others being farther out? Would these landforms likely be a good location for a permanent residence? Explain.

EXAMPLES OF STRONG ENVIRONMENTAL IMPACT— OTHER AGRICULTURE: FOUR LOCATIONS

Background. In North America there are several factors that have an impact upon size, layout, and appearance of farms, as well as on the types of crops and products harvested. These factors help to identify the general location of certain types of farming, and that knowledge leads to additional identification.

If a person were to examine a map showing the average size of farm by county across the United States, a quite distinctive pattern is apparent. (There is such a map in the U.S. National Atlas that was created by the U.S. Census Bureau.) The general change in farm size is an east–west relationship with the farms on the east coast and in Appalachia being small, those of the central corn belt area larger, the wheat farms more to the west larger still, and the farms and ranches of the intermontane areas the biggest of all.

These variations in size are a response to changing climate conditions primarily, but are also related to topography, soil quality, and economic demand. The east coast farms are mostly intensive truck cropping, while the farms of Appalachia and much of the southeastern states are small subsistence activities or cash crop operations such as tobacco. Soil fertility here is generally not as good as farther west.

The thousands of farms of the corn belt region, roughly centered on central Illinois, reflect the fact that it is the single largest area in the world of prime farmland. Temperature range, precipitation amount, soil quality, and flat to gently rolling topography allow farmers in this region to grow a wide variety of crops and most of them select the most profitable, field corn and soybeans. As a result these farms tend to be distinctive in their appearance.

These are all influencing factors affecting the size, type, number, and profitability of farms. When attempting to interpret airphotos or satellite images of agricultural landscapes, personal knowledge of these points will increase the likelihood of an in depth image analysis.

Lancaster County, PA. Having said all that above, the area shown in Figure 12–11 represents a somewhat anomalous situation. Lancaster County is an agricultural area in northeastern United States in which Corn Belt farming practices are carried on. This area is unique in that the conditions are more like those of Iowa than they are of New England. Corn Belt farming is typified by field corn as the major crop and its use in fattening cattle and other livestock. This stereogram acquired on 5/23/64 is typical of the area of southeastern Pennsylvania.

Sunflower County, MS. Figure 12–12 is a stereogram of a part of the Mississippi River floodplain where **old age river** features are the background for a superimposed agricultural land use. The date of the photo is 10/28/66. Compare this large scale view with the following TM image of the same region. Figures 12–13 and 12–14 are USGS topographic coverage for the preceding sites.

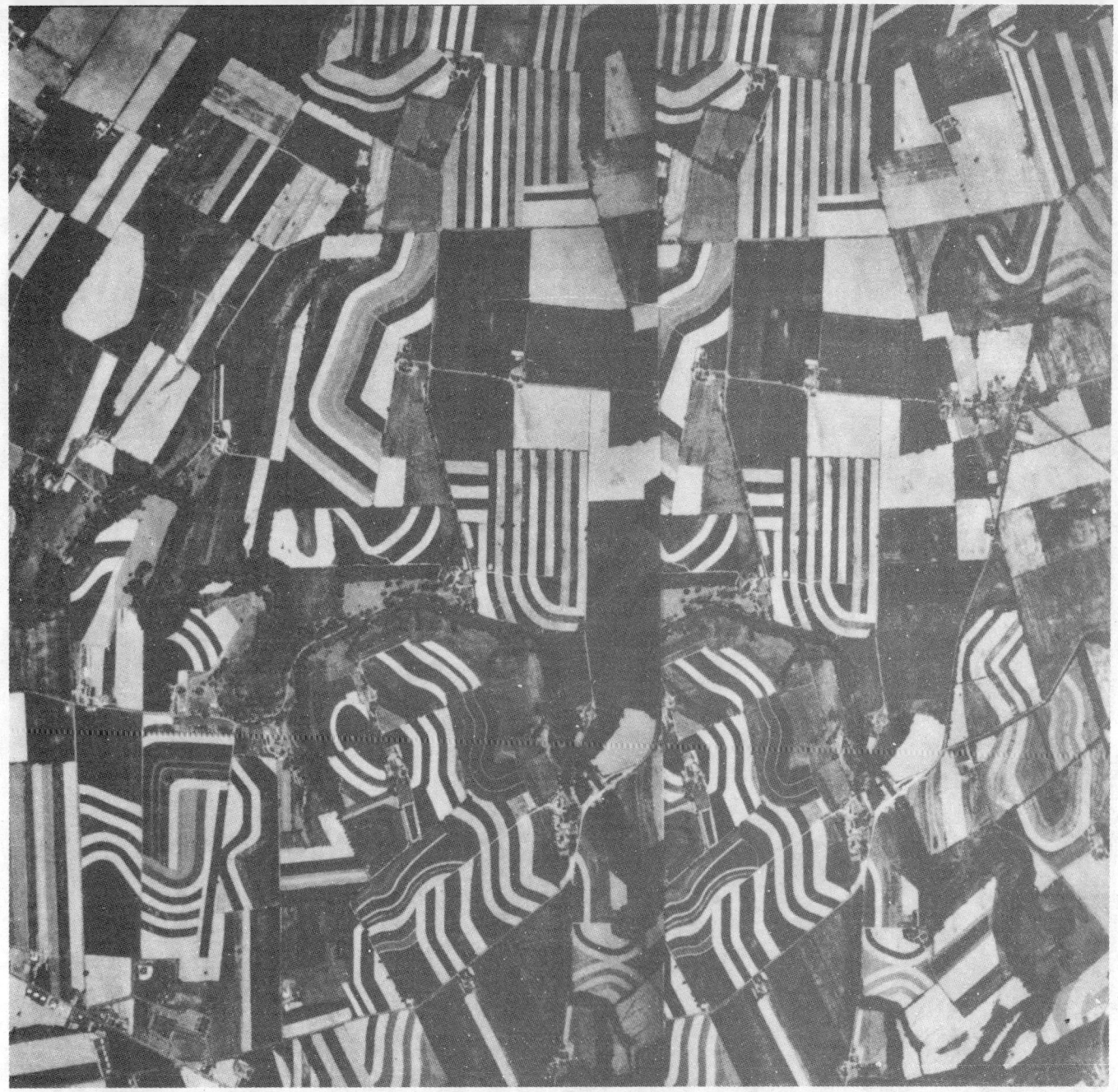

Figure 12–11. Black and white stereogram. Lancaster County, PA. Acquired 5/23/64. ID AHG5EE96,97. Courtesy USDA, ASCS.

Greenville, MS. This Landsat TM image was acquired 9/23/82 and is a black and white print of a composite of bands 2, 3, and 4 shown in Figure 12–15. Figure 12–16 is USGS topographic coverage of part of the scene and Figure 12–17 is a diagram that illustrates landform features of old age river systems.

Garden City, KS. Figure 12–18 is a TM image over the western part of Kansas and was obtained towards the end of the growing season in 1983. Figure 12–19 presents topographic information for the area. The circular fields are examples of **central pivot agriculture.** This is a type of farming that allows for controlling of plant growth through irrigating quarter-section fields by means of a mechanized, rotating sprinkler system.

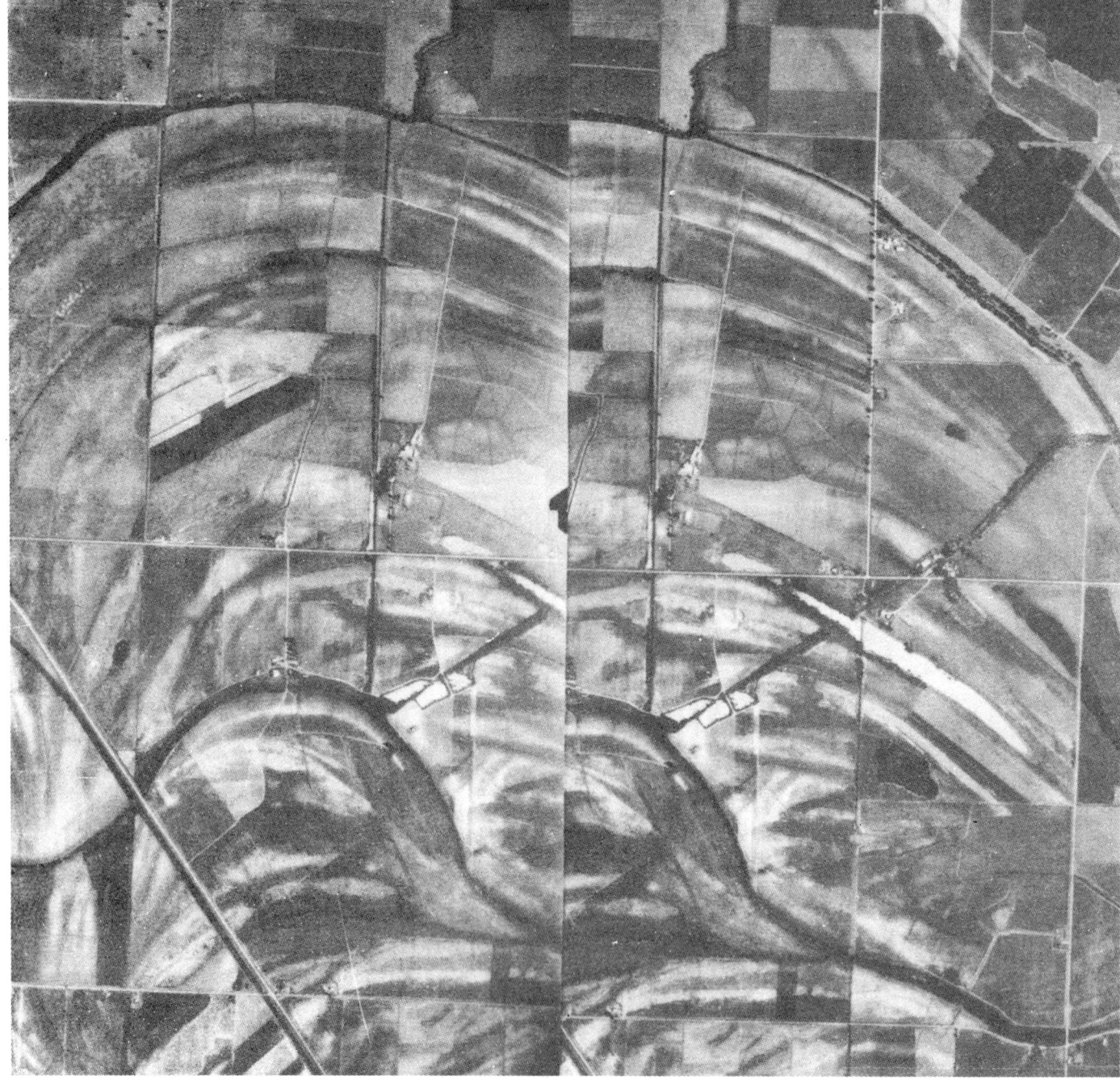

Figure 12–12. Black and white stereogram. Sunflower County, MS. Acquired 10/28/66. ID AVL1HH57,58. Courtesy USDA, ASCS.

Exercise 12.4 Interpretation of Selected Earth Surface Phenomena: Areas of Strong Environmental Impact—Agriculture

A. Lancaster County, PA

Compare the stereogram in Figure 12–11 with the contour map coverage in Figure 12–13. If one were to examine a map of Pennsylvania's overall topography and then be informed that this stereogram is typical of a relatively large area in the state, the facts seem contradictory.

1. What evidence is there of the local topographic relief?

2. How is the field pattern related to the topography? What might be the reason for such irregular plowing patterns?

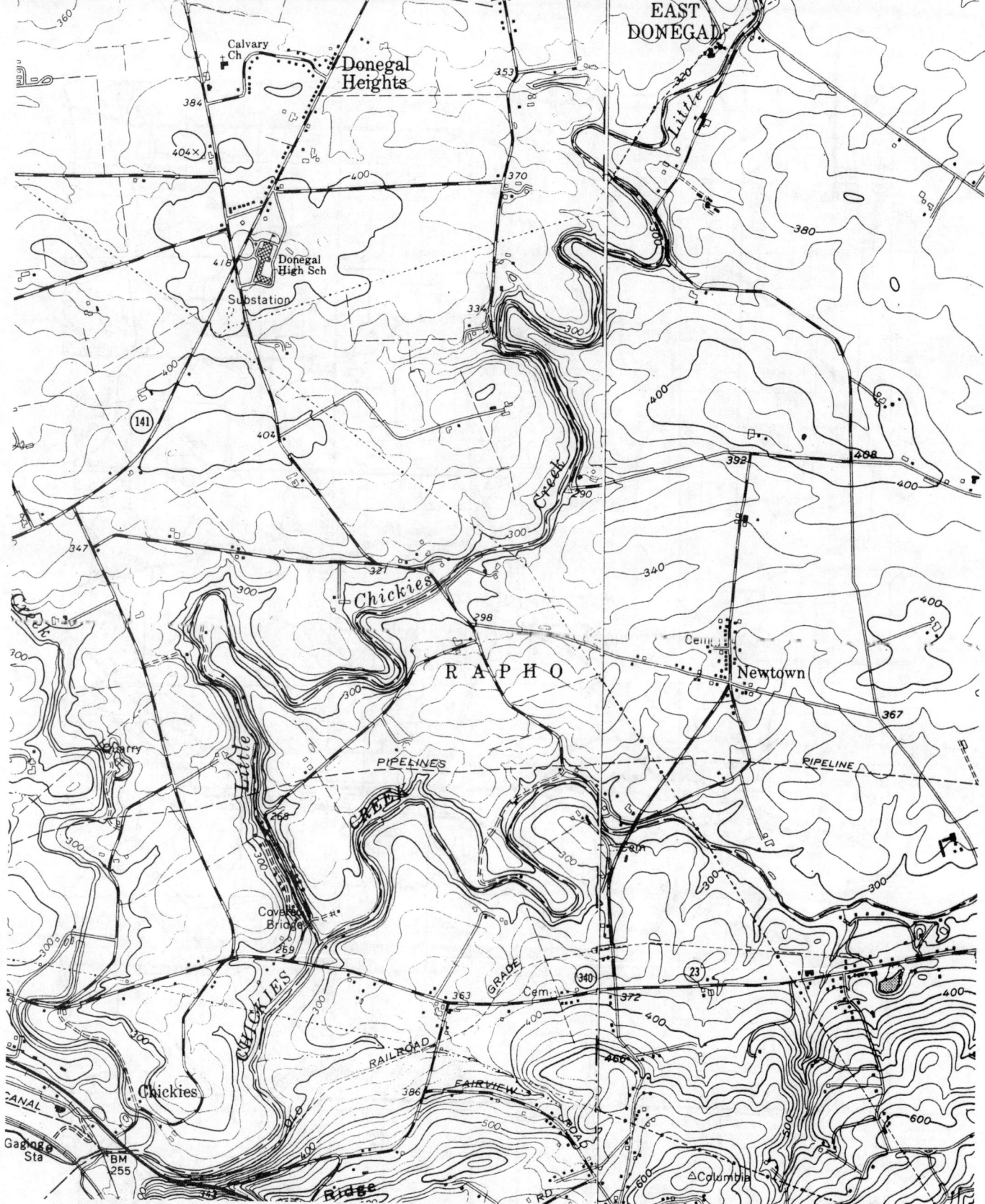

Figure 12–13. USGS topographic map (part). Lancaster County, PA. RF 1:24,000.

3. How much land is not being used for agriculture? Is this typical of New England? of Pennsylvania?

4. How many pieces of evidence can you list using the map and aerials that indicate this is a prosperous farming area?

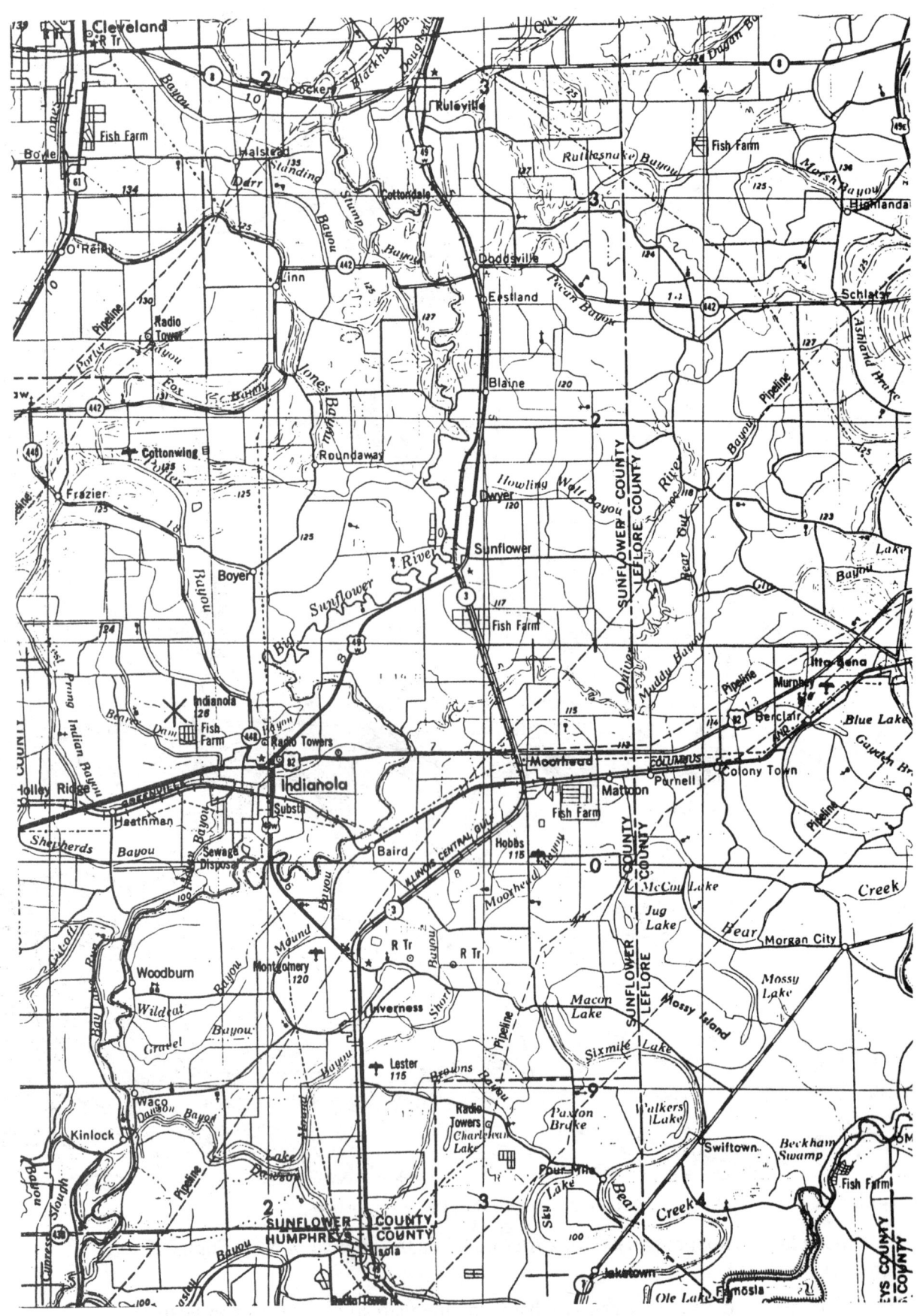

Figure 12–14. USGS topographic map (part). Moorhead Quadrangle, MS. RF 1:24,000

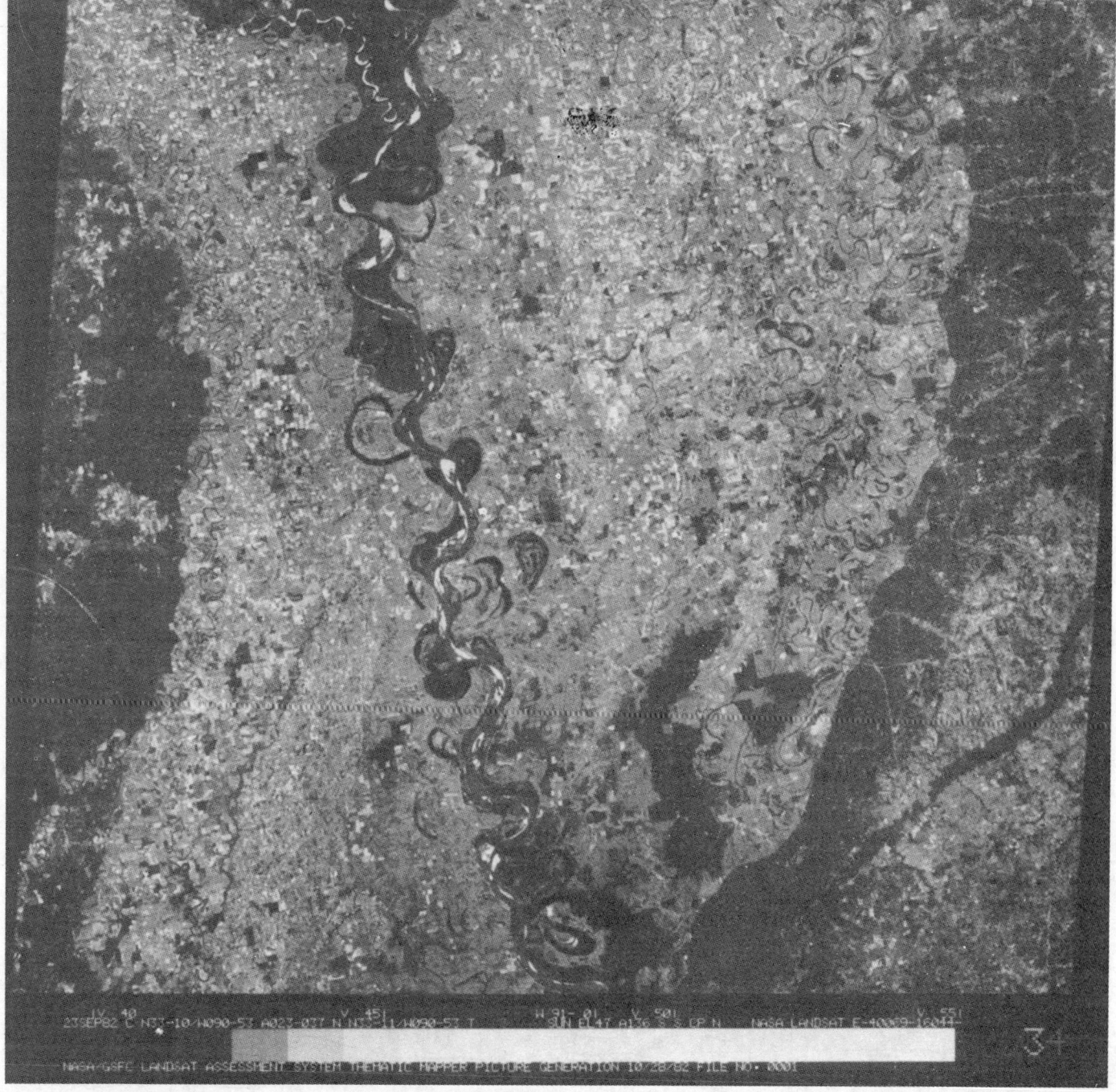

Figure 12–15. Landsat 4 TM bands 2,3,4, Greenville, MS. Acquired 9/23/82. Image ID 4006916044. Courtesy NASA/GSFC.

B. Greenville, MS and Sunflower County, MS

Compare figures 12–12, 12–14, 12–15, 12–16, and 12–17 and answer the following questions.

1. Note that the TM image includes a section across the valley of the Mississippi River from one valley bluff to the other. The bluffs rise 150 to 170 feet above the average elevation of the floodplain. What evidence is present on this image to indicate that the main course of the river has migrated back and forth across the valley?

2. Does the pattern of land use on the TM image give an indication of the limits of the floodplain? Explain.

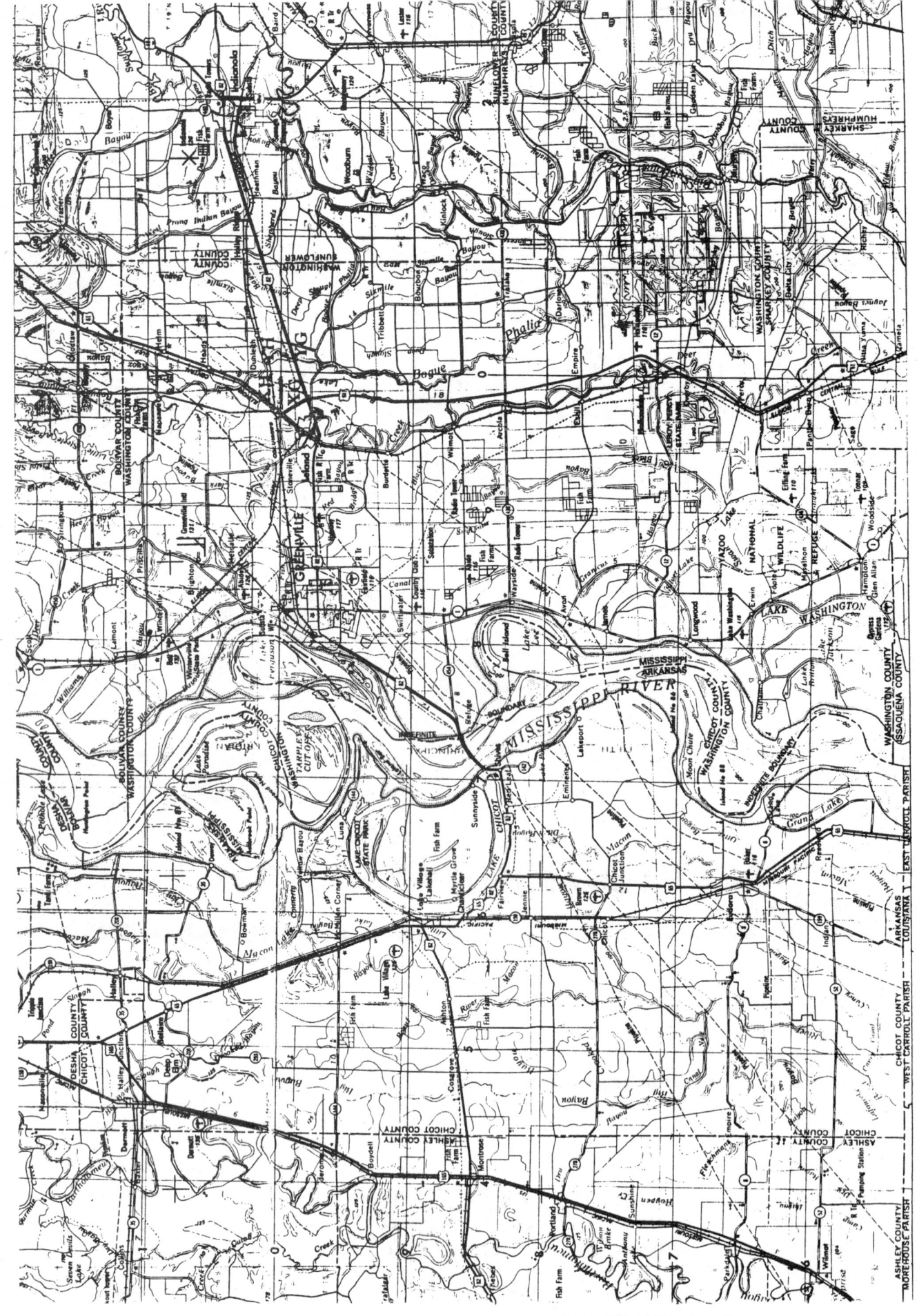

Figure 12–16. USGS topographic map (part). Greenwood, MS. RF 1:250,000 (reduced).

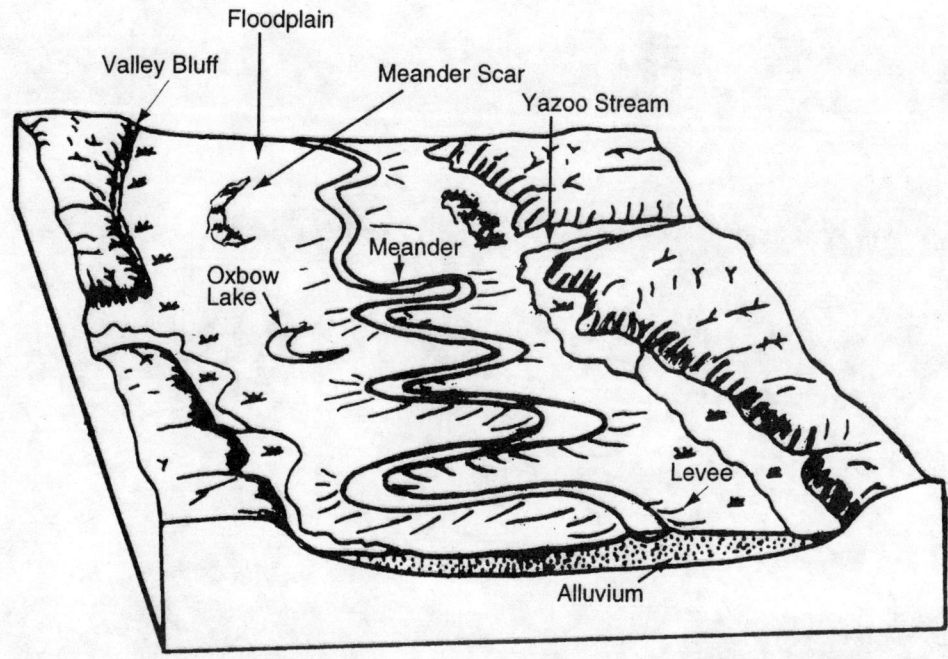

Figure 12–17. Diagram of old age riverine features.

3. Figure 12–12 is a stereogram of a small area that lies in the northeast corner of B-2 on the TM image. What evidence is present in the stereogram that indicates the main channel of the Mississippi River might have flowed through here?

4. What evidence is there in any of these figures that gives indication of how the physical and economic landscapes have interacted? Explain.

5. How many features of old age river development can you locate? Give feature names and specific locations.

C. Garden City, KS

Figures 12–18 and 12–19 are TM image and map, respectively, of an area in southwestern Kansas west of Dodge City and lying on the Arkansas River. This locale receives approximately 10–20 inches of precipitation and wheat is the major crop as compared to corn in the more humid eastern part of the state. Garden City is located north of the river in A-3. The topographic map of this area indicates that there are numerous gas wells and many windmills throughout the area to the south and west of Garden City.

1. Note that there is a considerable amount of central pivot agriculture. In comparing the map and the image do you see evidence of a source of irrigation water for these fields? What is one possible source?

2. Note that there are large areas in the B-1, B-2, and A-3 parts of the image that are not under cultivation. What map and image evidence might explain this circumstance when most of the area is being cropped?

3. What evidence is present on the map and image that makes it possible to guess the size of the central pivot fields?

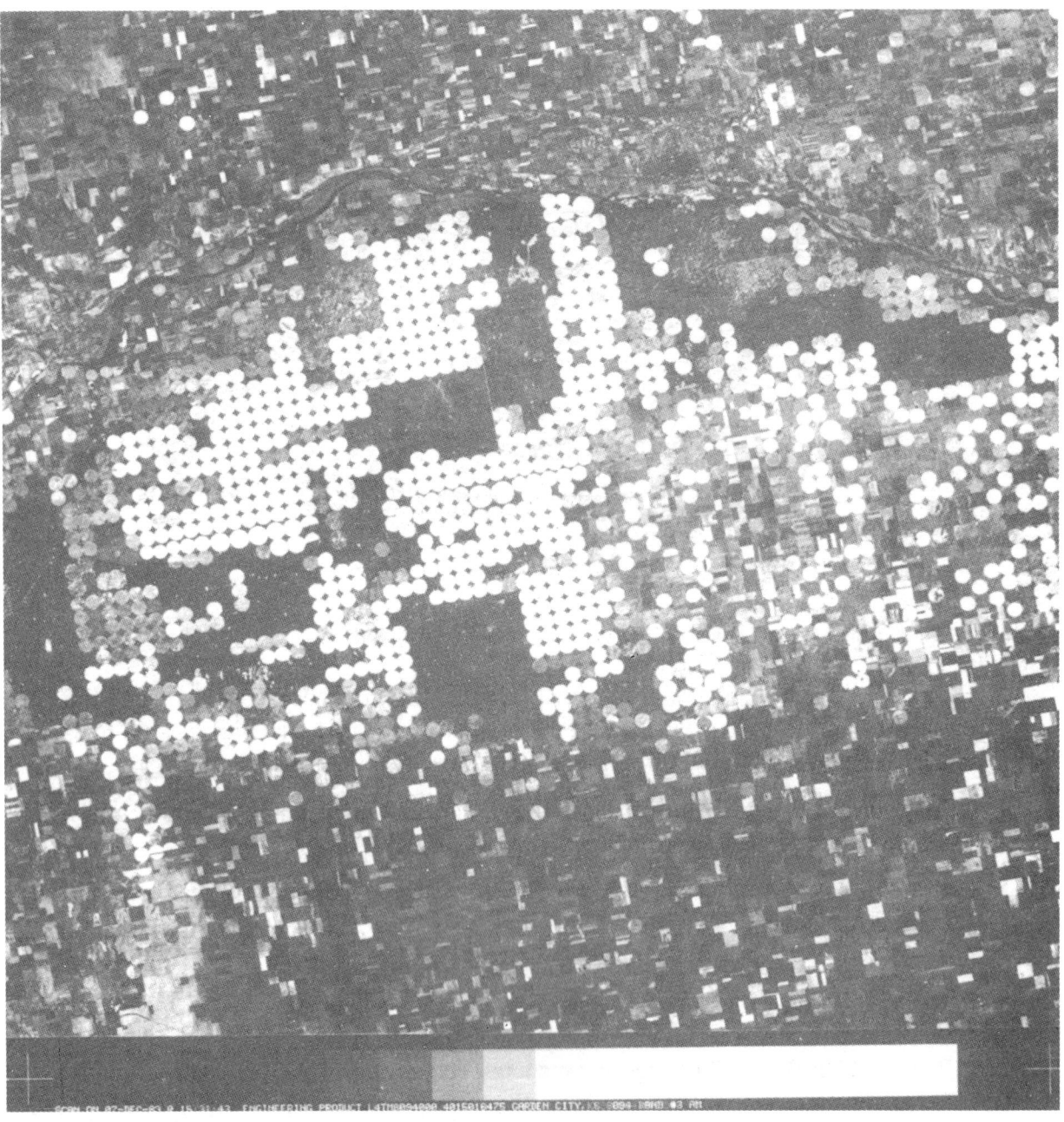

Figure 12–18. Landsat 4 TM bands 2,3,4, Garden City, KS. Acquired Fall, 1983. Image ID 4015016475. Courtesy NASA/GSFC.

4. Can you find any golf courses near Garden City? What are the signature characteristics of a golf course on this image? How might they vary if this were in color?

5. What are the factors of this image location that exert a strong influence on the use of the land? Explain.

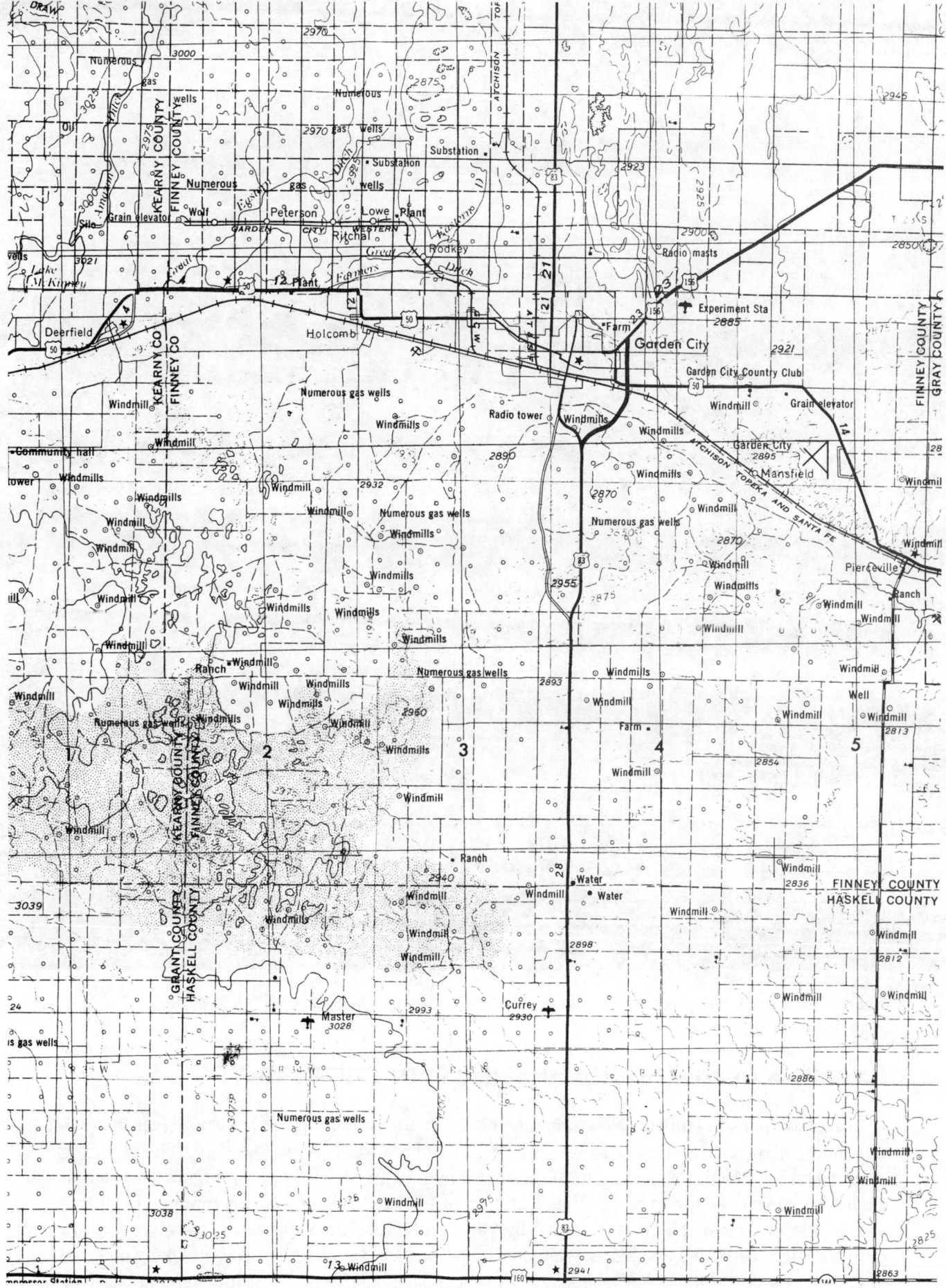

Figure 12–19. USGS topographic map (part), Dodge City, Scott City, KS. RF 1:250,000. 1975.

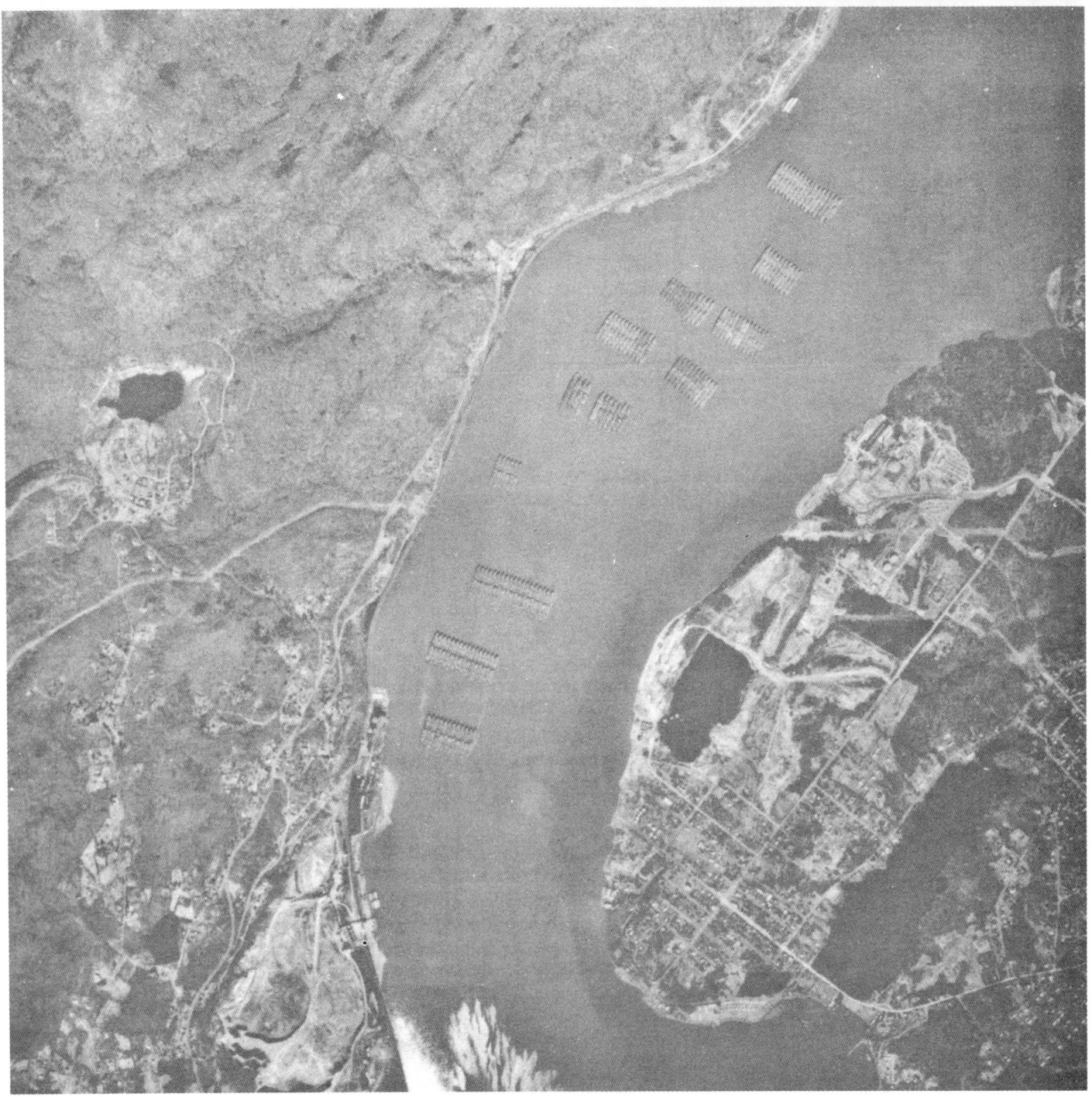

Figure 12–20. Black and white airphoto, Hudson River, NY. Acquired 3/27/68. ID NY-9-1491-1063. Courtesy Lockwood Mapping Inc.

EXAMPLES OF STRONG HUMAN IMPRINT ON THE LAND

Haverstraw Bay, Hudson River. Figure 12–20 is an airphoto of the Hudson River immediately north of Haverstraw Bay and approximately 10 miles north of New York City. The aerial was acquired on 3/27/68 and displays part of the area shown in the SLAR image in Figure 11–9 and on the map, Figure 11–10. Figure 12–21 is a portion of the Peekskill, NY USGS topographic map.

The left side of the photo is occupied by part of the **Taconic Mountains,** locally called the **Hudson Highlands,** and is sparsely populated. The surface is forest-covered with many outcrops of granite bedrock. Outliers of igneous rock in the adjacent **Triassic Lowland** stand up vertically on the west side of the Hudson River forming the **Palisades** which are represented on this photo in C-2. Local relief is about 1000 feet from ridge top to water.

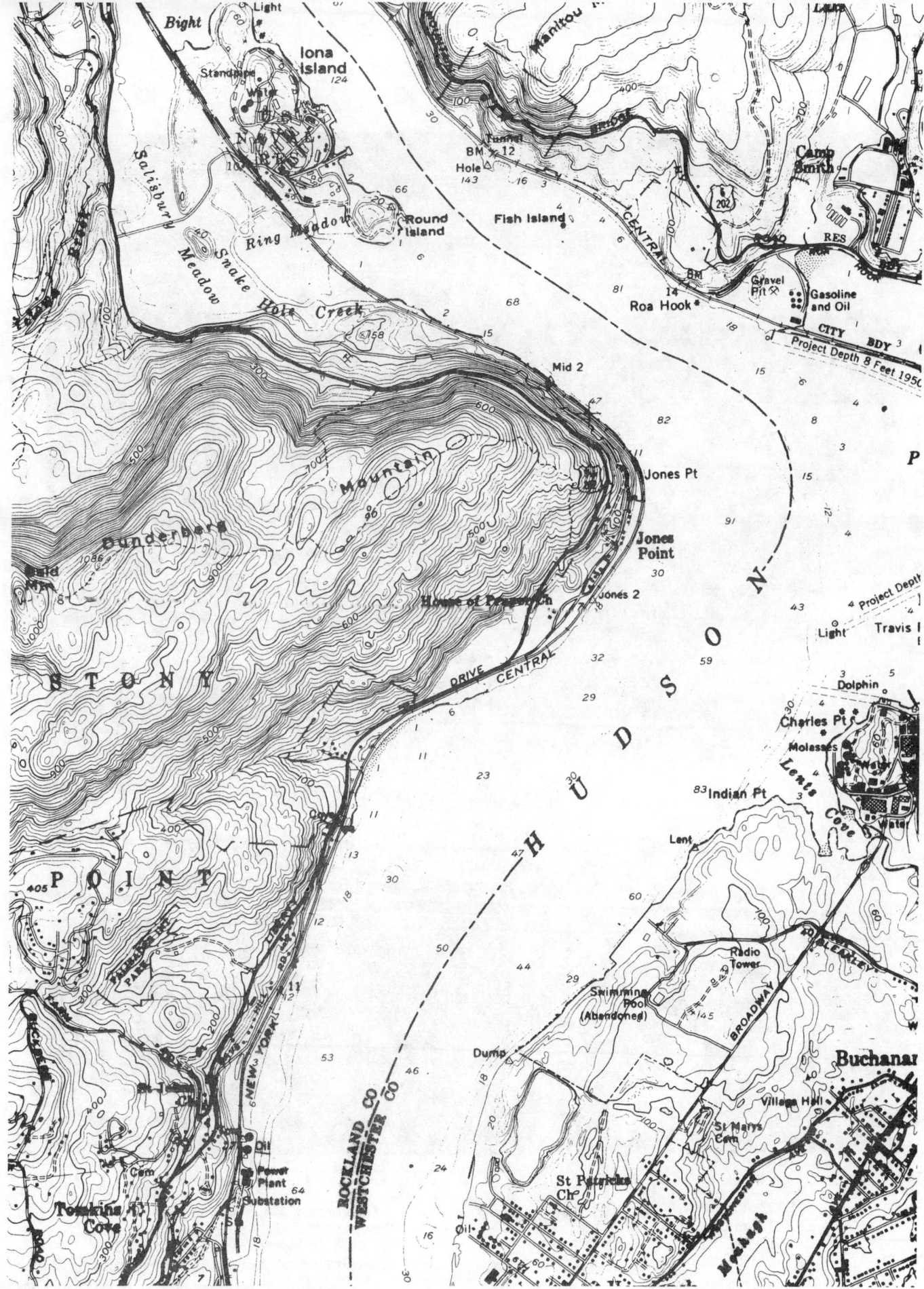

Figure 12–21. USGS topographic map, Peekskill, NY (part). RF 1:24,000.

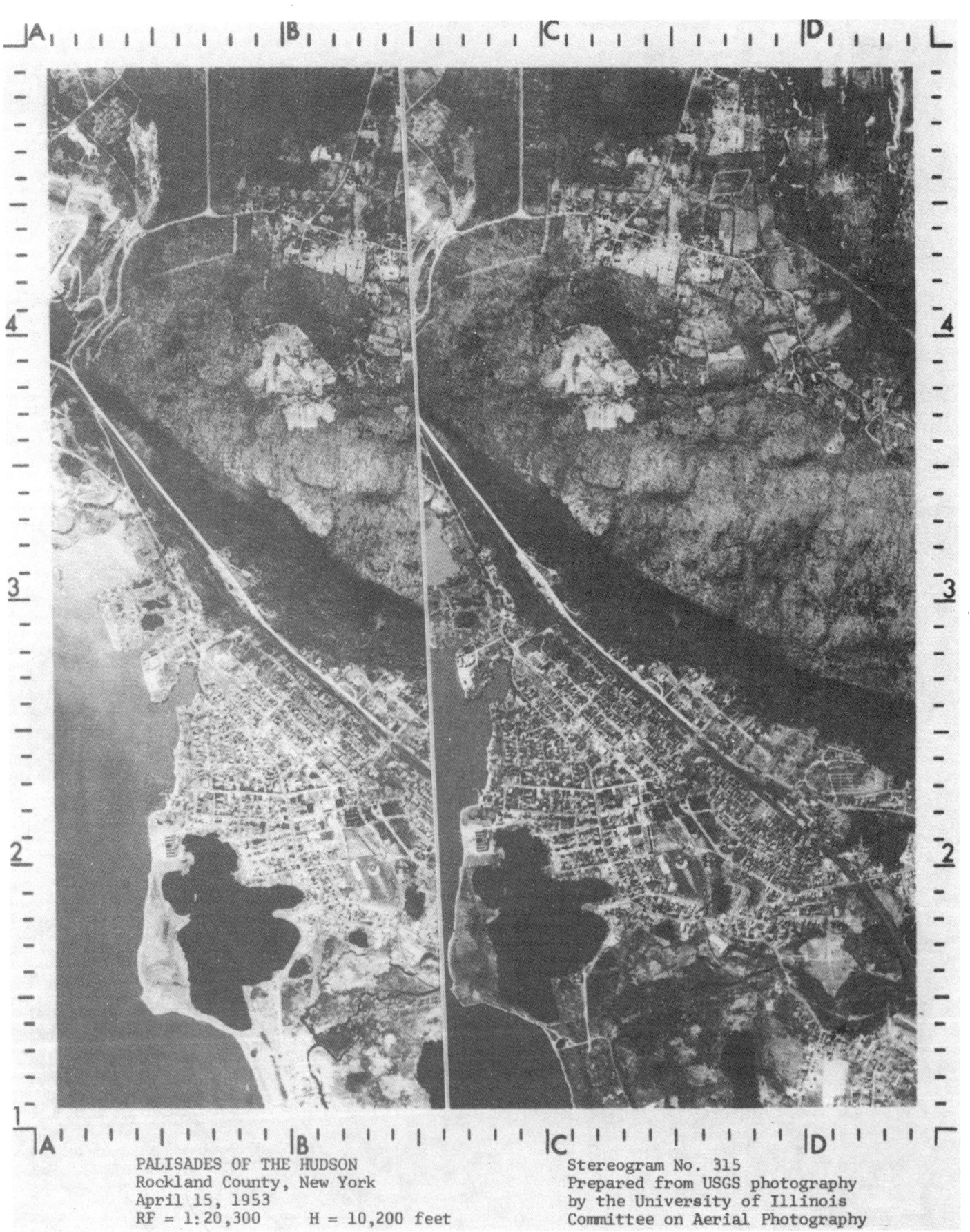

PALISADES OF THE HUDSON
Rockland County, New York
April 15, 1953
RF = 1:20,300   H = 10,200 feet

Stereogram No. 315
Prepared from USGS photography
by the University of Illinois
Committee on Aerial Photography

Figure 12–22. Courtesy University of Illinois GIS Laboratory.

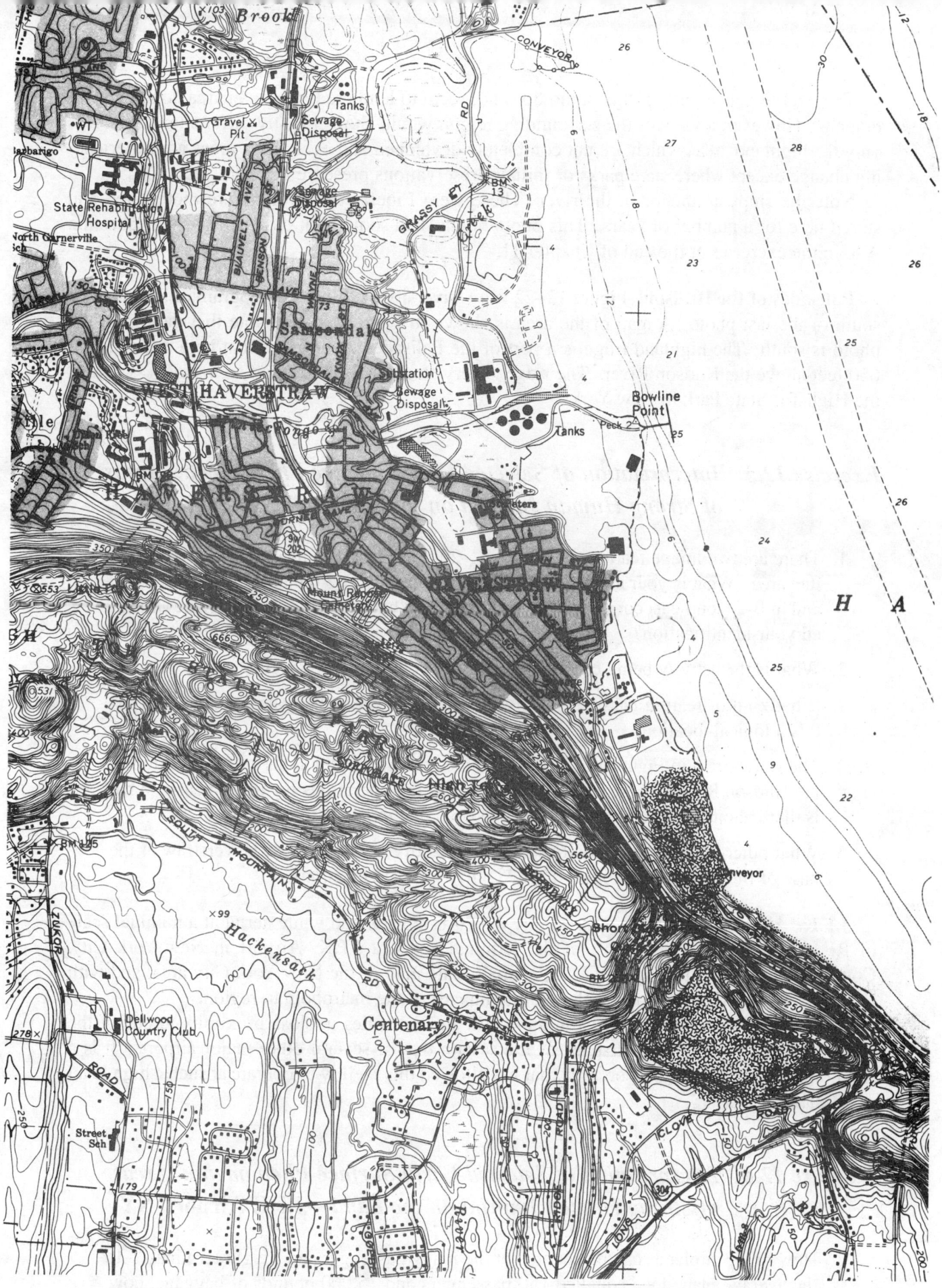

Figure 12–23. USGS topographic map (part). Haverstraw Bay, NY. RF 1:24,000.

The overall appearance of this photo does not seem to lend credence to an area dominated by humans. However, a view of the surrounding region would show a vastly different picture. It is only the highland areas which are not completely developed and even these areas are experiencing change except where state parks or military reservations preserve the status quo.

Note the ships at anchor in the river. These were Liberty Ships that were mothballed and stored here for a number of years. This photo might shed some light on the answer for question A.8. in the exercises at the end of Chapter 11.

Palisades of the Hudson. Figure 12–22 is a University of Illinois stereogram of the area just south of the last photo. A map of the area is shown in Figure 12–23. Note that the top of the photo is south. The highland ridge is a part of the Palisades escarpment that here stands about 650 feet above the Hudson River. The ridge is very steep and the forested area shown is part of the High Tor State Park in New York.

Exercise 12.5   *Interpretation of Selected Earth Surface Phenomena: Areas of Strong Human Imprint on the Land—Haverstraw Bay*

1. There are two image features in Figure 12–20 that hint at the nearness of a metropolitan area. What is your identification of the facilities in B-3 (center) on the east bank, and in B-2 (southeast corner) on the west bank? What signature characteristics assisted your identification?

2. What is the activity being carried on in C-1 next to the river? Evidence?

3. What kind of feature parallels the river's west bank near the water's edge? Why is this a logical location?

4. The stereogram in Figure 12–22 shows some enclosed water areas at the elevation of the Hudson River. These are sites of brick making operations for many years. What is there about the geologic history of this area that explains that activity?

5. What other land uses or activities can you identify in the town of Haverstraw at the base of the Palisades in Figure 12–22? Give specific locations.

J.F.K. Airport, New York, NY. Figure 12–24 is a black and white aerial at a nominal scale of RF 1:12,000 that covers most of the terminal areas at J.F. Airport in New York City. Compare this photo with Figure 12–25, a small scale map of the New York City area, and with Figures 9–4 through 9–7, small scale satellite images of the metropolitan region.

This airport is one of the busiest in the world and yet it lies in close proximity to one of the world's largest cities and its airspace overlaps with that of several other major airports. It is a technological marvel of transportation that has allowed this facility to operate at such a high level of efficiency.

Exercise 12.6   *Interpretation of Selected Earth Surface Phenomena: Areas of Strong Human Imprint on the Land—J.F.K Airport, NY*

1. How many airplanes can be counted at the many passenger and freight terminals? If the average plane load equalled 150 passengers and 15,000 pounds of baggage, how many people and tons of freight do these planes represent?

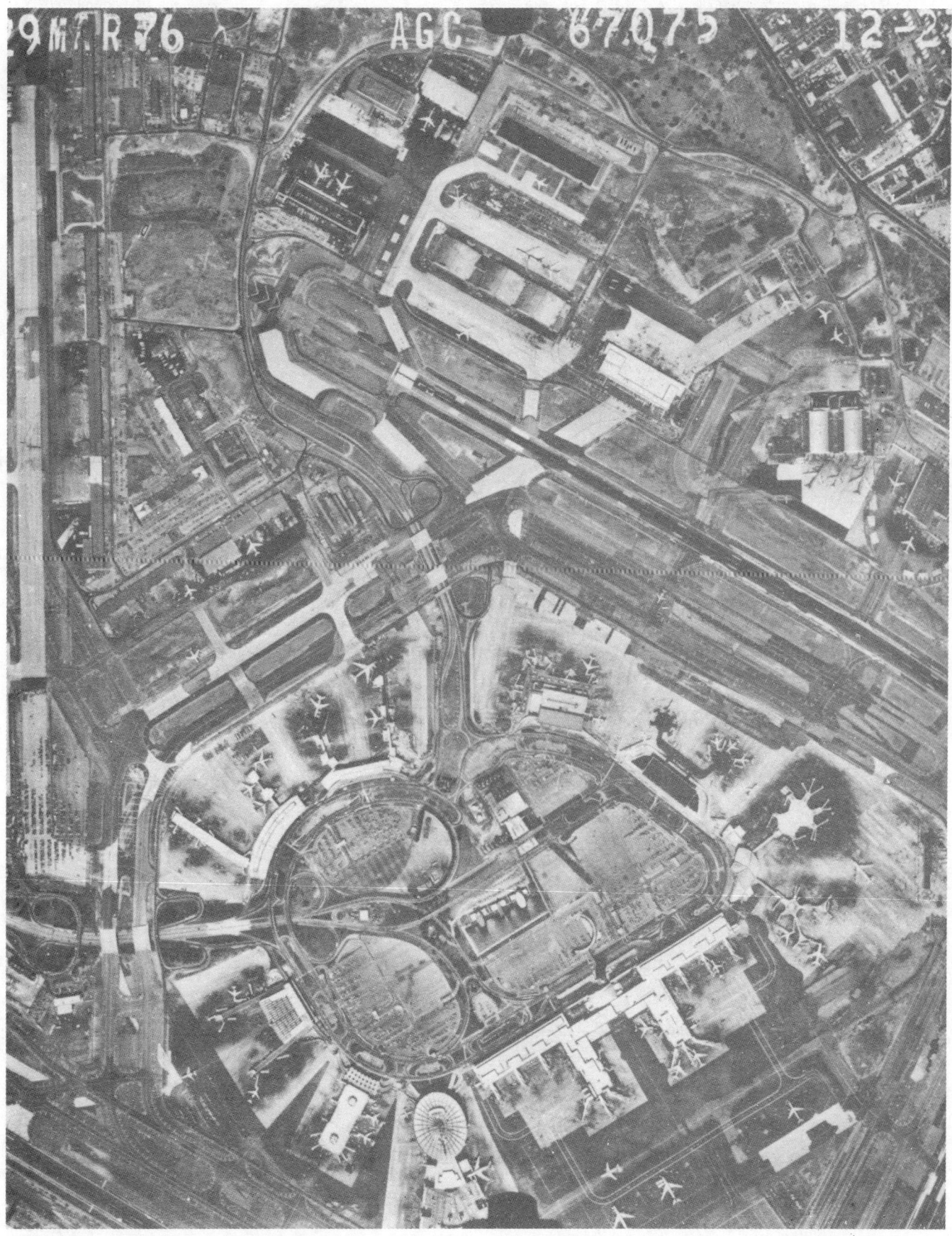

Figure 12–24. Black and white airphoto JFK Airport, NY. Acquired 3/29/76. ID AGC 67075 12-279. Courtesy AeroGraphics Corp.

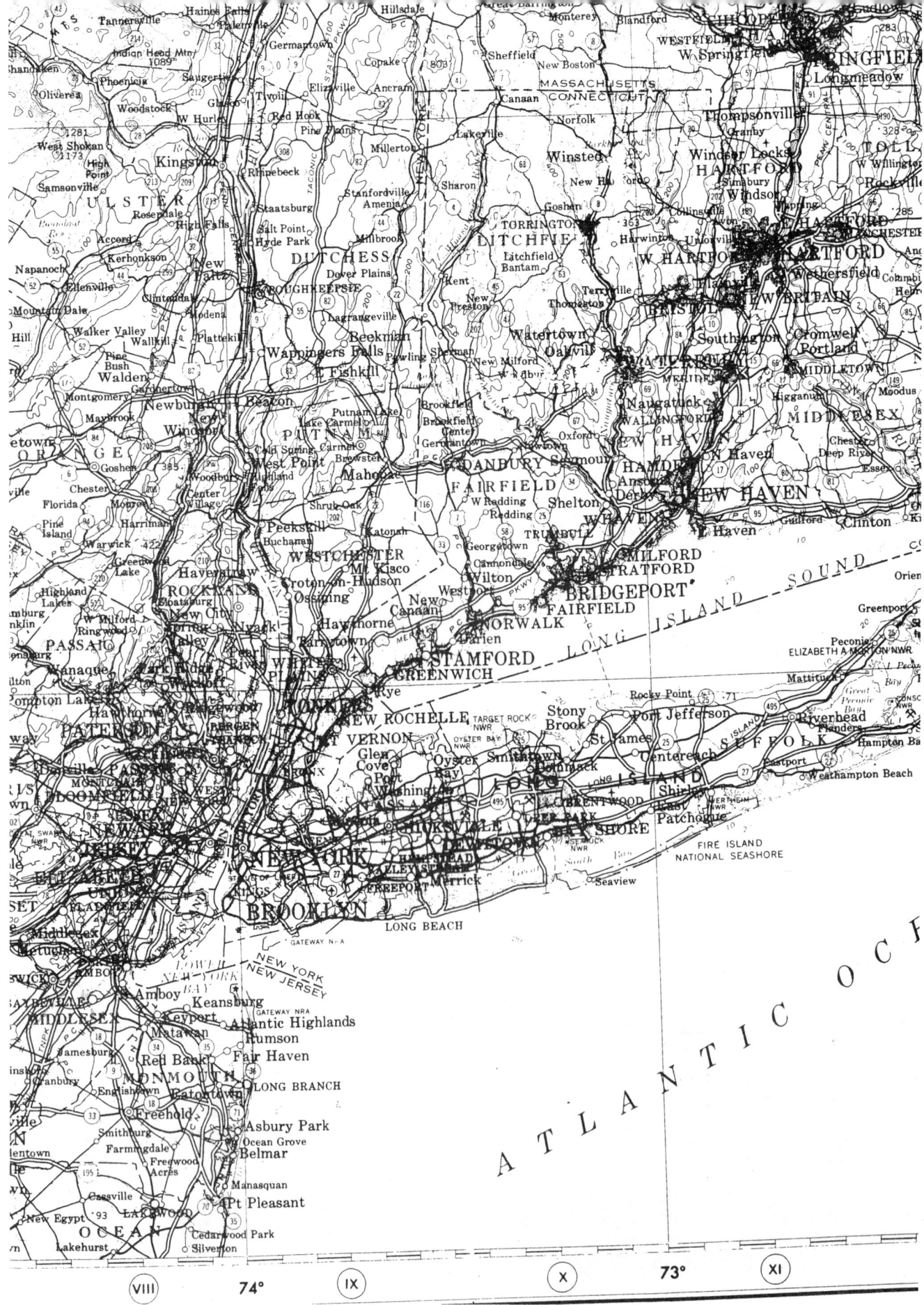

Figure 12–25. USGS topographic map (part), Hudson River. RF 1:1,000,000.

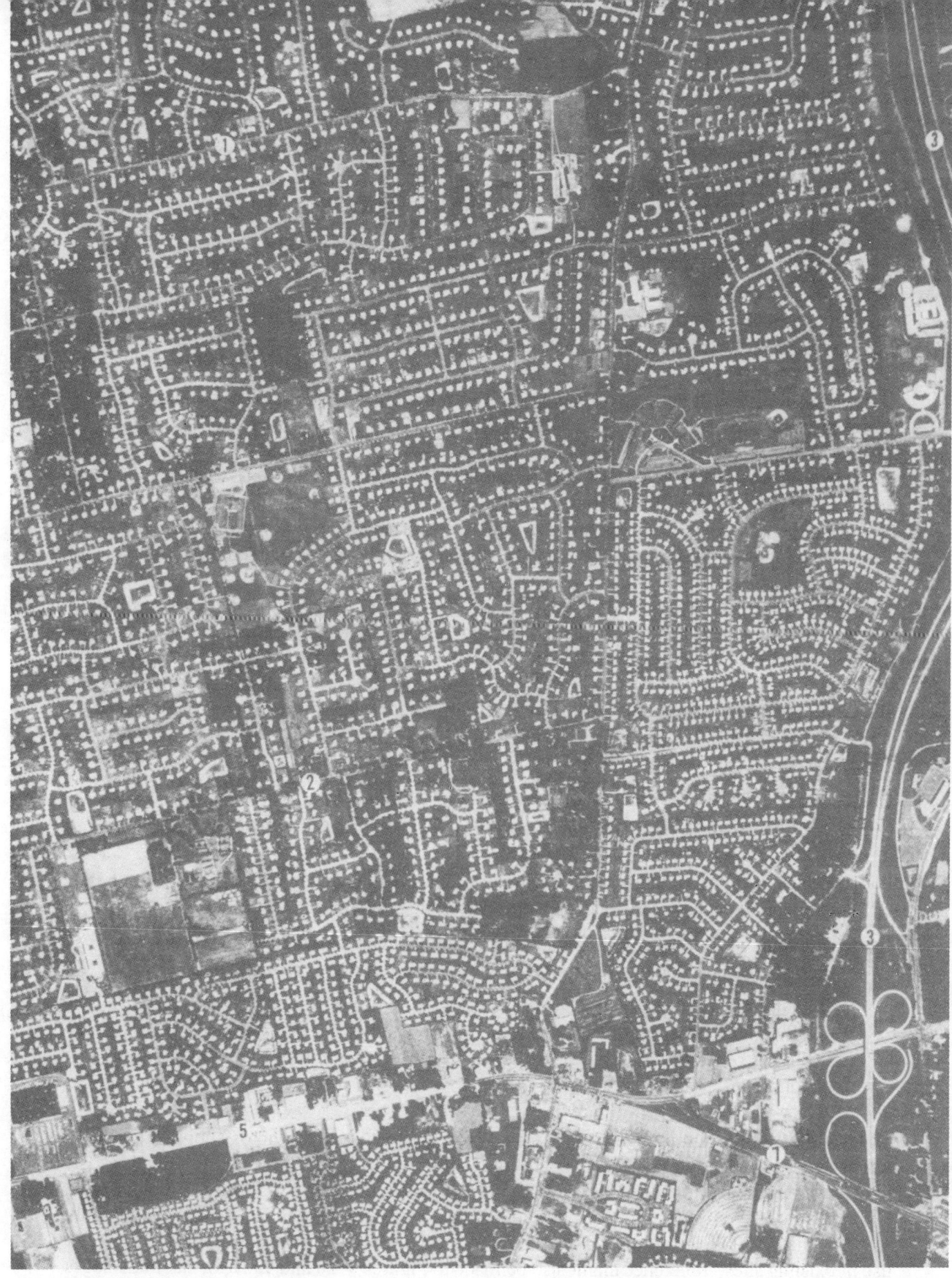

Figure 12–26. Black and white airphoto, Huntington, NY. Acquired Summer, 1970. Courtesy TRW REDI Property Data.

2. How many uncovered parking lots are visible in the photo? If all these parking areas were filled, how many cars would that represent? How did you arrive at your answer?

3. How many square feet of runway and taxiway areas are visible in this photo? If each square foot covered the equivalent of two cubic feet of paving and subsurface material, what volume of paving/building materials would be needed for eventual repair and rebuilding of these surfaces?

4. What types of **land use** would likely be **compatible** with this airport use? What types of land uses seem to be found adjacent to the airport? Are they economically and socially compatible? What kinds of problems are presented to airport managers when faced with these competing land use types?

Huntington, NY. Figure 12–26 is a black and white airphoto taken in the summer of 1970 of a small area in Suffolk County on Long Island, NY. The photo was graciously supplied by a Real Estate development firm, TRW-REDI, which is fitting. The location shown is only a few miles from Levittown, NY where a construction firm, Levitt & Sons, Inc., started a building boom at the close of World War II. It began to build small, residential homes with their own plot of land to serve the needs of United States' military personnel who returned from the war. These development ideas spread across the country rapidly and changed the landscape of America permanently.

The desire to own land has been a very strong force in most societies throughout the world for a very long time. That yearning and the mustering out pay to veterans at the end of the war came together with Levitt's ideas to unleash a transformation of land use in the United States that is still ongoing. Figure 12–27 is a portion of the USGS Greenlawn, NY topographic quadrangle of the area. It has an overall appearance that is matched by a thousand such maps all over the country.

Exercise 12.7   Interpretation of Selected Earth Surface Phenomena: Areas of Strong Human Imprint on the Land—Huntington, NY

1. How many types of land use are illustrated on the photo and map? Do you think these uses are all economically and socially compatible? Explain.

2. The scale of the map is RF 1:24,000. Compare the map and photo to determine how many houses per square mile would be built if this density were maintained.

3. How does that density compare with the use of urban land for high rise apartment or condominium buildings? What are some of the positive and negative aspects of the two types of residential density for the health of a community? Make a list of these factors.

Orange County, CA. Figures 12–28 and 12–29 are two more airphotos supplied by TRW-REDI. These are located in Orange County in California just to the east of Los Angeles. The photos were acquired in 1971. Figure 12–30 is a 1981 medium scale USGS map of Long Beach, CA that includes both sites. Note that this area of the Los Angeles Basin is developed to virtually 100% of that possible for the area. Future building will have to intensify construction for greater use of existing sites. Reference might also be made to Figures 9–20 and 9–21 for satellite images in the same area, as well as Figure 9–27 for a small scale map of the basin.

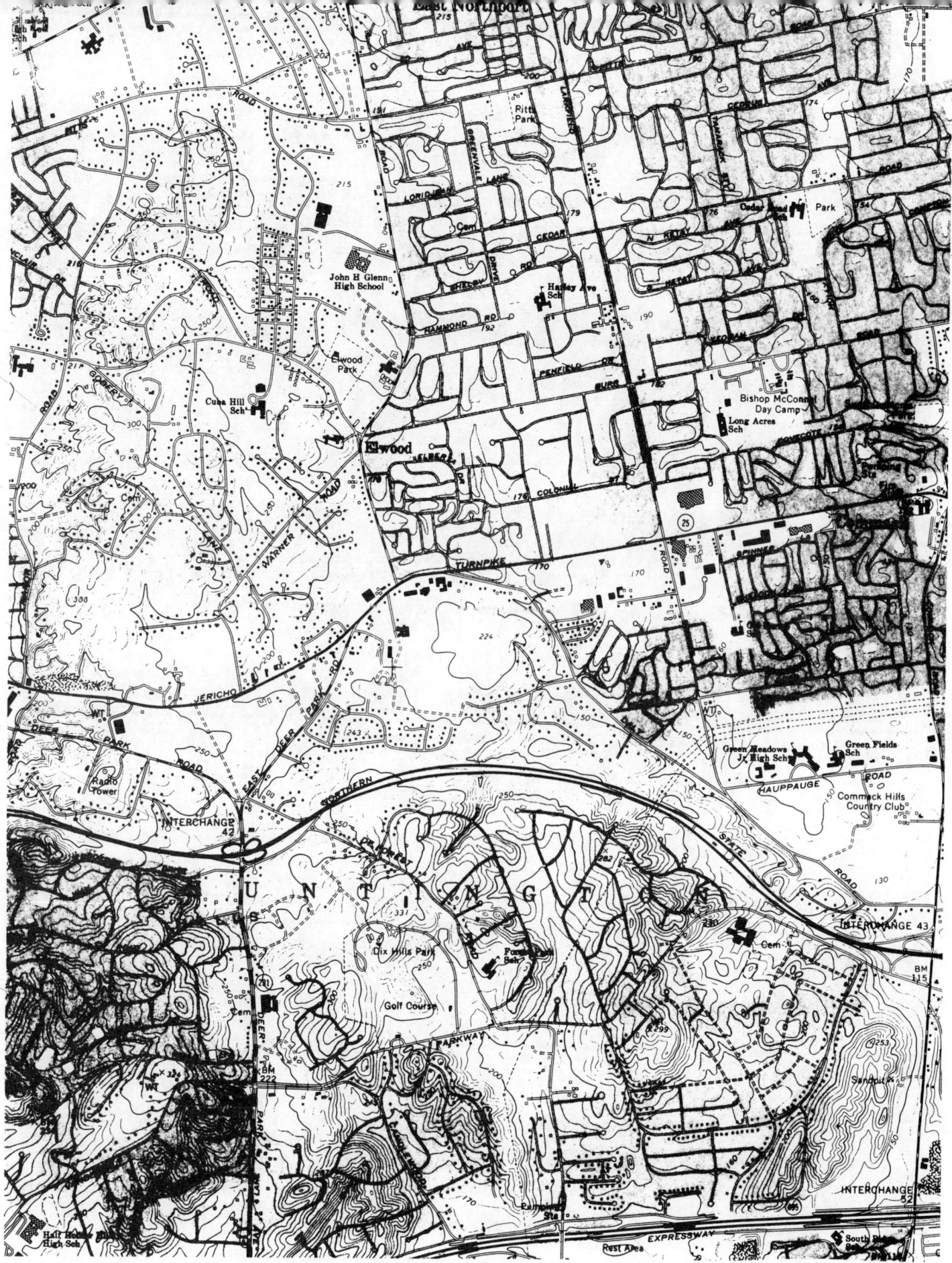

Figure 12–27. USGS topographic map (part). RF 1:24,000. Greenlawn, NY, 1979.

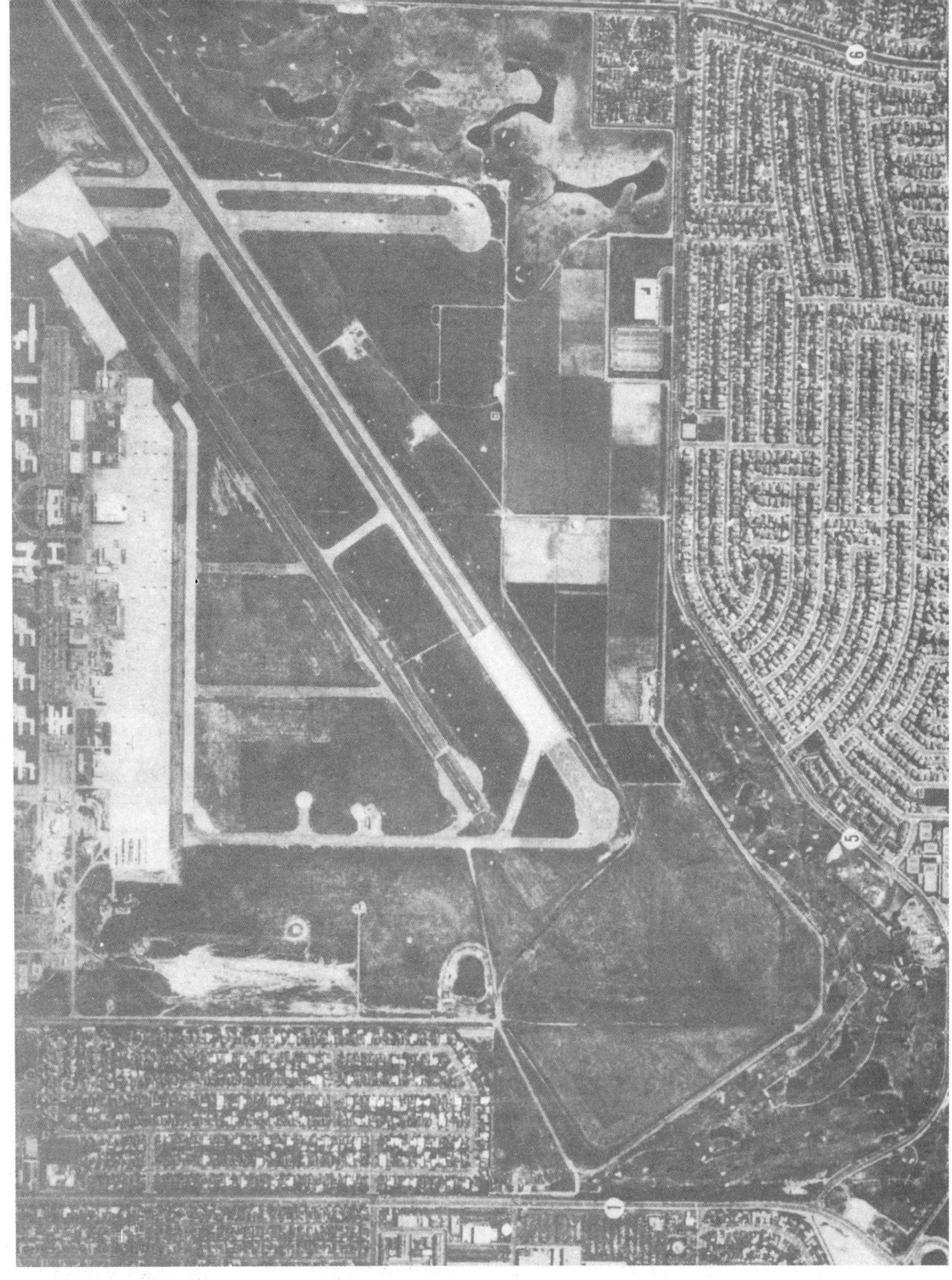

Figure 12–28. Black and white airphoto Los Alamitos, CA. Acquired March, 1976.
Courtesy TRW-REDI Property Data.

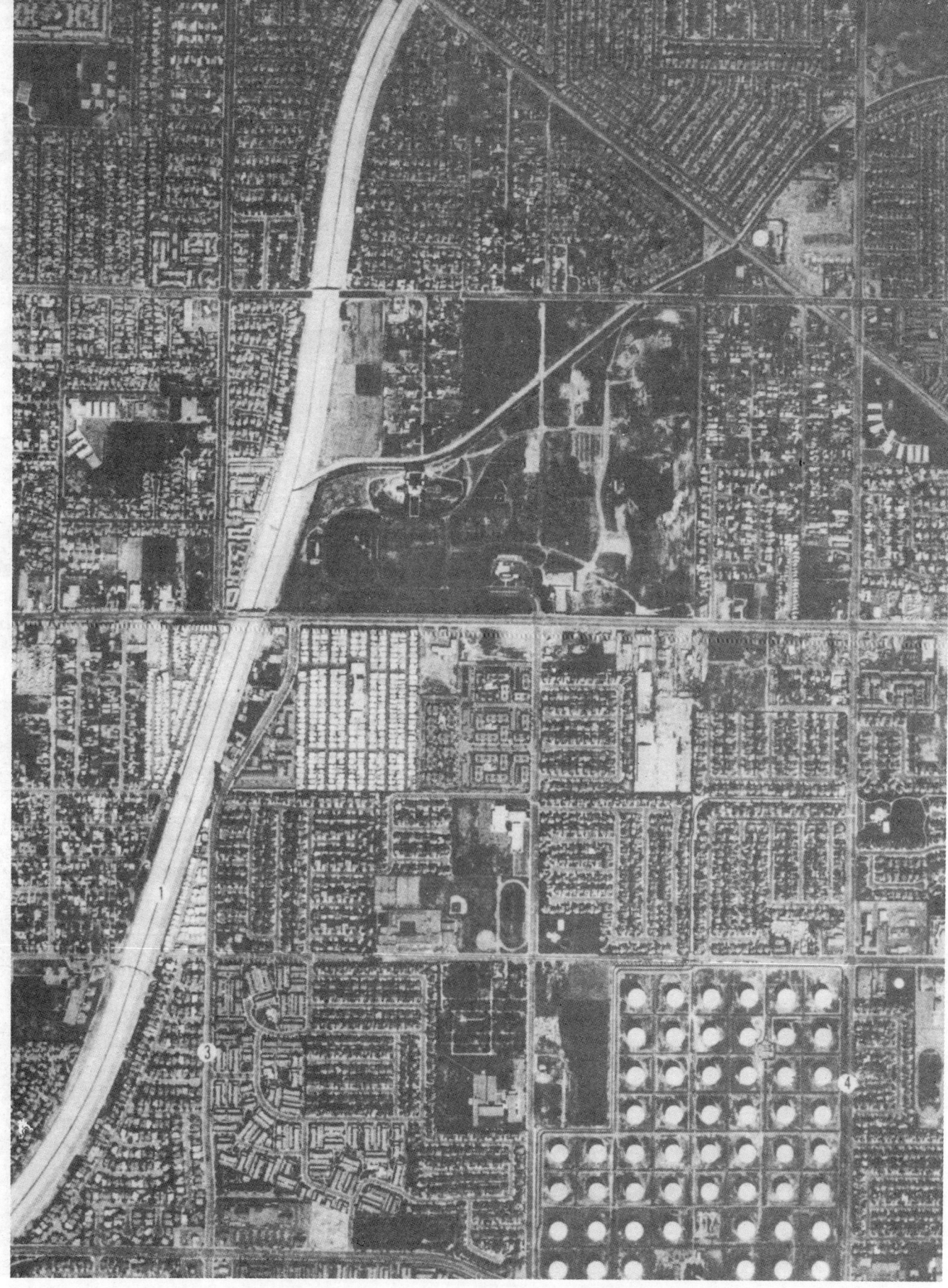

Figure 12–29. Black & white airphoto La Palma and Cypress, CA. Acquired March, 1976.
Courtesy TRW-REDI Property Data.

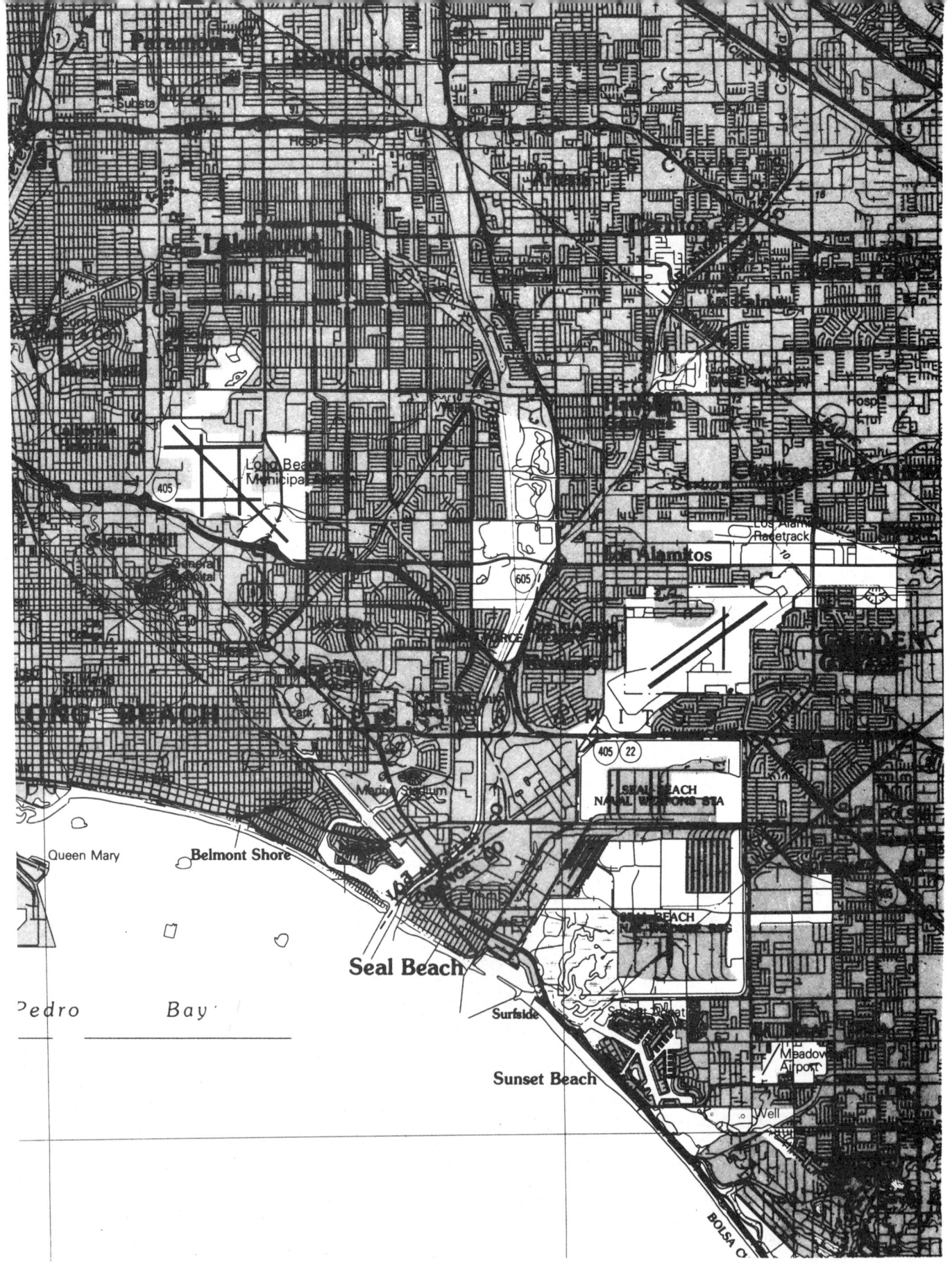

Figure 12–30. USGS map Long Beach, CA. RF 1:100,000.

Figure 12–28 displays part of the City of Los Alamitos and the U.S. Naval Reserve Air Station. Figure 12–29 shows parts of the cities of La Palma and Cypress, as well as the Coyote Creek open culvert, storm drain and its tributary, Moody Creek.

Exercise 12.8   Interpretation of Selected Earth Surface Phenomena: Areas of Strong Human Imprint on the Land—Orange Co., CA

1. The Los Angeles Basin displays an even higher level of development than the East Coast Suffolk County location. It is possible to see some similarities, however. What types of land uses surround the Naval Reserve Air Station in Figure 12–28? Do these uses seem more or less compatible than those around J.F.K. Airport?

2. How is the open space around the airport runways being utilized? Does this seem to make sense?

3. Do you believe the airport or the intense residential development was present first at this site? Why did you reach that conclusion?

4. Figure 12–29 is notable for the amount of space devoted to an open culvert, i.e., storm drain. Why are such features found throughout the Los Angeles basin and rarely in places such as New York City? What characteristics of the two locations make such features logical in southern California but not in downstate New York?

5. What types of land uses are present in Figure 12–29? Give examples of adjacent uses that seem to be compatible with each other, e.g., residential land and grammar school. Give examples of uses that do not appear to be compatible, e.g., railroad line (A-2, A-3) and residential use.

6. Elaborate on why such compatible and incompatible uses play a role in the economy and development of an area?

EXAMPLES OF PAST LANDSCAPES—ARCHAEOLOGY

Archaeology, as investigated through airphotos and satellite imagery, is an academic field that is still largely undeveloped in comparison to what could be accomplished using these sources. In earlier chapters it was stated that concepts such as **change through time** and **sequent occupance** are some of the most powerful frameworks for study of the Earth's surface.

In this section we have evidence that the manifestations of earlier activities are available for all to see, if only the vantage point is appropriate. It is possible to discover evidence of earlier civilizations through examination of images from above the Earth when such information is hidden from view at the surface itself.

The activities of former inhabitants of an area often leave evidence such as **soil marks** where underlying materials, roads, or building foundations leave their imprint in the form of varying agricultural productivity at the surface where it may be recorded by airphotos. Previous civilizations of present day desert areas have been discovered by the use of radar to penetrate to evidence of old watercourses. Here are two examples that give some hint of the treasure of information that awaits further study.

St. Lawrence River—Island of Orleans. Figure 12–31 is an airphoto acquired by the Canadian Government in 1960 of an area northeast of Quebec. Figure 12–32 is a USGS topographic map of the Quebec area that illustrates the general layout of land near Quebec. The photo includes part of the North Channel of the St. Lawrence River and a portion of the Island of Orleans. Note the layout of the fields at right angles to the river. This is an example of the **French long lot** land division system that marked off fields in land holdings, **rotures,** that were long and narrow. Owning frontage on a river was the norm as the rivers served as the main means of transportation and communication. Examples of this pattern can be found along the St. Lawrence River, as shown, and in parts of the state of Louisiana, e.g., along the Red River near Natchitoches, LA.

Mekong Delta, Vietnam. Figure 12–33 is an airphoto of another area where the French influence is visible. This airphoto was acquired over the Mekong Delta area prior to 1975. The photo contains some very interesting artifacts of previous landscapes.

The round area in A-2, A-3 is a granite outcrop that is not being farmed. The objects at the lower right corner of A-1 are part of the ruins of the ancient capital of Fu Nan. Most of the area of the photo is devoted to intensive agriculture from several periods of development.

In the B-2, C-1, and C-2 areas are remnants of the French plantation agriculture. The French arpent system of long lots is represented in the A-3, B-3, and C-3 areas along the present irrigation canal that runs almost vertically. Now examine Figure 12–34 for an enlargement of the top of the B-3 area.

Figure 12–34 displays a part of the present canal in the upper right corner of A-3. An ancient canal runs across the photo horizontally from B-1 to B-3. Note that in this enlargement it is possible to see vestiges of an ancient **ridge and furrow** agriculture lying at right angles to the ancient canal. In this form of intensive agriculture the irrigation ditches were excavated by hand and whatever soil was unearthed was used to build up the ridges between the ditches. Wet rice was grown under water and crops such as bananas were grown on the ridges.

Exercise 12.9   Interpretation of Selected Earth Surface Phenomena: Examples of Past Landscapes—Archaeology

1. Examine Figures 12–31 and 12–32. What other patterns of agricultural development can you recall that would leave a distinctive imprint or pattern on the land? What evidence is there that might indicate the size of a settlement or the intensity of development of an area based solely upon the agricultural imprint?

2. Consider Figures 12–9, 12–10, 12–12, 12–14, 12–15, 12–16, 12–18, 12–19, and 12–28 to 12–30. Each of these areas has the U.S. Public Land Survey superimposed on the surface. Notice the effects of this system on the layout of fields, roads, and settlements. If that survey system was discarded by a future group of inhabitants, would some aspects of it be visible at the surface years later? What kinds of features might show?

   Would the future use still be characterized by squares and rectangles? Would topography or climate be a factor in retaining vestiges of the earlier use? Explain.

3. Figures 12–33 and 12–34 display evidence of three cultural adaptations to farming in the Mekong Delta. What are the likely factors operating in each of those cultures that resulted in the patterns that are visible? Consider the levels of technology. Explain.

4. Can you think of another human imprint on the surface of the Earth that would be likely to be revealed by future airphotos or satellite imagery, even though the imprint had been covered over by soil and vegetation.

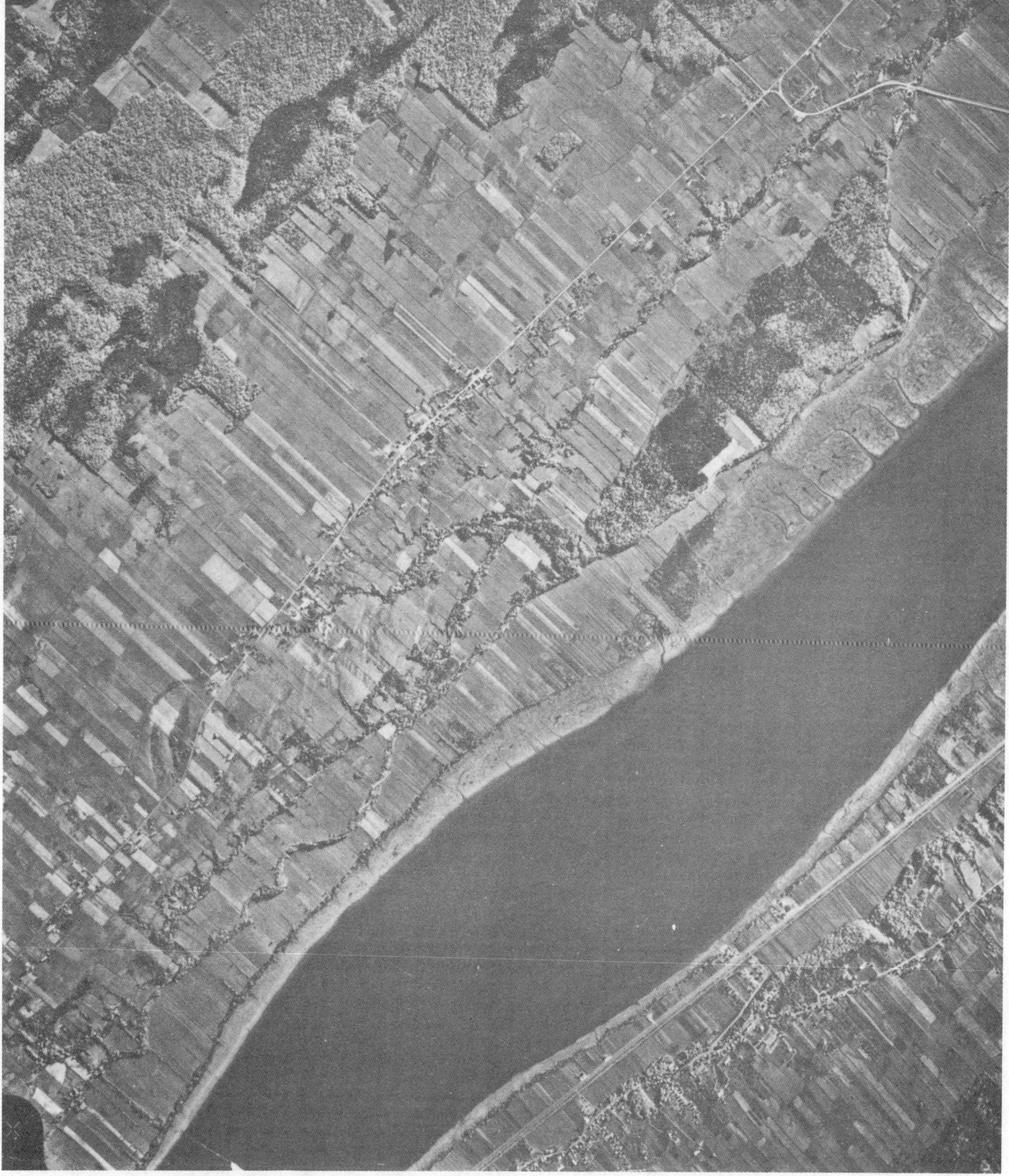

Figure 12–31. Black and white airphoto Ile D'Orleans, Quebec. Acquired 1960. Canadian Govt., Dept. Energy, Mines, & Resources. Courtesy Professor Mildred Berman. (North is toward bottom of page).

5. How does this line of thinking bear upon the use of airphotos and satellite imagery in geological exploration?

6. Might this approach to image analysis be utilized in the identification and mapping of wetland types?

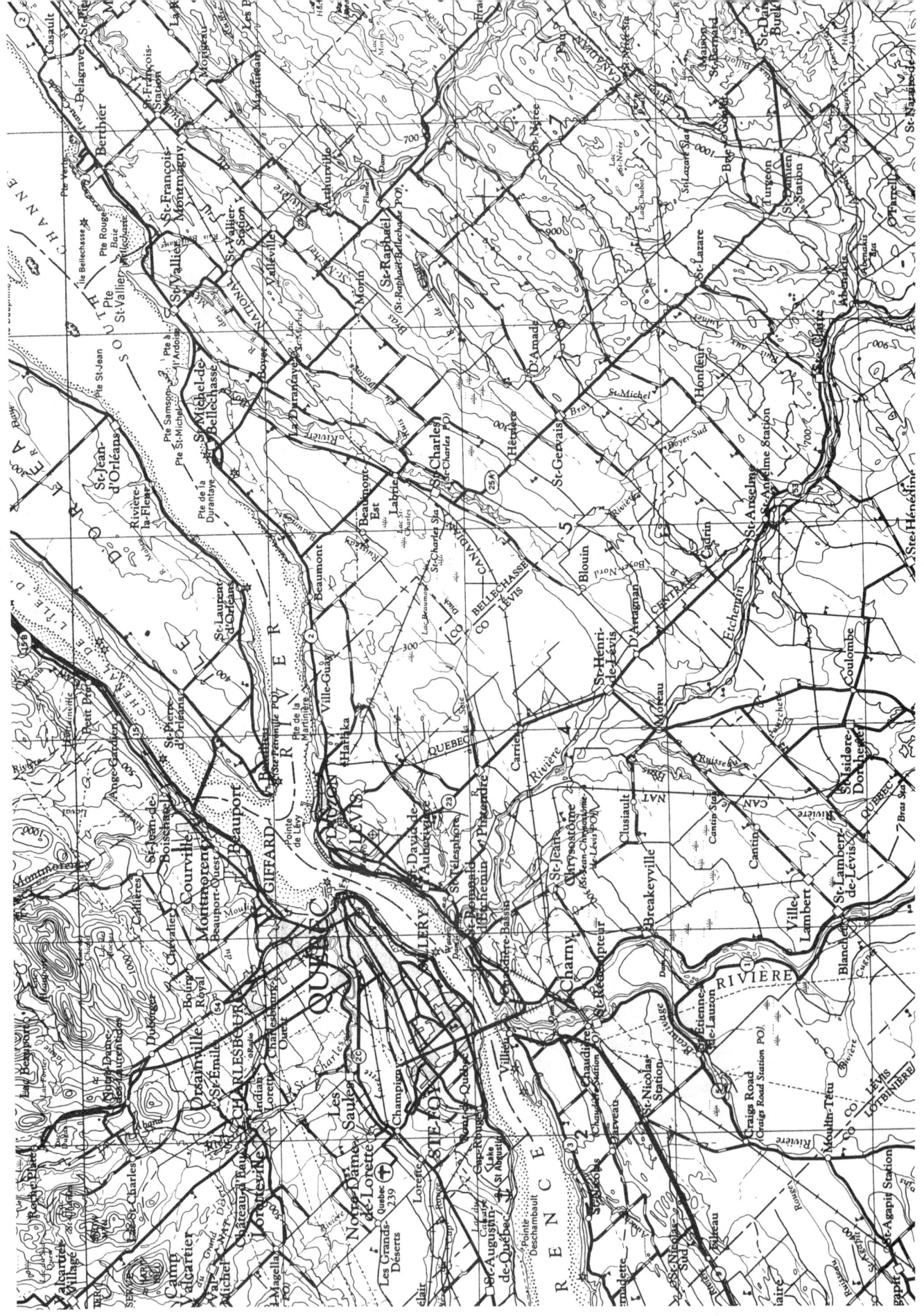

Figure 12–32. USGS topographic map (part).. Quebec. RF 1:250,000

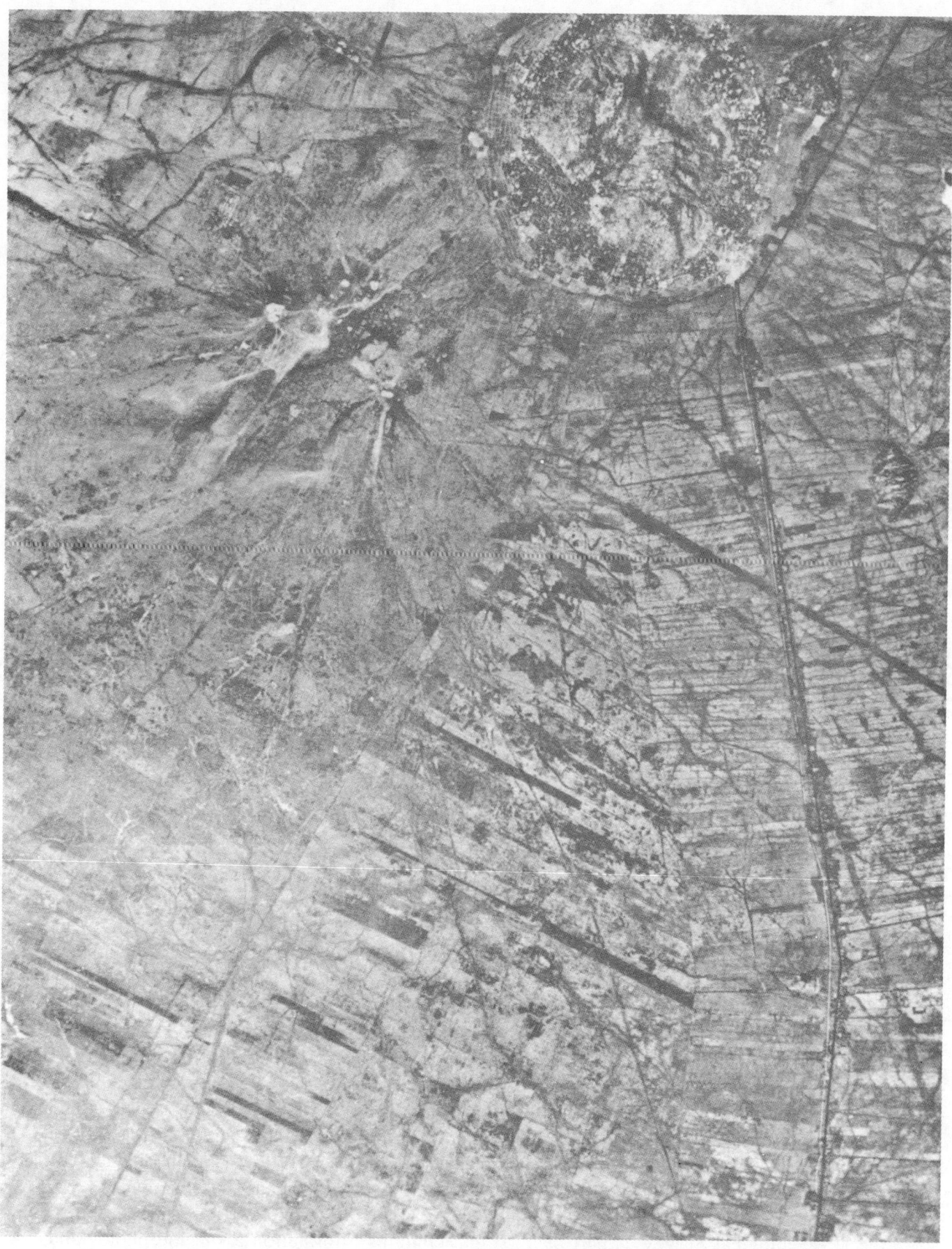

Figure 12–33. Black and white airphoto Mekong Delta, Vietnam. Acquired pre-1975. Courtesy Noel Ring.

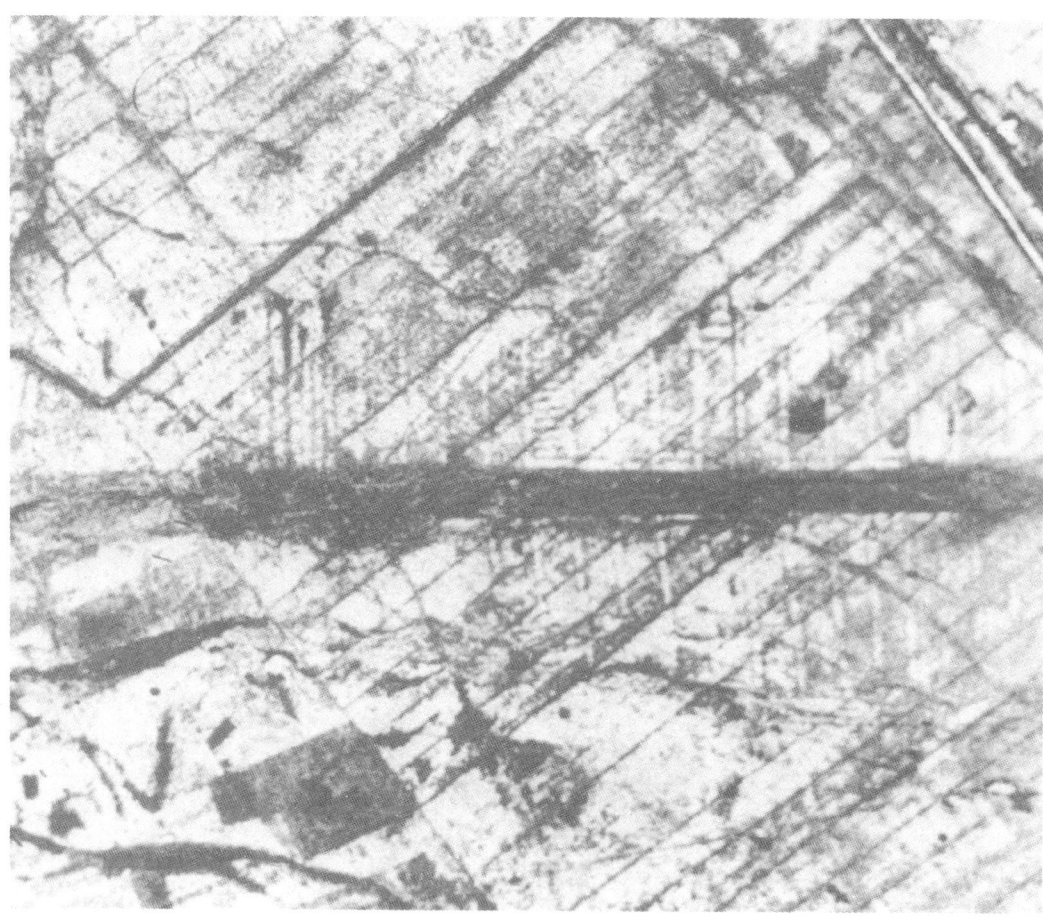

Figure 12–34. Black and white airphoto. Enlargement of Figure 12–33. Courtesy Noel Ring.

7. How many different academic/professional fields that deal with patterns at the Earth's surface might benefit by training in identification of archaeological artifacts from airphotos and satellite imagery? List several fields with some of the types of patterns upon which attention would likely be focused.

DECLASSIFIED SATELLITE PHOTOGRAPHY—A "NEW SOURCE" OF IMAGERY IN EARTH SURFACE INVESTIGATION

On February 23, 1995 the President of the United States issued an Executive Order authorizing the declassification of satellite photographs collected by the U.S. Intelligence Community in the 1960s and early 1970s. There are almost one million of the photographs which were acquired during the CORONA, ARGON, and LANYARD Missions. Currently released images were collected during the CORONA Mission which extended from August 1960 to May 1972.

All of this photography should be declassified and available to the public at the cost of reproduction by February of 2000, and much of it will likely be processed by November of 1996. It represents a source of information that has not been accessible to many scientists and should prove very fruitful. Studies of Earth surface features over a considerable period of time will be enhanced and changes in the size and shape of water bodies, volcanic flows, and the like may be documented. Perhaps the greatest value will be realized in environmental investigations.

Figure 12–35 is one of four photographs that are available from the EROS Data Center of the US Geological Survey as part of a kit that contains information about the photos and future releases. Images will be available in hard copy as prints, film positives, or film

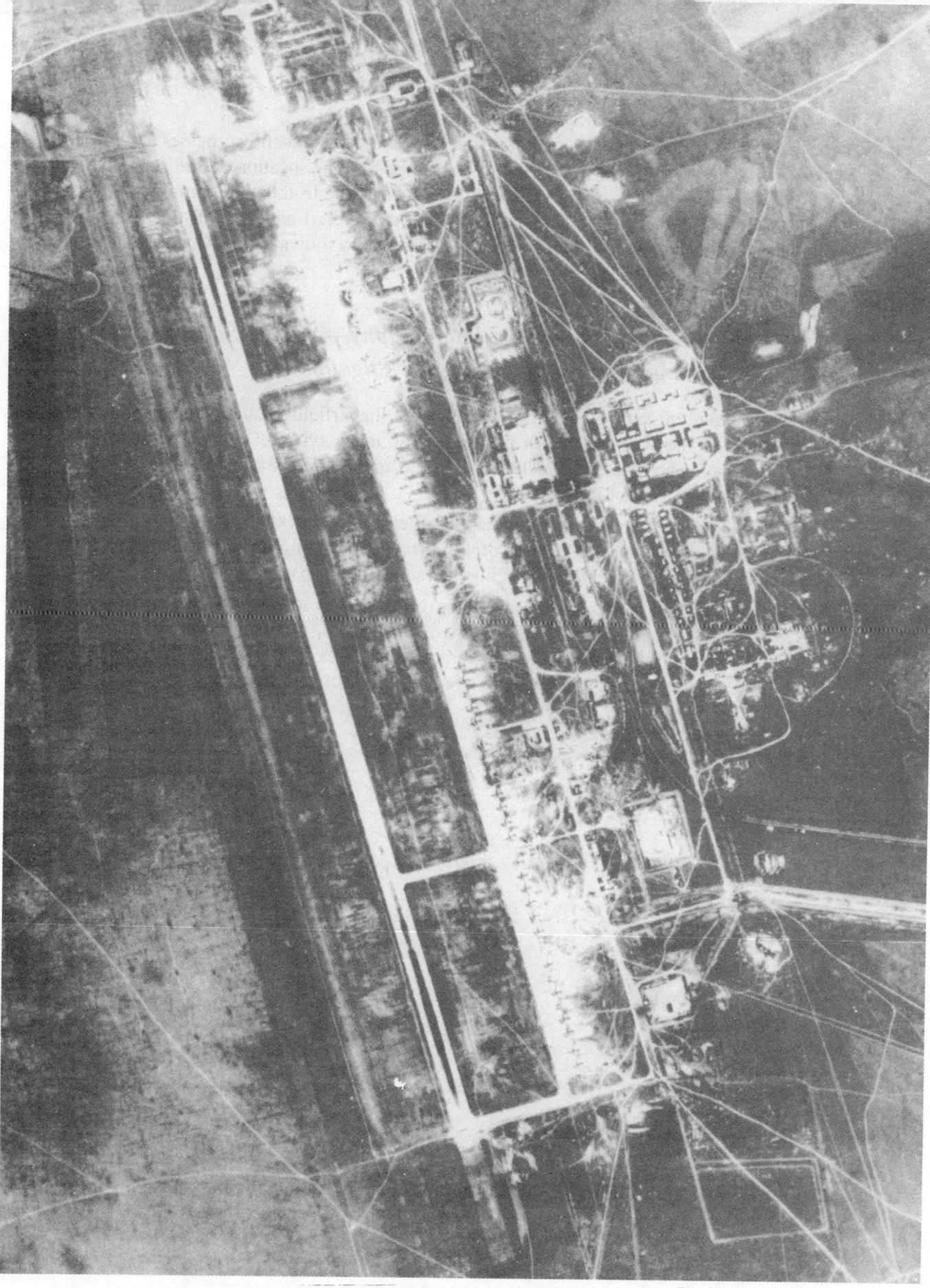

Figure 12–35. Satellite Reconnaissance Photo. Soviet Long-Range Airfield. Dolon. Kazakhstan. Acquired 8/20/66. Courtesy USGS EROS Data Center.

negatives, and may also be accessed through the World Wide Web of the Internet. Address is http://edcwww.cr.usgs.gov/dclass/dclass.html.

The resolution of Figure 12–35 is approximately 2 meters and that still surpasses any satellite system currently producing images for public consumption. The location of the scene is near Dolon, Kazakhstan on the Irtysh River north of Semipalatinsk. In this photograph of a Soviet Long-Range strategic bomber base, it is possible to count aircraft and even distinguish differences between heavy bombers and transport planes. The photo shown is a 30X enlargement of a portion of the original image.

Exercise 12.10   Declassified Satellite Photography—A "New Source" of Imagery in Earth Surface Investigation

1. How many aircraft can you count on the apron of the airfield? Can you discern any differences in the types of planes present? How many types of planes are there? How many of each type are present? Use low magnification (2× or 3×) or none on this published version of the photo.

2. Find this location in an atlas or on a map of at least RF 1:18,000,000, larger scale is preferable. Note the general location of this airbase. What might have been the logic regarding strategy and security in placing the facility in this part of the former USSR?

3. If this represents a 30× enlargement of the original photograph, how might these archived photographs be utilized in your particular area of academic or professional interest? What are the beneficial characteristics of the photos? What are the limitations for your use? How will scale and resolution impact your needs?

chapter *13*

DIGITAL IMAGE PROCESSING OF REMOTELY SENSED DATA

TERMINOLOGY

picture elements	histogram equalization
pixels	spatial filtering
digital numbers (DN)	high-frequency areas
raster image	low-frequency areas
vector image	low-pass filters
binary digits (bits)	high-pass filters
8-bit, 256 levels	directional filters
medical imaging	linears (lineaments)
desktop publishing	edge enhancement
pre-press image processing	density slicing
image restoration	seamless mosaicing
image "noise" or static	principal components analysis
banding (striping)	ratio images
line dropouts	multispectral classification
sun angle correction	false-color composites
haze correction	unsupervised classification
histogram	supervised classification
image enhancement	nearest neighbor analysis
contrast stretching	cluster analysis
linear stretch	training sites
non-linear stretch	Macintosh environment
area-specific stretch	IBM-compatible, PC environment
sinusoidal stretch	modem
Gaussian stretch	Internet
uniform distribution stretch	CD-ROM
scanner	video disc reader

BACKGROUND OF DIGITAL IMAGING

A digital image is one in which the **picture elements,** the smallest areal units of information, are characterized by numerical values. The units are usually called **pixels** and the values are referred to as **DNs,** or **Digital Numbers.** The values might be an indication of level of reflectance, temperature, or level of density or intensity.

The pixels are usually arranged in **rows** and **columns** to produce a **raster** image, but some digital images are created by the layout of **vectors** or lines of direction such as a road. These two forms of digital images have not been particularly compatible so that they could be utilized together, but rapid advances are being made in that direction. The emphasis here will be on raster type images.

The values assigned to a pixel must be in the form of **binary digits,** or **bits,** so that they can be read by a computer. The most common bit scale of values has become an **8-bit** which allows for a range in DNs of 0 (usually black) to 255 (usually white). This range of **256 levels** allows for considerable contrast and therefore visual differentiation in a typical digital image.

The people who are engaged in digital image processing come from many different fields. Remote sensing began to receive data and images from space in digital form in the 1960s, but the onset of the **LANDSAT** program produced voluminous digital images of the Earth starting in 1972. These images contained millions of pieces of data and since the satellite could return to the same ground location, a sequence of images through time could be recorded. This capability made these images very attractive to remote sensing scientists and their use increased dramatically.

Medical imaging is another field that utilizes image processing techniques. Early processing borrowed some ideas from remote sensing, but current analysis of **CAT** (computer axial tomography) scans and images produced by **MRI** (magnetic resonance imaging) have become highly sophisticated and now are governed by medical concepts and technology. There is not much overlap between remote sensing and medical imaging at present.

One other area of image processing should be mentioned here because it presents some pitfalls for the remote sensing scientist. **Desktop publishing** and **pre-press image processing** are areas that are heavily involved in manipulating images. It is important to note, however, that the focus of attention in these endeavors is the appearance of the final image, and not its faithful representation of some Earth surface feature or atmospheric phenomenon. Techniques that have been developed for these fields may seem to produce the same image or data product, but they frequently do not.

THE VALUE OF DIGITAL IMAGE PROCESSING

In earlier chapters the focus has been directed towards the visual perception of airphotos and satellite images. By working on "hard copy" prints, one could make direct measurements of linear and areal values, carry out the direct measurement of object heights, gain a perception of surface texture, and develop procedures to isolate phenomena based upon signature identification. All of these approaches have been the foundation of "hands on," visual analysis of airphotos and satellite images for years, and will likely continue in importance.

Being able to take the same "hard copy" photo and, using a relatively inexpensive **scanner,** produce a digital version of the image which can be displayed on the CRT screen (monitor) of a PC has opened up a much broader field of image analysis. Perhaps of even greater importance is the fact that all of the needed hardware and much software is easily affordable for virtually any remote sensing scientist.

Almost all of the techniques that could be used with photographic prints can still be applied to such images in digital form, but now more powerful methods of identification and analysis are available. One false-color composite Landsat MSS scene is comprised of three bands that have been combined, each band consisting of approximately 7.5 million pixels, or more than 22 million pieces of data, ready for mathematical analysis. Not only can these digital numbers be subjected to many statistical procedures to extract information, they can be modified in a variety of ways to enhance the original image.

Figures 9–4 through 9–7 are examples of hard copy versions of three Landsat MSS bands and a false-color composite. Figures 9–10 through 9–17 represent the similar form of data for a

Landsat TM product. All of these printed versions could be scanned and presented in digital form on a computer. Various combinations of bands could be used to enhance certain types of surface phenomena. A large scale example will be presented later in this chapter.

DIGITAL IMAGE PROCESSING PROCEDURES

Image restoration. All images recorded by some sort of sensing system, whether film, video tube, charge-coupled device, or radiometer, are subject to some distortion because of **"noise"** or **static.** Some sensors are more susceptible to this than others, but it is a common problem, somewhat analogous to static on a radio transmission or glare from the sun in a snapshot photo.

Since working with an image free of interference is desired, restoring image quality by eliminating noise is a common procedure and is carried out first in the image processing sequence. It is fortunate for the typical end user of digital image products that those who produce digital image products customarily remove most noise by means of computer algorithms.

Other types of image distortions may be produced by a sensor that has gone bad or that is no longer calibrated correctly. In a multi-sensor system, this can produce **banding** or **striping.** This is also normally corrected before the user receives the product and is accomplished by one of several procedures that corrects pixel values by applying average DNs of nearby pixels.

Another type problem that affects image quality is termed **line dropouts.** This can be caused by a failed sensor or sticking scanning mirror and it is corrected by assuming the missing values are equal to an average of pixel values above and below the line in question. Other corrections are made when scenes are to be joined in a mosaic and were recorded on different dates so that the **sun angle correction** must be made mathematically. **Haze correction** is often made by comparing a near infrared histogram with visible bands that are more susceptible to scattering. The ability to view the distribution of pixels on a **histogram** where the X axis is range of DNs and the Y axis is number of pixels occurring at a given DN, is the basis of many image processing procedures. See Figures 13–1 through 13–7. These are histograms of the seven TM bands of a Landsat subscene located in northeastern Massachusetts and acquired on 6/14/88. Another histogram of TM band 4 acquired over the same area on 5/11/93 is also shown.

Image enhancement. Once the raw data have been restored to a distortion free image, the matrix of pixel values may be manipulated to gather increased information about some aspect of the digital scene. It is a truism that whenever the DN values are modified in some form of enhancement, other data are lost. For this reason it is important to maintain a copy of the original digital file when carrying out image enhancement.

Contrast stretching is one of the common enhancement techniques wherein the histogram of DN values is modified to produce greater contrast in the viewed scene. This is accomplished by a number of **linear** and **non-linear** procedures. One of the linear methods of stretching is accomplished by trimming the scattered, less significant DN values on the tails of the histogram so that the remaining DN observations can then be spread over a larger range of values, thereby increasing contrast. Note, for example, that Figures 13–1 through 13–5, and 13–7 all have small numbers of pixels out at the tails of the histogram allowing for less contrast in the main body of pixel values (the tallest columns).

There are a number of statistical procedures that may be employed to produce stretching. Other linear techniques allow for the DN values in a particular area of an image to be stretched to provide more detail or to divide an image into homogeneous areas based upon vegetation, soil, rock units, etc., and enhance contrast within those units. Examples of such enhancement are **area-specific stretch** and **sinusoidal stretch.**

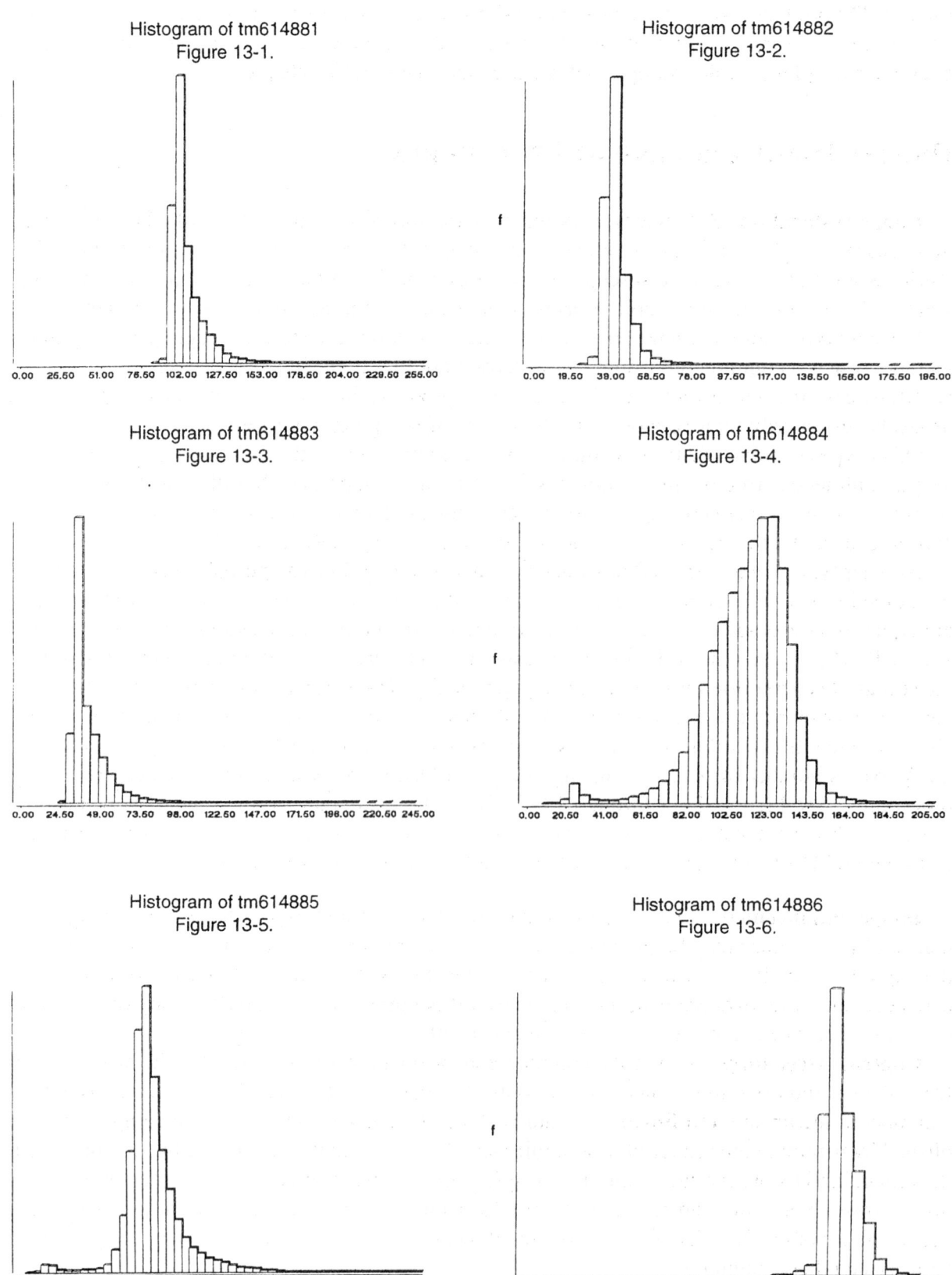

Figures 13–1 through 13–6. Histograms of TM bands 1–6. Subscene in Topsfield and Danvers, MA. Acquired 6/14/88. Generated by Idrisi software. Data Courtesy EOSAT Co.

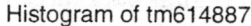

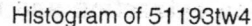

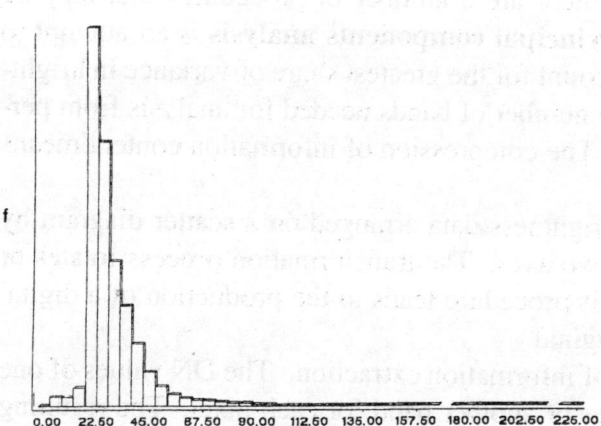

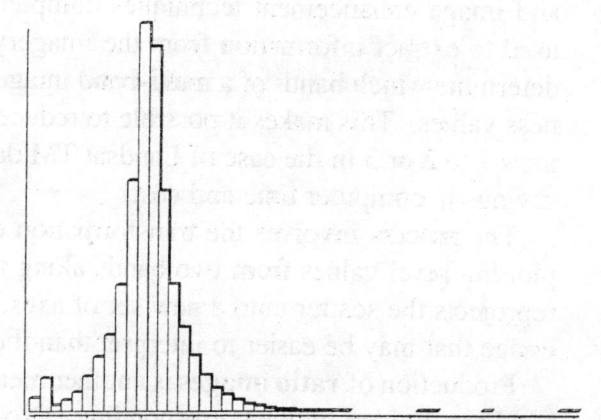

Figure 13–7. TM band 7. Acquired 6/14/88.
Generated by Idrisi software.
Data Courtesy EOSAT Co.

Figure 13–8. TM band 4. Acquired 5/11/93.
Data Courtesy EOSAT Co.

Non-linear techniques also modify the histogram and therefore the image by utilizing a number of statistical methods. The histogram values can be modified to conform to the normal curve in a **Gaussian stretch** if a scene is characterized by areas too bright or too dark and detail is lost visually. Another version is a **uniform distribution stretch** or **histogram equalization** in which pixels are reassigned values so that a uniform density of pixels is distributed along the X axis. The greatest stretching in this procedure occurs in parts of the image where the largest number of pixels fall in a small range of DN values.

Another form of image enhancement is **spatial filtering.** This procedure is performed in order to regionalize areas of an image that exhibit many changes in brightness values, **high-frequency areas,** and areas that exhibit very few changes in brightness level, **low-frequency areas.** This procedure may be used to sharpen edges in a scene or to reduce the effect of noise. The algorithms used to perform these operations are termed **low-pass filters,** low frequency or smooth areas are passed, and **high-pass filters,** high frequency or rough areas are kept by the filtering.

Directional filters represent another form of filtering that is designed to enhance linear features. This procedure is especially useful applied to radar imagery as it increases the visibility of **linears** or **lineaments** such as fault lines, roads, railroads, etc. There are other types of filters that enhance linear features through **edge enhancement.**

Density slicing is another form of enhancement in which the continuum of gray tones in a scene is divided into a number of intervals that are demarcated by specific digital numbers. DN's of different numbers are lumped together to emphasize the changes of values displayed in a scene. This approach has proven particularly useful in interpretation of GOES weather satellite imagery. Figures 10–17 and 10–19 are thermal IR images that display temperature levels as steps or density slices.

Mosaicing of multiple images makes use of the fact that the pixel values can be mathematically manipulated so that **seamless mosaics** can be produced from several digital images. Figure 13–9 (see Plate IX) is an example of such a mosaic produced by IBM with Landsat MSS data for an area of Montana and Wyoming in the Yellowstone River watershed. The image shown is part of an 8 scene mosaic false color composite of bands 4, 5, and 7. Note that the only difference visible along seam lines is the roughly north–south seam between the July 30th and July 31st images in 1973. The variation is due to differences in cloud cover and atmospheric conditions on the two days.

Information extraction and image classification. Once the digital data have been restored and image enhancement techniques completed, there are a number of procedures that may be used to extract information from the imagery. **Principal components analysis** is an attempt to determine which bands of a multi-band image account for the greatest share of variance in brightness values. This makes it possible to reduce the number of bands needed for analysis from perhaps 7 to 2 or 3 in the case of Landsat TM data. The compression of information content means savings in computer time and cost.

The process involves the transformation of brightness data arranged on a scatter diagram by plotting pixel values from two bands along the two axes. The transformation process rotates or reprojects the scatter onto a new set of axes. This procedure leads to the production of a digital image that may be easier to interpret than the original.

Production of **ratio images** is another means of information extraction. The DN values of one band are divided by the corresponding DN values of another band for each pixel. The resulting image shows areas of dominance by the numerator band as bright pixels and areas where the denominator of the ratio is greater as dark areas. Pixels of roughly equal values at the same locations in each band are unchanged. This procedure is useful when examining areas where vegetation or rock type are the same on two sides of a ridge, but the light and shadow values vary because of sun angle. The ratio values will indicate that the spectral curves of the two areas are similar.

Multispectral classification is another information extraction process. Once the **PCA** process has been completed, the three bands that account for the lion's share of variation in pixel values can be utilized in one or more statistical procedures that are carried out by the computer.

Prior to the classification step, it is often valuable to create **false-color composites** of various combinations of bands. This allows the analyst to see the area as though it had been recorded in natural color, in color infrared, or various other color combinations. By using this approach it is possible to isolate or enhance specific elements of the image area. Separating urban and suburban use from forest, agriculture, or wetlands is made a lot easier visually and the specific bands and precise pixel values in the core and on the margins of land use/land cover areas may be identified so that their values can be used in classification processes.

The images shown on Plate X are examples of this approach and represent various false-color results for a Massachusetts Audubon Society Wildlife Sanctuary in Topsfield, MA. The images are from a Landsat TM subscene acquired on 6/14/88 and are presented here through the courtesy of the **Earth Observation Satellite Company, EOSAT.** Figure 13–10 (see Plate X) is a natural color representation, 13–11 (see Plate X) is a replication of a color infrared version, and 13–12 (see Plate X) and 13–13 (see Plate X) are examples of near-middle infrared dominance to enhance many specific features not brought out in the other renditions.

Following PCA analysis and false-color production a decision can be made as to which TM bands to use for the multispectral classifications. These can take the form of an **unsupervised classification** in which the computer classifies the scene without direction from the analyst or as **supervised classifications** wherein control is exercised.

The former procedure uses **nearest neighbor** and similar spatial concepts to perform **cluster analysis** so that pixels that are similar and seem to represent homogeneous regions can be grouped together. The resulting image may be interpreted by comparing the clusters produced with maps and airphotos and other materials related to the image site. The nature of the ground surface where each cluster is located can then be used to aid understanding of the classified image produced.

Supervised classifications call for the interpreter to identify **training sites** where groups of pixels are a good representation of a particular type of land use, land cover, or other surface phenomenon. Once these sites have been located the computer can be given instructions as to the pixel values in those areas and these values are then used in a number of statistical manipulations.

Each statistical procedure will result in a different classified image and one or more of these may accurately represent a logical breakdown of the features that are under study. The following chapter deals with environmental monitoring and assessment and classified images will be presented in that context.

GETTING STARTED IN DIGITAL IMAGE PROCESSING

Hardware and software needs. Considerable image processing is being carried out in a **Macintosh** environment, especially in desktop publishing, advertising, and artistic endeavors. There are a number of excellent software imaging programs that serve these interests. Some of the better known are **Adobe's Photoshop, Microgafx' Picture Publisher,** and **Aldus' Pagemaker.** These are all excellent for processing and enhancing images for the needs mentioned above. As was mentioned earlier, the focus is usually on resulting appearance and not necessarily to faithfully represent some aspect of the Earth's surface or environment.

Most of the scientific image processing seems to be taking place at present in a **PC environment** with **IBM-compatible** hardware and software. The image processing literature indicates that this situation will become more pronounced in the future. Some of the most popular software in this realm is relatively expensive for the individual user. Packages such as **ESRI's Arc/Info, Macdonnel-Douglas' GSI, ERDAS** products, **Intergraph's MGE,** and many other image processing and GIS packages have excellent reputations, but are costly.

While there is considerable sophisticated image processing taking place on mainframes, mini computers, and network systems, it is quite possible to gain access to this activity without buying expensive hardware or software. The basic hardware needed is likely to be the largest expense, but even here the individual is faced with a wide variety of choices and costs.

Perhaps a suggested minimum initial outlay would require obtaining a PC at least with 80386 CPU, although an 80486 is preferable and the faster it operates the better. Memory should be a minimum of 4 megabytes (MB), but 8 or 16 MB is more suitable depending upon the size of images that will be manipulated at one time. The hard drive should be 300 MB or larger, and again more is better. It is surprising how quickly you can outgrow your system when your main interest is image processing. An important consideration when obtaining hardware is to acquire a system that is easily expandable.

A peripheral device that is often included when buying a PC today is a **modem.** The cost of this item can be repaid many times over as it can assist you in acquiring free or very inexpensive software and imagery. Several Federal agencies, including NASA, NOAA, USGS, and the USDA, have made imagery and some software available to anyone who wants it. There are also a number of private companies who offer startup or demonstration software for little or no cost. To take advantage of this situation you must gain access to the **Internet** via a state government educational network, local college or university computer, or one of the commercial on-line services such as **America On Line** or **Prodigy.**

There are some not-for-profit organizations which make some types of software available at very reasonable prices. One very good example is **IDRISI** which is a GIS/Image Processing package that is produced by the Graduate School of Geography at Clark University in Worchester, MA (see Appendix E). This software comes with manuals and some images to practice with. The tutorial section leads you through all the procedures that have been mentioned above and the material is upgraded on a regular basis.

If you are working or affiliated with an educational institution, a **video disc reader** is a very useful piece of hardware as it is possible to buy video discs that contain thousands of handheld photographs taken from space at a cost of pennies per image. These can be viewed in any sequence and selected images can be downloaded in digital form to use in a system such as IDRISI.

The Federal Government has begun producing **CD-ROM** disks that contain a wide variety of images. This source of images seems to be on the rise, so a CD-ROM drive in your computer is a good investment.

Exercise 13   *Digital Image Processing of Remotely Sensed Data*

1. Examine the histograms in Figures 13–1 through 13–7. The subscene was acquired well within the growing season, so the data portray reflectance from vigorous vegetative growth. As can be observed in Figures 7–3, 7–4, 7–7, and 7–8, the area that is covered by the subscene is heavily vegetated throughout. Figure 9–10 included the spectral ranges for each of the TM bands.

   Compare the histograms to each other. Which ones seem to be characterized by a smaller range in DN values for the bulk of the pixels (ignore smaller numbers of pixels in tails)? Would that be likely to produce more or less contrast in an image display of those bands? If you were told that band 4 exhibited the greatest amount of contrast and DN range (for the bulk of pixels), how would that influence your answer in the previous question?

   Note that each band records different types of information and that lower DN numbers are darker areas and higher DN numbers are lighter image areas. Band 7, Figure 13–7, is sensitive to water, soil moisture, and moisture in vegetation. The image of band 7 is the darkest of the seven and shows the least contrast. If you were told that grass areas and agricultural fields are lighter and surface water areas darker on the image, what would you predict for the gray tone for wetlands vegetation?

2. Figures 13–10 through 13–13 display several different band combinations of TM imagery. Compare these images of the MAS Wildlife Sanctuary with Figures 5–1, 5–2, 7–3, 7–4, 7–6 through 7–8, and the maps in Chapter 14. Can you determine that some false-color composites identify some surface features better than others? Give examples.

   How might the date of image acquisition impact the information presented in these false color composites? Trees usually reach full leaf-out condition approximately one month before the date of this Landsat image at this location. How might these false-color composites differ for imagery that was acquired on May 14th? Compare Figure 13–7 with 13–8, the band 4 histogram for June 14, 1988 and May 13, 1993. There is some similarity in the shapes of the histograms, but the mean DN value for June band 4 is 114 and for May band 4 is only 63. If you knew that this band shows peak reflectance for vegetative vigor, how might that affect your answer?

   What types of map information or airphoto information that is easily seen on the items listed in #2 above is recognizable on the Landsat false-color composites? What is the most important characteristic of these features that allows you to identify them? For example, is a highway picked up by its shape, size, contrast with surroundings, or effect on land use patterns, etc.?

3. Compare the image in Figure 13–9 with an atlas map of the Yellowstone River basin. What evidence can you draw from the image as to areas that seem to be wetter or drier in these July 30, 31 scenes? If your map gives indications of relief, how does the orientation of mountain ranges seem to affect the distribution of moisture?

4. A useful exercise related to this chapter is to draw up a "wish list" of hardware and software needs to fulfill your goals in digital image processing. Refer to the sources in Appendix E, computer magazines available in any magazine outlet, and local newspaper advertisements. There are a number of chain stores that specialize in computers where lower prices are usually available. What is the total money amount that will start you on your way? Are there grants or assistance programs in your workplace, educational institutions, or city/state governments that would help you defray the initial costs? It should prove pleasantly surprising to you how inexpensively one can pursue a sophisticated digital image processing program.

chapter 14

ENVIRONMENTAL MONITORING AND ASSESSMENT

TERMINOLOGY

Earth as closed system
USGS Topographic Maps
orthophotos
slope maps
land use/land cover maps
digital terrain maps
hydrologic maps
multi-sensor approach
multi-band sensing
real-time CIR video

RASCAL, Flight Landata, Inc.
VIFIS system
histogram peaks
broad clusters
fine clusters
homogeneous sites
LU/LC signatures
maximum likelihood
minimum distance to means
standard deviation units

THE SCIENTIFIC AND POLITICAL SETTING

If one is willing to accept the view that **the Earth is a closed system,** then some conclusions are inescapable. It is not possible to pollute the atmosphere, land surfaces, ground water, oceans, and other water bodies without eventually paying a price in a degraded quality of life on Earth. The questions that then must be addressed are: **What is the acceptable level of pollution** in a given environmental setting? **How do we detect and measure the amounts of pollution? What steps must be taken to ameliorate a degraded environment?**

Unfortunately these questions are not easily answered and because there is a cost factor, the political interest is raised and the issues are debated in some forums that are unusual. The objective here is not to debate the need for environmental monitoring and assessment, that is taken as a given. Discussion in this chapter will focus upon a few selected measuring techniques that fall into the realm of image interpretation and analysis.

MATERIALS OF ASSESSMENT IN EVERY LOCALITY

Perhaps the examination of the problem should start with materials that are easily acquired and highly useful. Virtually all individuals have access to **USGS Topographic Maps** and other similar materials. Such maps are suitable for studying environmental problems on the Earth's surface, for mapping the area of concern, and for acquiring at least general information about slopes, relief, vegetative cover, surface water, and wetlands. Oftentimes these maps are overlooked because they are too small scale and therefore don't show enough detail. Nevertheless, they represent an excellent starting point for investigation as they are available for several different years and were compiled by experts employing analysis of airphotos of the local area.

Other maps that are available for many areas in the United States are **orthophotos** printed as black and white or as false color products. These maps offer the added data that are obtained when the Earth's surface is photographed at a particular moment in time. The condition of agricultural fields, drainage courses, vegetative cover, and settlement patterns are all revealed. Any knowledge that can be gleaned about an area of interest is useful when the approach to investigation is being formulated.

The maps mentioned above are usually already at hand nearby. There are many other maps that also might be useful for a particular project such as **slope maps, land use and land cover maps, digital terrain maps,** and **hydrologic maps.** These can all be ordered from any of the **National Cartographic Information Centers (NCIC)** as well as individual **separations of a topographic map,** e.g., just the hydrology or just the vegetation, etc. (see Appendix E).

The Multi-sensor Approach to Monitoring & Assessment

With the launch of the first Landsat vehicle in 1972 there instantly became available in large volume a host of image data for most of the Earth's surface. Perhaps most importantly these data were recorded in several bands that were spatially registered so that the same pixel could be evaluated from one band to another. Having several wavelengths of energy recorded for the same specific Earth surface location allowed for comparison of energy that was reflected or emitted. This led to identification of surface materials that often reflect or emit at different levels in the various wavelengths. Figures 14–1 and 14–2 are graphs depicting the spectral responses for several types of Earth surface materials and general categories of vegetation.

Landsat vehicles have produced many thousands of scenes of the Earth in multi-band form. More recently the French satellite, SPOT, has also produced coverage in several registered bands. Now Japan, India, Russia and Canada are also recording multi-band satellite imagery or are in the final stages of such projects. There are a multitude of such images from past years and the promise is for even more with greater wavelength coverage in the future.

As useful as the satellite coverage has been, once a vehicle is launched and placed in orbit its sensors have been calibrated for bands of specific widths. This, together with the small scale of such imagery, makes the products less responsive to the need to monitor changing conditions in particular circumstances at the Earth's surface. This has led to the development of multi-band sensing systems that are designed to meet these demands.

Developments in Large Scale Multi-band Sensing

There has been considerable experimenting with the use of multi-band video systems for a number of years. Early designs attempted to replicate CIR video through the use of filters with good results. Now **real-time CIR video** is available to be carried on small fixed-wing aircraft and helicopters. This has resulted in a rapid response to a localized environmental problem by consulting and aerial surveillance firms. The fact that some of the systems utilized can be "tweaked" so that precise band widths are recorded makes these systems all the more practical and useful.

Figures 14–3 (see Plate XI) through 14–6 are products of a sensing system called **RASCAL** (Remote Airborne Sensor Computer Analysis Link) that is flown by **Flight Landata, Inc.** of Lawrence, MA. The examples shown are video frames over Newbury, MA along a salt marshupland boundary. Figure 14–3 is a false-color CIR image which was produced by combining Figures 14–4 (0.545–0.555), 14–5 (0.644–0.656), and 14-6 (0.846–0.857). These band widths given in micrometers are the green, red, and near IR bands recorded as blue, green, and red images respectively in the hard copy product.

The false color image in Figure 14–3 portrays essentially the same information as in color infrared film, Landsat MSS bands 4, 5, and 7, or Landsat TM bands 2, 3, and 4. The defining dif-

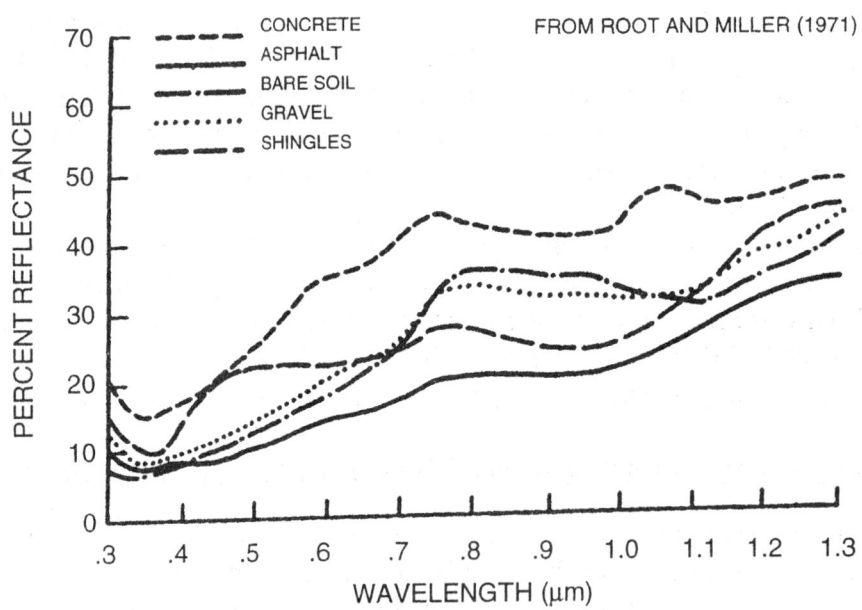

Figure 14–1. Spectral response curves for non-vegetated earth surface materials.

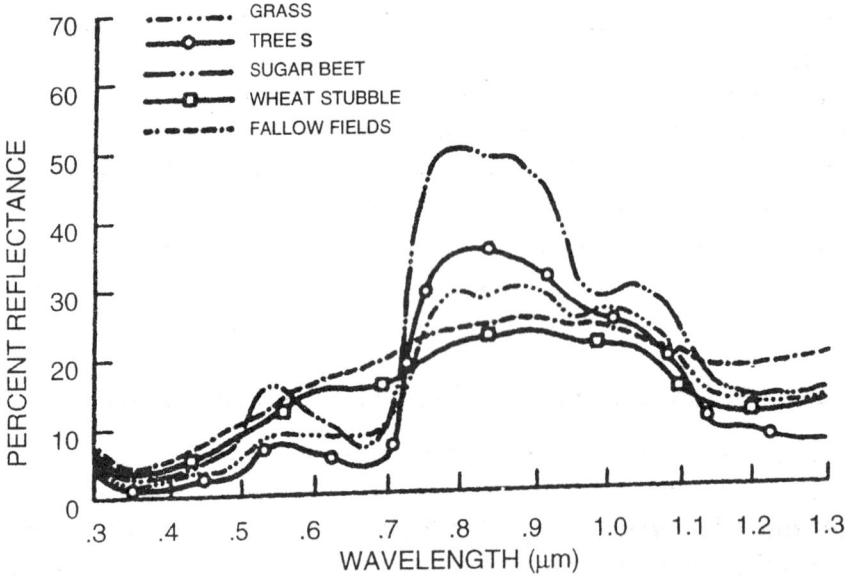

Figure 14–2. Spectral response curves for selected types of vegetation.

ference in this case is that the band widths of Figures 14–4 through 14–6 cover *small portions* of the Landsat TM bands 2, 3, and 4. The spectral information revealed is essentially the same as was shown in Figure 9–10, except that the narrower bands of RASCAL allow for more specific identification of vegetative types. Comparison of the three bands with on-site field investigation will allow more precise identification than is the case with the more encompassing band widths of the satellite imagery.

This system is a real-time video, multi-band unit that can be mounted in a plane or helicopter and acquires three narrow band video image tapes in the visible and near IR. These bands are encoded into a high resolution, false-color composite that is monitored on board as it is being recorded. The video recordings are digitized and registered to ground control points and then imported into a GIS system. Thermal IR and ultraviolet bands can be substituted for any of the bands listed above.

The most important features of the system are one meter resolution, the ability to see the imagery as it is being acquired, being able to make accurate scale changes by adjusting aircraft height, and the capability to fine tune individual bands to very narrow widths to serve a specific imaging task. All of these data may then be introduced into the GIS system for merging with other kinds of information.

The scene shown in Figures 14–3 to 14–6 is a small portion of the coastal salt marsh where it meets the upland boundary. The interface between these two disparate areas is easily recognized on site, on topographic maps, on black and white airphotos, and on large and small scale CIR aerials. It is even identifiable on photography taken from Skylab 270 miles above the Earth. Obviously it stands out strongly on large scale, low level videography.

Figure 14–4, the green band, shows turgidity in water, some pollution plumes in water, and the peak of green reflectance from vegetation. The darkest areas are water or strong green reflectance. Figure 14–5, the red band, gives dark values in areas of water and chlorophyll absorption and reflects brightly from areas of thinly vegetated soil. Figure 14–6, near IR band, is brightest where the other bands are dark due to peak reflectance of vegetative vigor from the mesophyll layer. Combining the three bands produces a CIR product, but one with a potential to allow identification of plant species once the field comparison and verification is carried out.

It is obvious that this type of hardware is more responsive to local, large scale environmental investigation than could be the case with present satellite imagery. The disadvantage would be the cost of using such a system to cover a large area such as a state or large metropolis.

Another exciting development in multi-sensor remote sensing is the emergence of camera systems that can make one flyover of a target area and then many individual bands of data may be separated out from that one pass. Flight Landata is perfecting a **Variable Interference Filter Imaging Spectrometer (VIFIS)** system that can acquire high resolution, contiguous spectral band images in one pass.

This instrument can be configured in such a manner that it can produce continuous, spectral images of approximately 0.01 micrometers band width in a range from 0.40 to 1.10 micrometers using only a single flight pass. It is obvious that this type of system will have major impact on environmental surveillance and monitoring for projects of small to very large size due to the low cost of flight time. Other similar systems are being developed in several other countries.

APPLICATION OF MONITORING AND ASSESSMENT TECHNOLOGY TO A LOCAL STUDY AREA IN ESSEX COUNTY, MA

The Massachusetts Audubon Society operates a Wildlife Sanctuary in Topsfield, MA. The Sanctuary and surrounding lands have served as a training site for students of Salem State College at the undergraduate and graduate levels. This area has been studied on site and in the laboratory using a wide variety of investigative techniques. The intent is to educate, but also to create a variety of hard copy and digital classification products of the study area. It is hoped that the procedures used to develop these materials will be appropriate to apply to the monitoring, assessing, and managing of the entire watershed in which the Sanctuary lies. Figure 7–6 is a CIR photo taken from Skylab that portrays the area of the watershed of the Ipswich River. That land drained by the river extends from B-1 to B-2 where it empties into Plum Island Sound at the southern tip of the long north–south barrier island and north of Cape Ann in B-3.

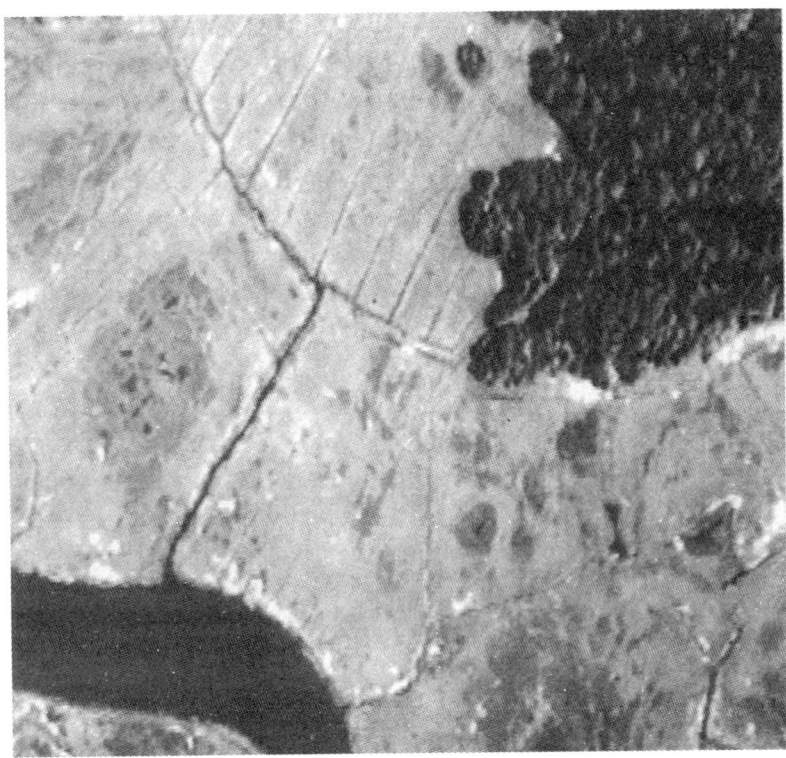

Figure 14–4. RASCAL band width 0.545–0.555 micrometers. Area in Newbury, MA 6/02/88 by Flight Landata, Inc. Flight Landata, Inc., P.O. Box 528, Newburyport, MA 01950.

The area has been examined and evaluated as to the existing land use/land cover, specifically in regard to the types of wetlands that exist along the Ipswich River and as a result of the glaciation of the area. A separate but related project is the creation of a large scale GIS database for the study area that may also have value to apply to the entire Ipswich River watershed.

The use of maps and airphotos. Several groups of students have made field trips to the Sanctuary to compare on-site views and perceptions with the information that can be gleaned from a variety of maps and airphotos. Topographic coverage of the site was produced by combining portions of the Salem, MA and Georgetown, MA USGS topographic quadrangles. Figure 14–7 is a black and white version that shows culture, wetland symbols, and contours. This was enlarged and wooded swamps were delineated vs. marshland for in-class study (see Figure 14–8). Figure 14–9 is an enlargement of part of the MAS map of the Sanctuary. It shows wooded wetlands in a light, dotted pattern and brush·or marsh areas as darker (compare with Figure 14–7).

Other maps of the study area were presented earlier as Figures 5–1 and 5–2. The former was a Soil Conservation Service map of soils distribution and the latter was a land use/land cover map overlaid on the local USGS quads.

Aerial photographs were available in several formats and all of these were made accessible to students yn both **hard copy and digital file formats.** In time sequence from oldest to most recent were a black and white aerial acquired 6/11/71 (see Figure 7–7), and the 1974 Skylab CIR photo mentioned above. Other photography included an NHAP CIR image of 4/1/86 (Figure 7–3), a black and white 70mm photo acquired on 4/9/90 (Figure 7–8), and the most recent was an NAPP CIR dated 4/13/92 (Figure 7–4).

Satellite imagery. All of these map and photo products were used for comparison with on-site conditions related to season, as well as to those satellite images that were also acquired for the area. Students were able to refer to Landsat Thematic Mapper images of 9/10/82 (hard copy and scanned RGB), 6/14/88, and 5/11/93 (the latter two subscenes as digital files for all seven bands).

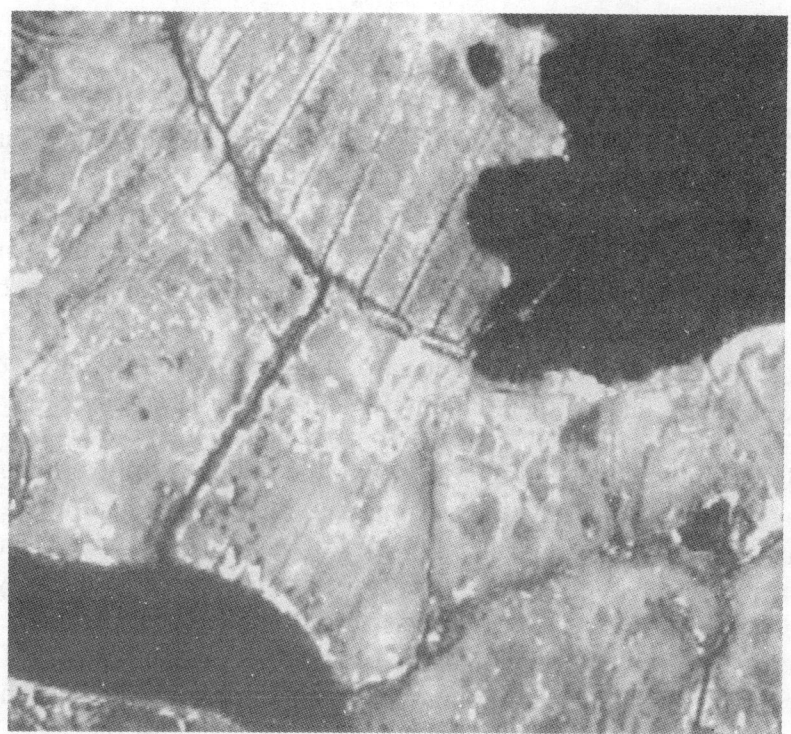

Figure 14–5. RASCAL band width 0.644–0.656 micrometers. Area in Newbury, MA 6/02/88 by Flight Landata, Inc. Flight Landata, Inc., P.O. Box 528, Newburyport, MA 01950.

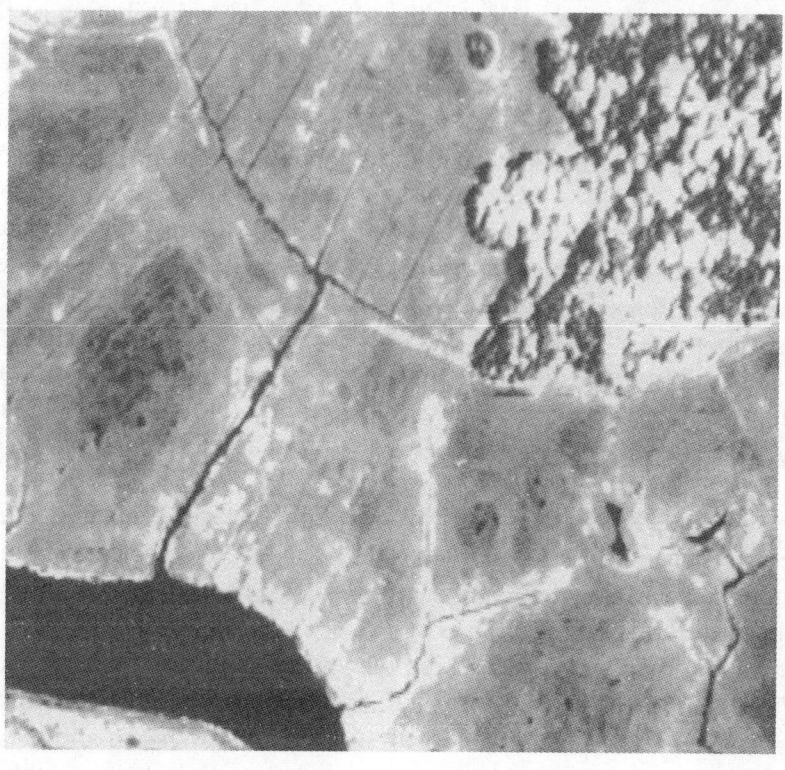

Figure 14–6. RASCAL band width 0.846–0.857 micrometers. Area in Newbury, MA 6/02/88 by Flight Landata, Inc. Flight Landata, Inc., P.O. Box 528, Newburyport, MA 01950.

Students were encouraged to study all of the image materials carefully and to take note of several characteristics. What were the dates of acquisition and how did that change the appearance of vegetation patterns in wetland and upland areas? Visits to the site at several times during the year produced information about the water table elevation and the amount of surface water present at different times.

Field visits focused upon vegetation on site in regard to the variety of species, heights, and vegetation density or biomass. Slopes, variation in soil types, area of ground covered by a pixel on each type of image, and signature characteristics that seemed to characterize the different wetland types were examined.

An inventory of vegetation species had been carried out for the entire Sanctuary area recently. Students reviewed the maps and compared them with on-site observations.

Products of digital image processing. Following the review of all the map, airphoto, satellite image, and field evidence, an attempt was made to classify the land use/land cover of the Sanctuary utilizing digital images and the IDRISI software. Chapter 13 discussed the use of principal components analysis and false color composite images as ways in which the image bands that displayed the most variance could be selected. Students experimented with different 3-band combinations and with the RGB layers of the CIR photography to find which ones seemed to work best in breaking out the categories of land use/land cover for the study site.

Having selected the image bands that students felt would be most revealing about relatively homogeneous patterns of land use and land cover, **unsupervised classification** of the digital data was undertaken using IDRISI software. The processing of the data by the computer was based upon **peaks of pixels** in the **histogram** distribution of reflection values. The computer processed the three-band composite that had been selected by the student. The resulting digital image was a display of **clusters** that represented dominant spectral response patterns. Students could also produce a second level of clusters using another option in the software. This distribution of **fine clusters** produced a greater number of identified areas of spectral dominance.

The two cluster images then needed to be compared with the maps, imagery, and information gleaned from field investigation to determine whether the statistically defined clusters in fact represented reality. At this point some decisions had to be made regarding whether to digitally combine some of the clusters.

Following these experiences the next procedure involved running a number of tests that came under the heading of **supervised classification.** In these activities, the computer is given guidance as to what spectral values coincide with certain classes of land use, land cover, vegetation types, or the like.

The guidance is based upon all the image information and the ground truthing carried out at the study site. This step is much more successful if detailed knowledge of the area under investigation is obtained. The next step involves identifying characteristic areas on site that may be used as **training sites** to guide the computer processing.

Figure 14–10 (see Plate XII) shows the training sites that were selected as representative of 7 identified use/cover types. These training site signatures were selected as being **homogeneous** in terms of ground cover and therefore in spectral values. The larger such a homogeneous area could be, the better the results were likely to be in supervised classification.

The several statistical manipulations carried out in the supervised classifications can be performed with all 7 TM bands or fewer. Once the training sites for **land use/land cover signatures** have been selected, they can be printed out as an image map as shown above. In addition, they can be printed out in graphic and numerical form so that spectral values of each signature can be compared with others. Areas of overlapping spectral values can be a problem and sometimes lead to refining or combining some of the signatures.

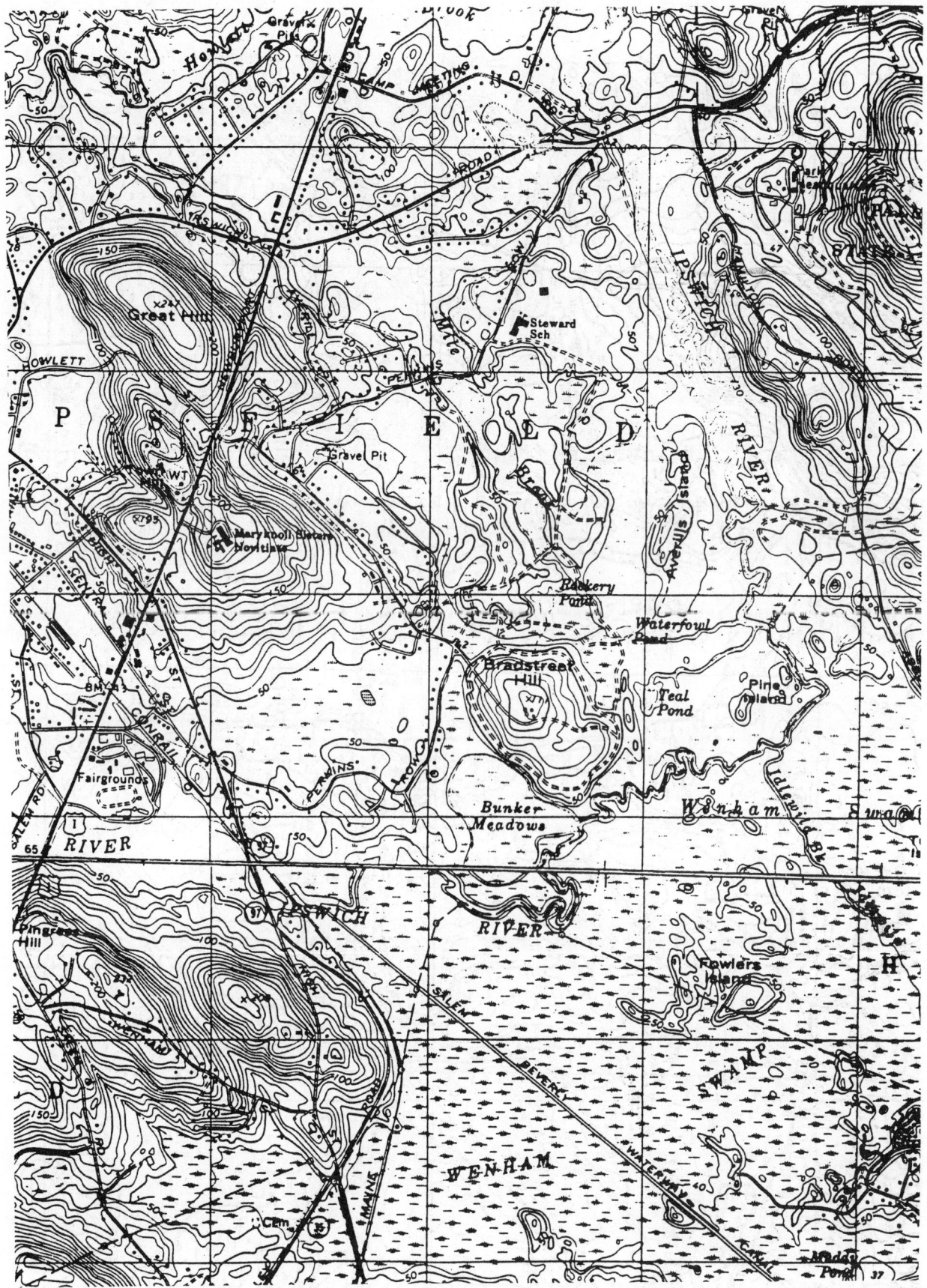

Figure 14–7. USGS quadrangles Salem & Georgetown, MA (part). Topsfield, MA. Massachusetts Audubon Society Sanctuary.

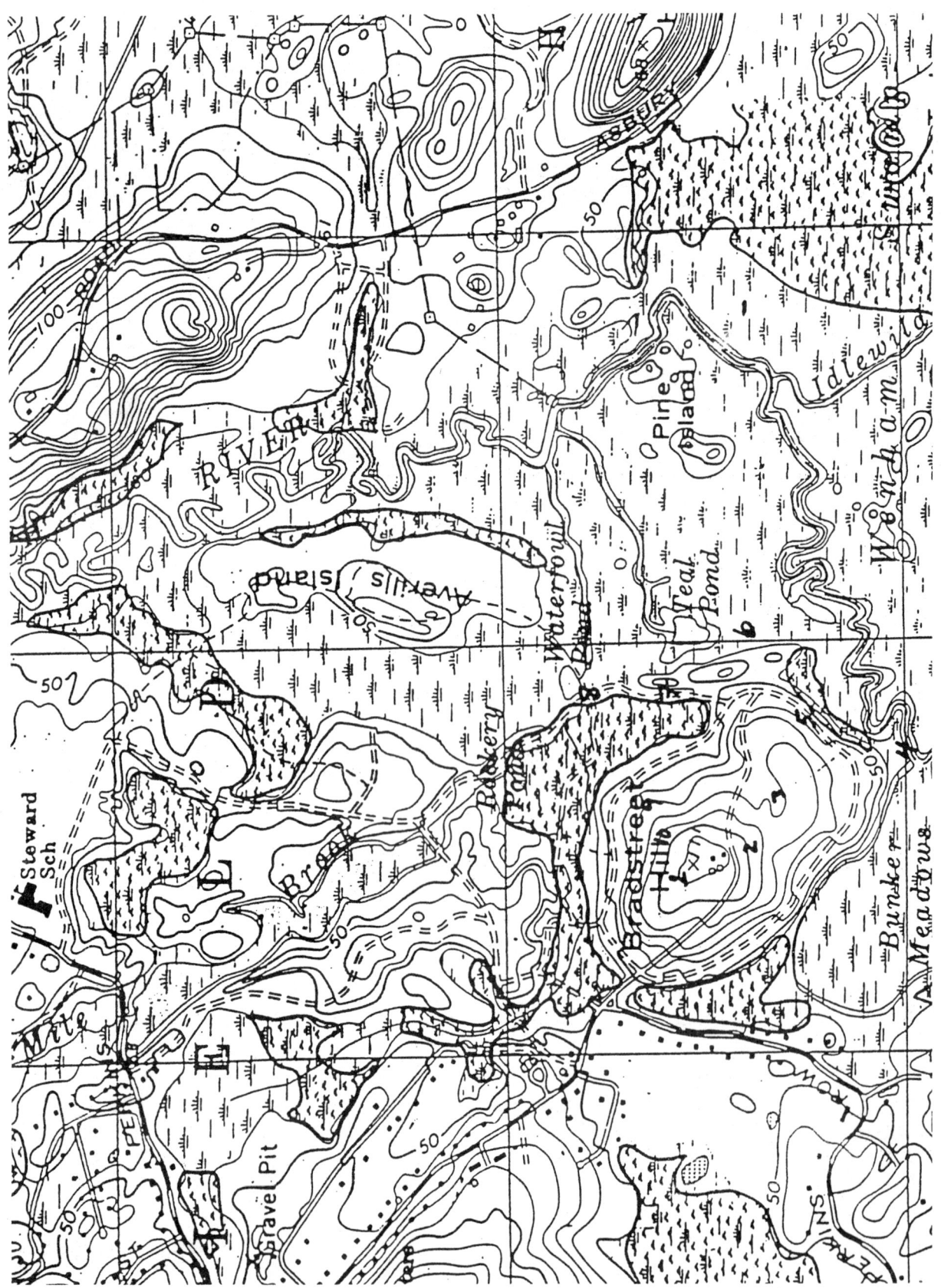

Figure 14–8. Enlarged version of part of Figure 14–7. Topsfield, MA. Wooded swamps outlined in bold black line.

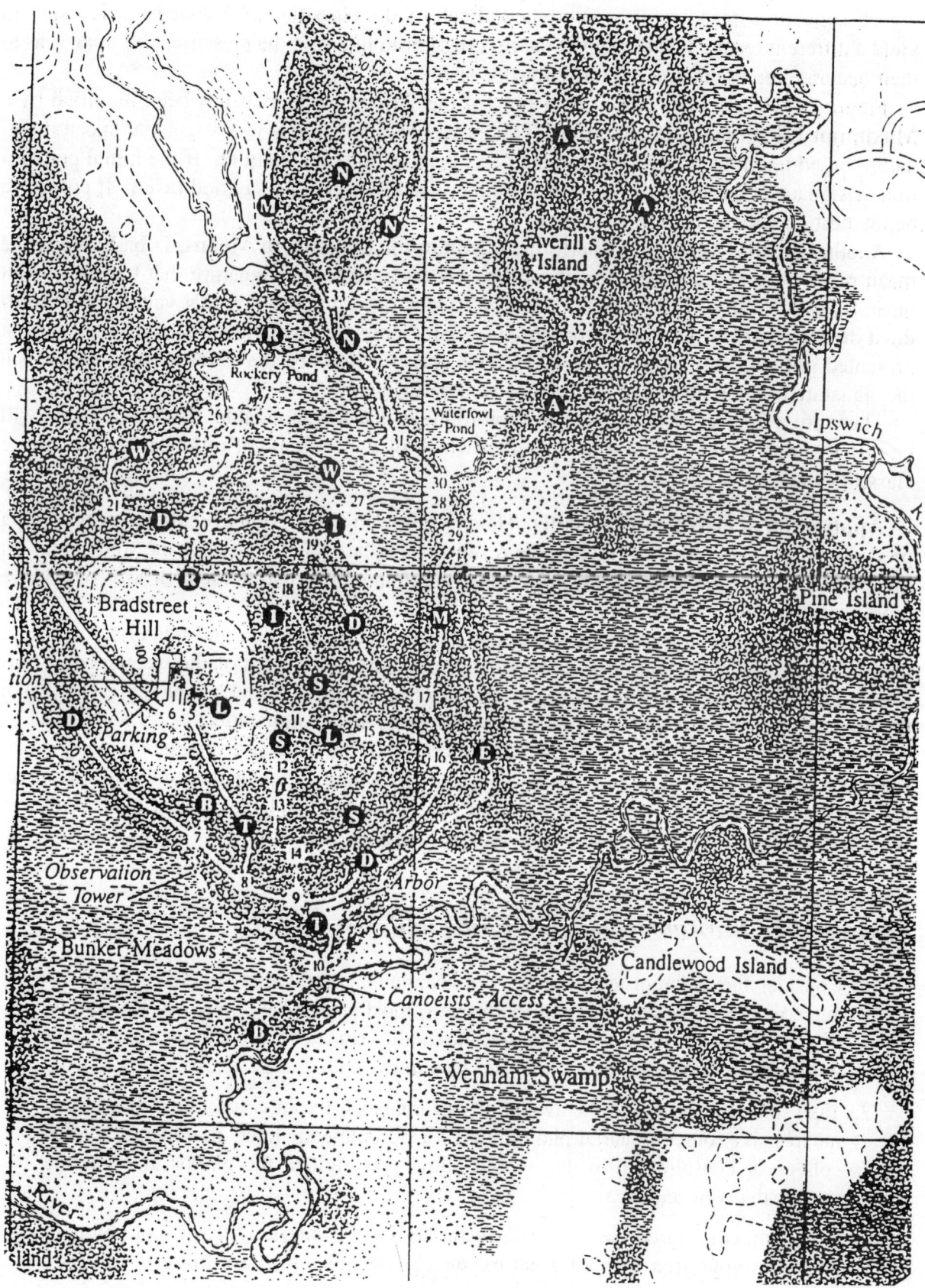

Figure 14–9. Map of MAS Wildlife Sanctuary. Topsfield, MA. Courtesy Massachusetts Audubon Society.

In this investigation five classification procedures were carried out, but only two of the resulting classification images will be presented. Each of the classification procedures is likely to yield a different result. It is then up to the investigator to interpret the classification images as to their accurate representation of reality for this site.

Figure 14–11 (see Plate XII) is a display of the TM subscene as it has been modified by a **Maximum Likelihood Classification.** This process assigns pixels to a particular spectral signature based upon the probability that the pixel belongs to that signature. If the initial preparation of signatures has been accurate, this routine is likely to produce a good result. It proved to be the best classification in this instance.

Another of the classifications performed, **Minimum Distance to Means,** is based upon the mean pixel value of each signature. Each pixel is assigned to the signature that has the closest mean reflection value. The procedure may be carried out using the raw DN values, or in **standard deviation units.** The latter procedure is usually more successful and that is the image result presented in Figure 14–12 (see Plate XII). This classification process tends to work well when the signature training sites have not been as well defined.

There are a number of other classification routines that are applicable. Each study area will be different and therefore one cannot predict which of the tests might produce the best image classification.

Application of classification results. Having followed these extensive and time-consuming procedures, it is appropriate to review the results as they are able to classify the study area. At this point it may be desirable to carry out some fine-tuning of signature identification to achieve more representative training sites.

Once a classification scheme has produced an image classification and numerical signature parameters for the study site, it is appropriate to examine other areas of the watershed (in this instance). Will the training sites and signatures developed for the initial site be representative of the broader area, or must they be modified by change or added signatures?

Eventually the classification of the entire watershed or the extensive area under study is carried out and this can serve as an important step in developing a management plan for the larger area or region. Sampling several representative sub-regions of this larger area is an efficient way to save time and reduce the cost of establishing a management plan.

Exercise 14   Environmental Monitoring and Assessment

1. Consider the different techniques of environmental evaluation that have been discussed in this chapter. Which of them seems most appropriate for your city, county, etc.? Why do you think you might have to modify some of these approaches because of the characteristics of your particular location?

2. Take some time to assess the availability of images for your community. Is there an inexpensive source of aerial photography? Does someone have a scanner at a local college or consulting firm that would allow you digitize the photos and satellite images that you acquire?

3. Have you considered the available resources on the internet? How can you gain access in your area without great expense? The resources are there if you look for them.

4. What map and photo resources can you obtain for an area of interest at low cost?

5. Draw up a plan to evaluate some area of interest in regard to environmental monitoring and assessment. Construct a plan to carry out ground truthing of the area to determine what type of land use/land cover or other landscape units are present in the area.

6. Having taken these steps, investigate the availability of support from local governmental, business, or educational organizations. Most people in the country want their quality of life to be enhanced by maintaining or improving environmental conditions. If you find ways of holding down costs, you will find the support to allow you to involve yourself in a productive way in environmental monitoring and assessment.

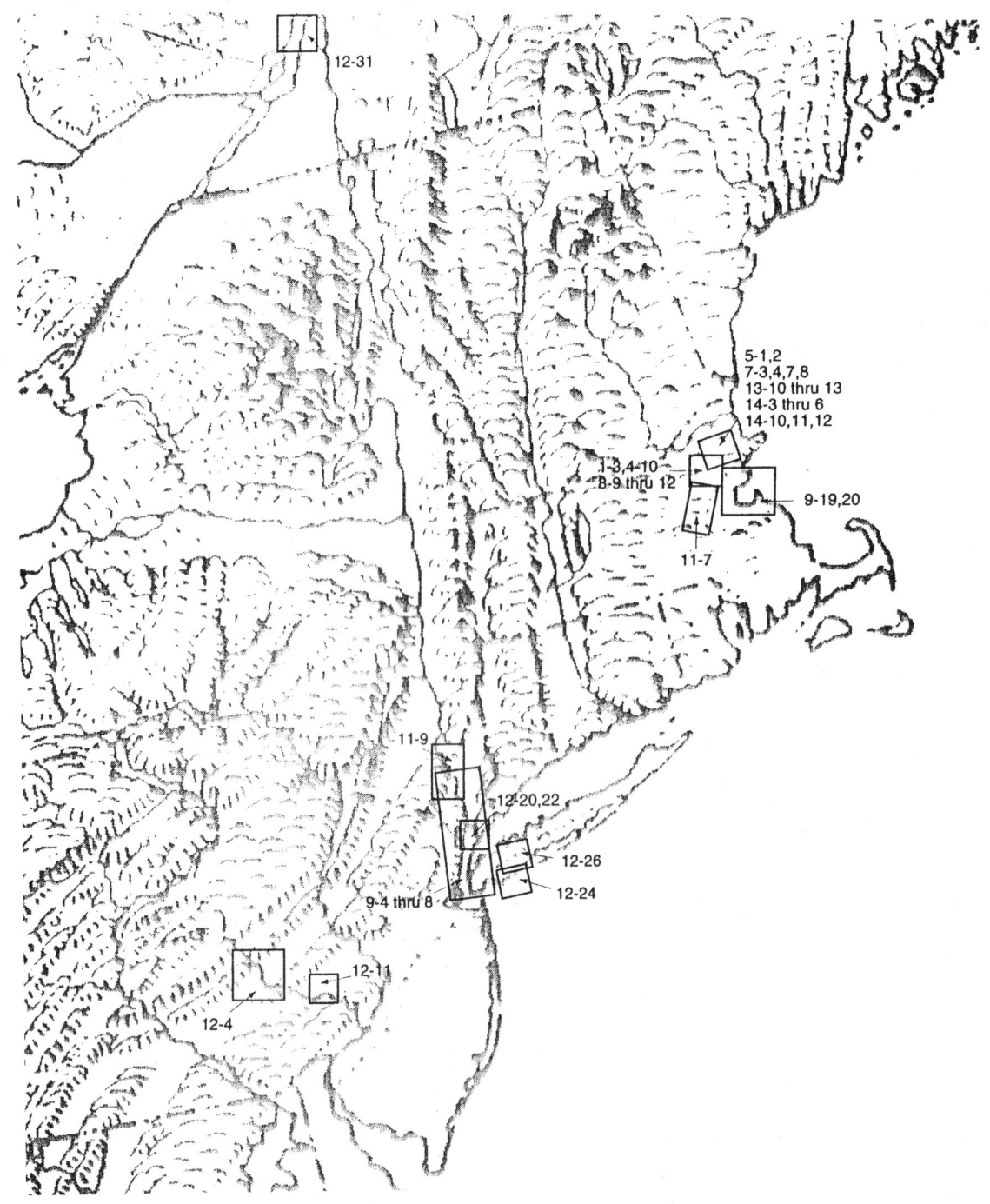

12-31

5-1,2
7-3,4,7,8
13-10 thru 13
14-3 thru 6
14-10,11,12

1-3,4-10
8-9 thru 12

9-19,20

11-7

11-9

12-20,22

12-26

9-4 thru 8

12-24

12-11

12-4

RAISZ LANDFORM MAP of the UNITED STATES (Image locations shown by chapter & figure number).
Copyright by Erwin Raisz 1957. Reprinted with permission by GEOPLUS, Danvers, MA.

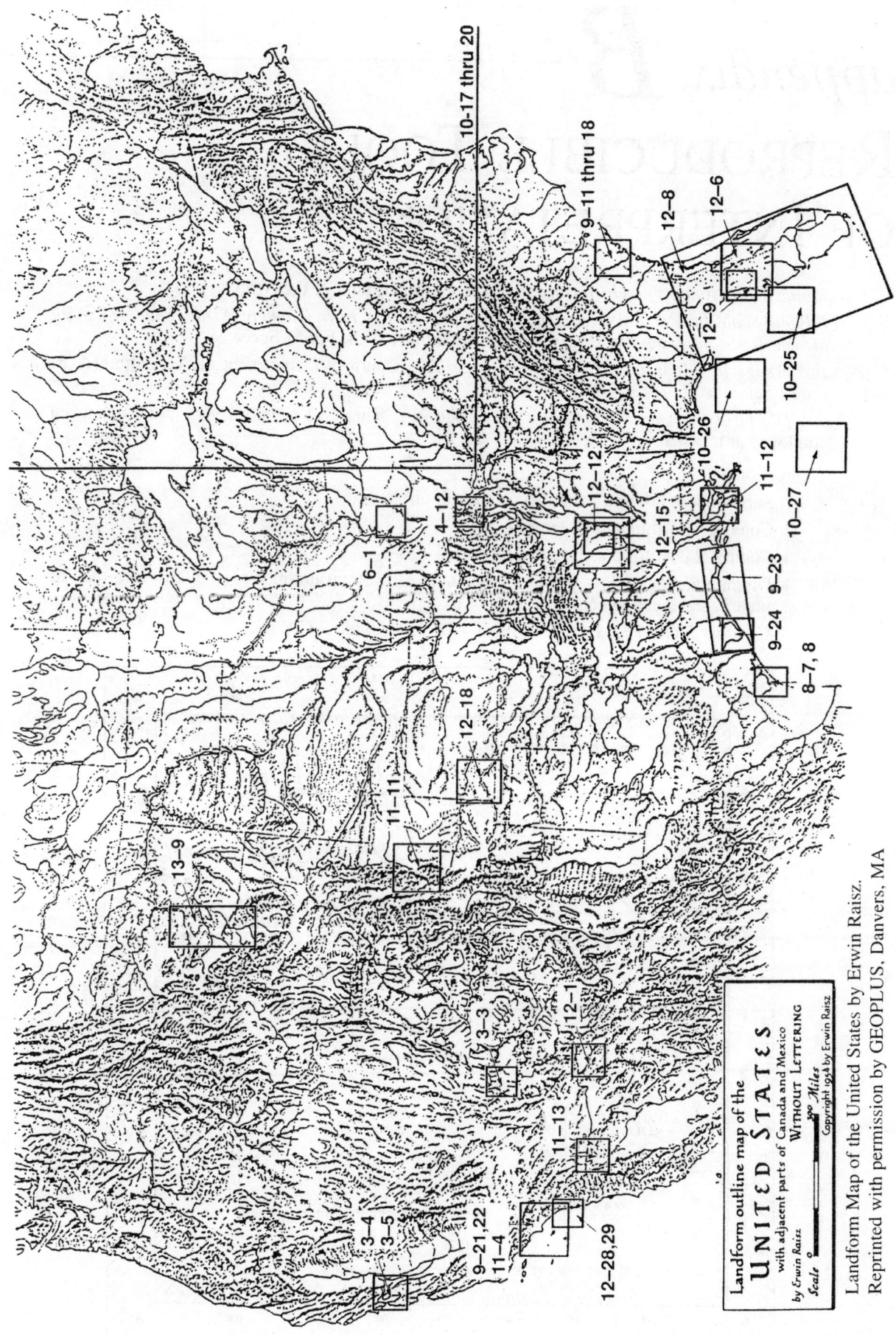

Landform Map of the United States by Erwin Raisz.
Reprinted with permission by GEOPLUS, Danvers, MA

appendix B

REPRODUCIBLE TOOLS
OF INTERPRETATION

1. *Coordinate Scale (Roamer) and Protractor.* Courtesy Department of the Army.
2. *Aerial Photo Scale-Protractor* (various scales) See page 219. *Crown Diameter Scale; Shadow Length Measuring Scale.* Courtesy Department of Agriculture, U.S. Forest Service.
3. *Parallax Wedge for Mountainous Areas.* See page 220. Courtesy Department of Agriculture, U.S. Forest Service.
4. *Random Dot Grid for Area Measurement.* See page 221. This grid provides 90% precision when employed at this size.

Directions:
1. Superimpose dot grid over area on map or airphoto.
2. Count dots within area of interest.
3. Count 1/2 dots for dots bisected by boundaries.
4. Divide number of dots within area of interest by total number of dots covering map or photo area to get percent.
Courtesy STRATEX INSTRUMENT CO., Los Angeles, CA.

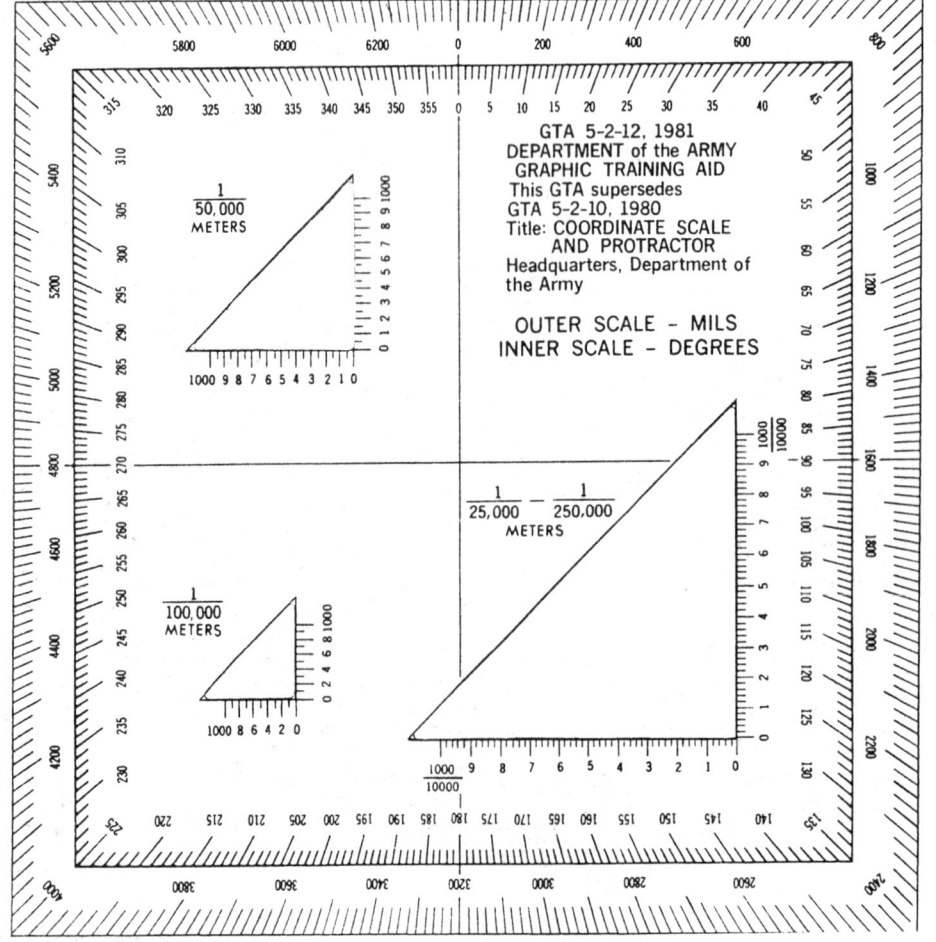

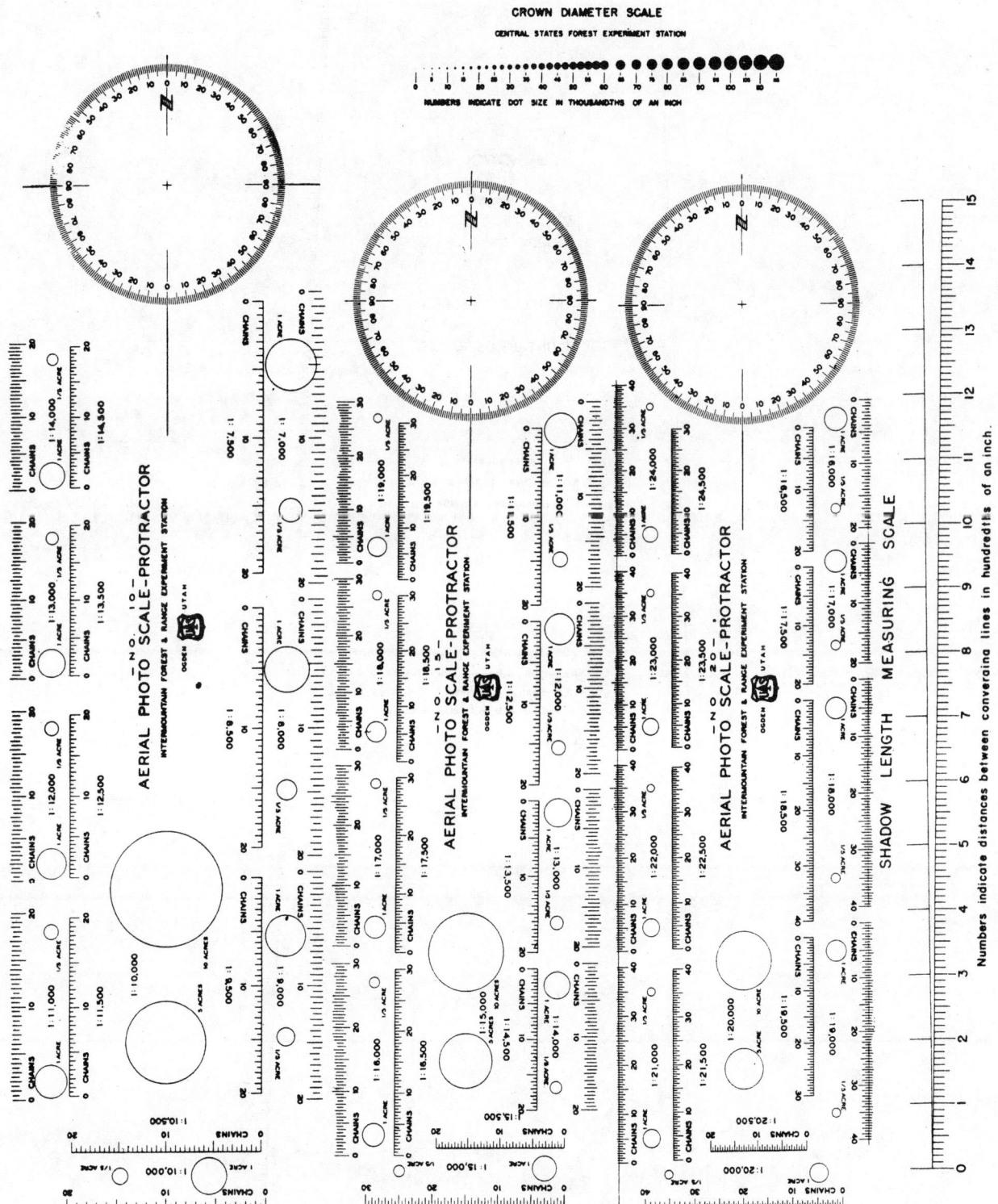

CROWN DIAMETER SCALE

CENTRAL STATES FOREST EXPERIMENT STATION

NUMBERS INDICATE DOT SIZE IN THOUSANDTHS OF AN INCH

AERIAL PHOTO SCALE-PROTRACTOR
— NO. 10 —
INTERMOUNTAIN FOREST & RANGE EXPERIMENT STATION
OGDEN UTAH

AERIAL PHOTO SCALE-PROTRACTOR
— NO. 15 —
INTERMOUNTAIN FOREST & RANGE EXPERIMENT STATION
OGDEN UTAH

AERIAL PHOTO SCALE-PROTRACTOR
— NO. 20 —
INTERMOUNTAIN FOREST & RANGE EXPERIMENT STATION
OGDEN UTAH

SHADOW LENGTH MEASURING SCALE

Numbers indicate distances between converging lines in hundredths of an inch

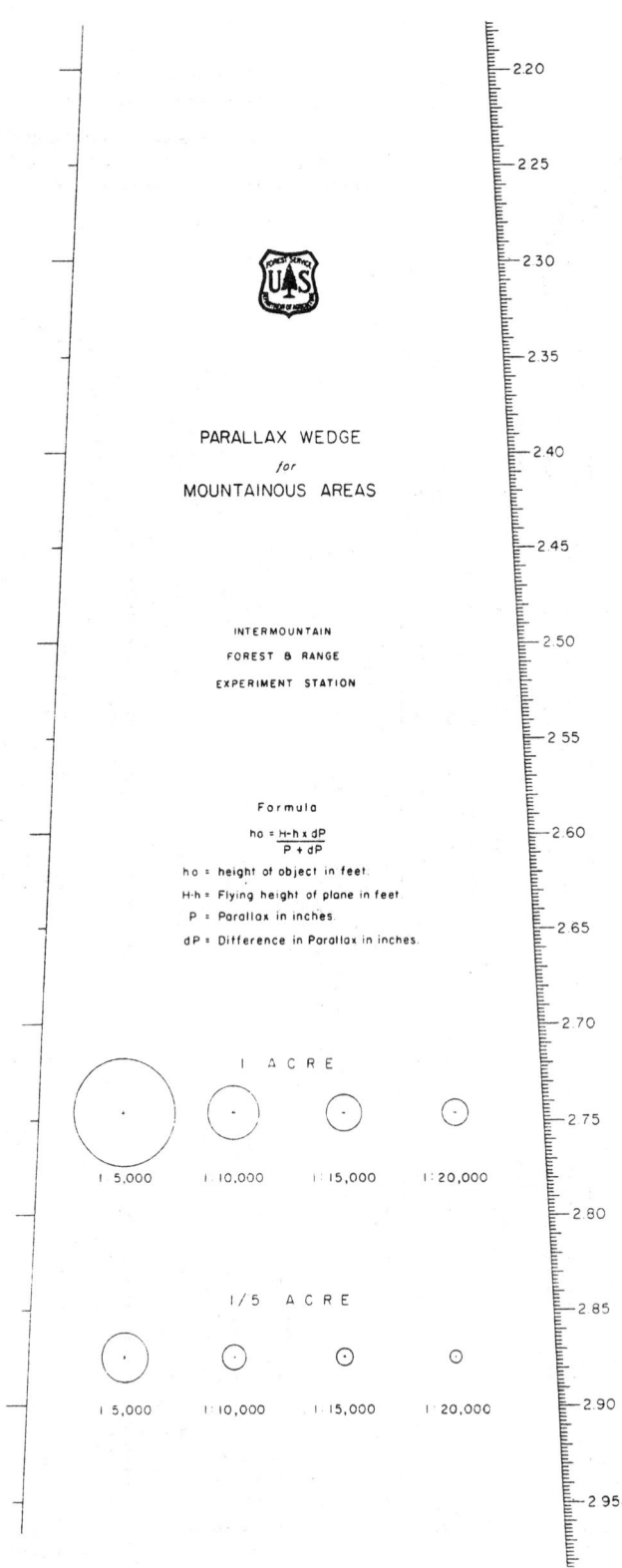

PARALLAX WEDGE
for
MOUNTAINOUS AREAS

INTERMOUNTAIN
FOREST & RANGE
EXPERIMENT STATION

Formula

$$ho = \frac{H-h \times dP}{P + dP}$$

ho = height of object in feet.

H-h = Flying height of plane in feet.

P = Parallax in inches.

dP = Difference in Parallax in inches.

I ACRE

1:5,000   1:10,000   1:15,000   1:20,000

1/5 ACRE

1:5,000   1:10,000   1:15,000   1:20,000

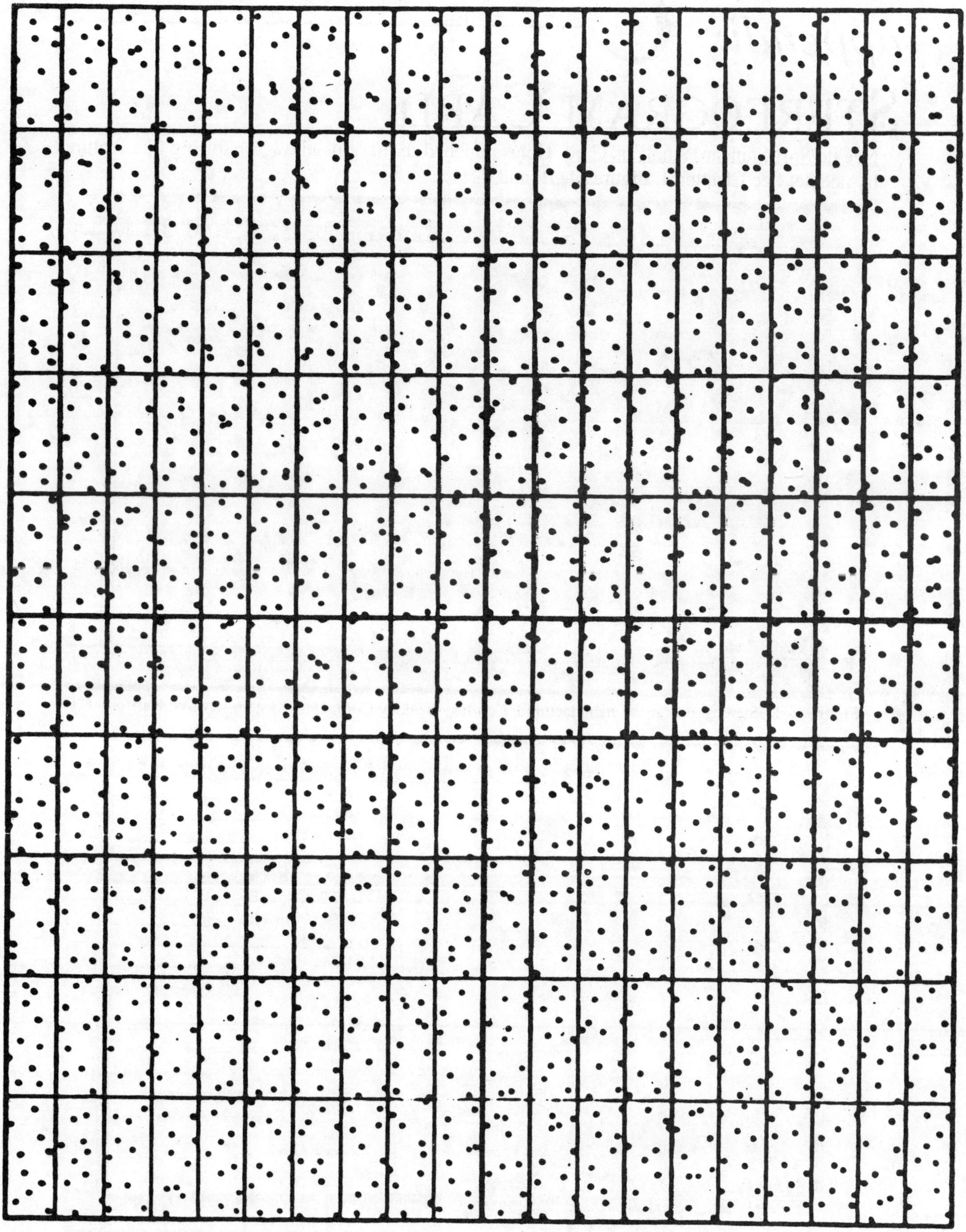

appendix C

STEREOGRAM CARD

<u>Needle/Sort Standard Analysis Card</u>. Holes punched in margins allow for sorting like features by needle. See Chapter 4, Figures 4–10 and 4–11.

Figure C–1. Stereogram card as manufactured. Courtesy Beekley Corp, Data System Div., W. Hartford, CT.

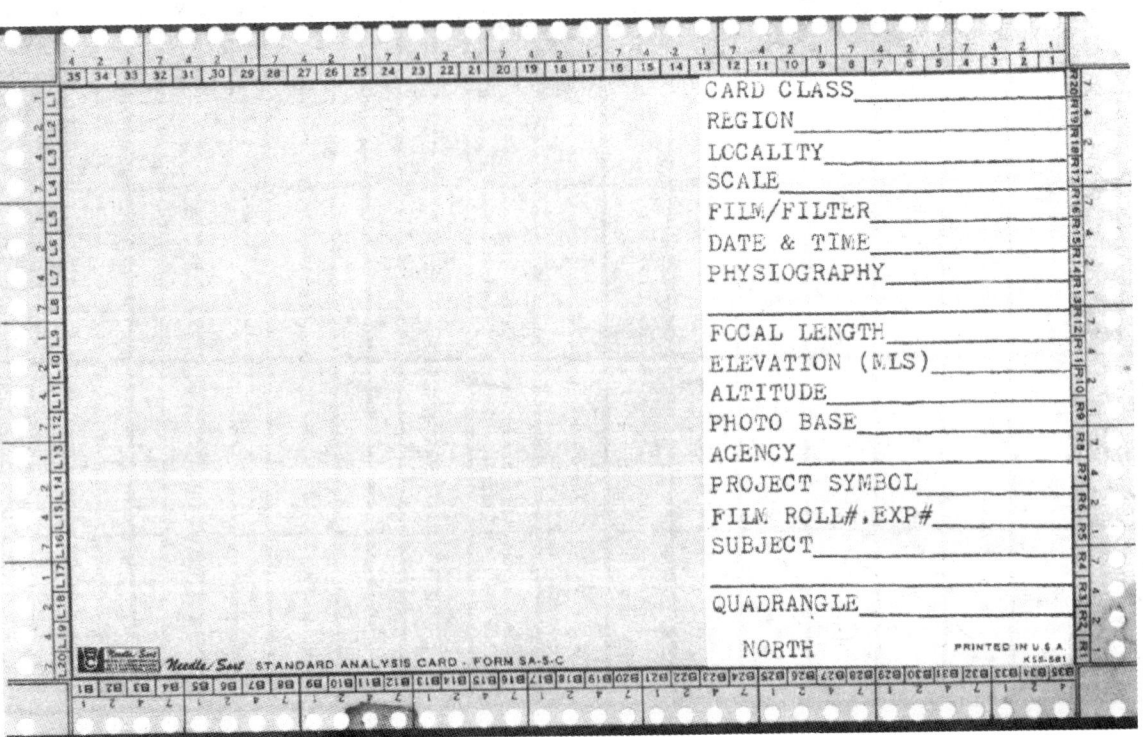

Figure C–2. Stereogram card with sample key identifying features of a stereogram. Courtesy Beekley Corp, Data System Div., W. Hartford, CT.

LAND USE AND LAND COVER CLASSIFICATIONS

Explanatory Note. Three classification schemes are presented here as examples of some of the approaches currently being utilized in various disciplines that focus upon land use and land cover. The version below was used by the Town of Ossining, NY and is typical of planning efforts at the local level. The USGS classification developed by Anderson and others is intended to be used with remote sensor data from small to large scale and additional levels of detail may be included. The classification produced for the NOAA Coastwatch Change Analysis Project was designed to aid study of the wetland/upland interface along the US Atlantic coast. One of these classifications should contain elements that might be used in new land use/land cover investigations and are included here to help in that effort.

EXISTING LAND USE

RESIDENTIAL	single family	915	lemon yellow*
	two family	917	yellow orange
	3 & 4 family	942	yellow ochre
	5 or more families (multiple)	943	burnt ochre
	mid to high rise housing	946	dark brown
COMMERCIAL	general business (single or strip)	929	pink
	local (neighborhood) commercial (center)	921	vermillion red
	central business district	923	scarlet lake
INDUSTRIAL	office research	936	slate grey
	industrial	964	light grey
	–light manufacturing	L	light grey
	–heavy manufacturing	H	light grey
	–warehousing	W	light grey
	–storage	S	light grey
	transportation/utility	931	purple
PUBLIC	schools	919	sky blue
	public buildings	902	ultramarine
QUASI PUBLIC	churches	905	aquamarine
	quasi-public buildings & institutions	901	indigo blue
	private schools (w/name)	903	true blue
	cemeteries	920	light green
OPEN SPACE	parks	910	true green
	–neighborhood park	N	true green
	–regional park	R	true green
	–playfield	P	true green
	commercial recreation	912	apple green
	vacant land	938	white

*Prismacolor Pencil code

LAND USE AND LAND COVER CLASSIFICATION SYSTEM AFTER ANDERSON, HARDY, ROACH, AND WITMER (1976) AND FLORIDA (1976)
As presented in F. F. Sabins Jr. *Remote Sensing, Principles and Interpretation.* 2nd ed. (New York: W. H. Freeman Co., 1986).

Level I	Level II	Level III
100 Urban or Built-up	110 Residential	111 Single Unit, Low Density (less than 2 dwelling units/acre DUPA)
		112 Single Unit, Medium Density (2-6 DUPA)
		113 Single Unit, High Denisty (greater than 6 DUPA)
		114 Mobil Homes
		115 Multiple Dwelling, Low-rise (2 stories or less)
		116 Multiple Dwelling, High-rise (3 stories or more)
		117 Mixed Residential
	120 Commercial and Services	121 Retail Sales and Services
		122 Wholesales and Services, including trucking and warehousing
		123 Offices and Professional Services
		124 Hotels and Motels
		125 Cultural and Entertainment
		126 Mixed Commercial and Services
	130 Industrial	131 Light Industrial
		132 Heavy Industrial
		133 Extractive
		134 Industrial Under Construction
	140 Transportation	141 Airports, Including Runways, Parking Areas, Hangers and Terminals
		142 Railroads, Including Yards and Terminals
		143 Bus and Truck Terminals
		144 Major Roads and Highways
		145 Port Facilities
		146 Auto Parking Facilities (where not directly related to another use)
	150 Communications and Utilities	151 Energy Facilities (Electrical and Gas)
		152 Water Supply Plants (Including Pumping Stations)
		153 Sewage Treatment Facilities
		154 Solid Waste Disposal Sites
	160 Institutional	161 Educational Facilities, Including Colleges, Universities, High Schools, and Elementary Schools
		162 Religious Facilities, Excluding Schools
		163 Medical and Health Care Facilities
		164 Correctional Facilities
		165 Military Facilities
		166 Governmental, Administration, and Service Service Facilities
		167 Cemeteries
	170 Recreational	171 Golf Courses
		172 Parks, Zoos
		173 Marinas
		174 Stadiums, Fairgrounds, Race Tracks
	180 Mixed-Urban	
	190 Open Land and Other	191 Undeveloped Land Within Urban Area
		192 Land Being Developed, Intended Use Not Known
200 Agriculture	210 Cropland and Pasture	211 Row Crops (e. g. corn)
		212 Field Crops (e. g. wheat)
		213 Pastures

Level I	Level II	Level III
	220 Orchards, Groves, Vineyards, Nurseries, Ornamental Horticultural Areas	221 Citrus Orchards 222 Noncitrus Orchards 223 Nurseries 224 Ornamental Horticultural 225 Vineyards
	230 Confined Feeding Operations	231 Cattle 232 Poultry 233 Hogs
	240 Other Agriculture	241 Inactive Agricultural Lands 242 Other
300 Rangeland	310 Grassland	
	320 Shrub and Brushland	321 Sagebrush Praries 322 Coastal Scrub 323 Chaparral 324 Second Growth Brushland
	330 Mixed Rangeland	
400 Forest Land	410 Evergreen Forest	411 Pine 412 Redwood 413 Other
	420 Deciduous Forest	421 Xeric Oak 422 Other Hardwood
	430 Mixed Forest	431 Mixed Forest
	440 Clearcut Areas	
	450 Burned Areas	
500 Water	510 Streams and Canals	
	520 Lakes and Ponds	
	530 Reservoirs	
	540 Bays and Estuaries	
	550 Open Marine Waters	
600 Wetland	610 Forested Wetland	611 Evergreen 612 Deciduous 613 Mangrove
	620 Nonforested Wetland	621 Herbaceous Vegetation 622 Nonvegetated
700 Barren Land	710 Dry Lake Beds	
	720 Beaches	
	730 Sand and Gravel Other than Beaches	
	740 Exposed Rock	
800 Tundra	810 Shrub and Brush Tundra	
	820 Herbaceous Tundra	
	830 Bare Ground Tundra	
	840 Wet Tundra	
900 Perennial Snow or Ice	910 Perennial Snowfields	
	920 Glaciers	

MODIFIED COASTAL LAND COVER CLASSIFICATION
SCHEME for USE with REMOTE SENSOR DATA
(after NOAA C-CAP STUDY)
Derived from USGS LU/LC of Anderson et. al. for Upland Areas
and from USF and WS of Cowardin et. al. for Wetlands and Deep Water Habitats

Level 0	Level 1	Level 2	Level 3
Upland			
	1.0 Developed	1.1 High Intensity (solid cover)	1.11 Residential
			1.12 Commercial
			1.13 Industrial
			1.14 Transportation, Communications and Utilities
		1.2 Low Intensity	1.21 Residential
			1.21 Commerical
			1.23 Indutrial
			1.24 Transportation, Communications and Utilities
			1.25 Rural Development
	2.0 Cultivated	2.1 Woody	2.11 Orchards/Groves
			2.12 Vine/Bush
		2.2 Herbaceous	2.21 Cropland
	3.0 Grassland (Herbaceous)	3.1 Herbaceous	3.11 Unmanaged
			3.12 Pasture
			3.13 Groomed
	4.0 Woody	4.1 Deciduous	4.11 Forest
			4.12 Shrub
		4.2 Evergreen	4.21 Forest
			4.22 Shrub
		4.3 Mixed	4.31 Forest
			4.32 Shrub
	5.0 Exposed	5.1 Soil	5.11 Transitional Developed
		5.2 Sand	5.21 Beach/Dune Complex
			5.22 Sandy, other than beach
			5.23 Extraction Pits
		5.3 Rock	5.31 Outcrops
			5.32 Quarries/Mines
			5.33 Unconsolidated
		5.4 Evaporite Deposit	5.41 Dry Salt Flats
	6.0 Snow and Ice	6.1 Snow and Ice	6.11 Perennial Snowfields
			6.12 Glaciers
Wetland			
	7.0 Wetland	7.1 Marine (intertidal)	7.11 Aquatic Bed
			7.12 Reef
			7.13 Rocky Shore
			7.14 Unconsolidated Shore
		7.2 Estuarine (intertidal)	7.21 Aquatic Bed
			7.22 Reef
			7.23 Streambed
			7.24 Rocky Shore
			7.25 Unconsolidated Shore
			7.26 Emergent

Level 0	Level 1	Level 2	Level 3
			7.27 Scrub/Shrub
			7.28 Forested
		7.3 Riverine (tidal)	7.31 Rock Bottom
			7.32 Unconsolidated Bottom
			7.33 Streambed
			7.34 Aquatic Bed
			7.35 Rocky Shore
			7.36 Unconsolidated Shore
			7.37 Nonpersistent Emergent
			7.38 Open Water
		7.4 Riverine (lower perennial)	7.41 Unconsolidated Bottom
			7.42 Aquatic Bed
			7.43 Rocky Shore
			7.44 Unconsolidated Shore
			7.45 Nonpersistent Emergent
			7.46 Open Water
		7.5 Riverine (upper perennial)	7.51 Rock Bottom
			7.52 Unconsolidated Bottom
			7.53 Aquatic Bed
			7.54 Rocky Shore
			7.55 Unconsolidated Shore
			7.56 Nonpersistent Emergent
			7.57 Open Water
		7.6 Riverine (intermittent)	7.61 Streambed
		7.7 Riverine (unknown perennial)	7.71 Rock Bottom
			7.72 Unconsolidated Bottom
			7.73 Aquatic Bed
			7.74 Rocky Shore
			7.75 Unconsolidated Shore
			7.76 Nonpersistent Emergent
			7.77 Open Water
		7.8 Lacustrine (littoral)	7.81 Rock Bottom
			7.82 Unconsolidated Bottom
			7.83 Aquatic Bed
			7.84 Rocky Shore
			7.85 Unconsolidated Shore
			7.86 Nonpersistent Emergent
			7.87 Open Water
		7.9 Palustrine (wetland)	7.91 Rock Bottom
			7.92 Unconsolidated Bottom
			7.93 Aquatic Bed
			7.94 Unconsolidated Shore
			7.95 Moss-Lichen
			7.96 Emergent
			7.97 Scrub/Shrub
			7.98 Forested
			7.99 Open Water
Water and Submerged Land			
	8.0 Water and Submerged Land	8.1 Marine (subtidal)	8.11 Rockbottom
			8.12 Unconsolidated Bottom
			8.13 Aquatic Bed

Level 0	Level 1	Level 2	Level 3
			8.14 Reef
			8.15 Open Water
		8.2 Estaurine (subtidal)	8.21 Rock Bottom
			8.22 Unconsolidated Bottom
			8.23 Aquatic Bed
			8.24 Reef
			8.25 Open Water
		8.3 Riverine (tidal)	8.31 Rock Bottom
			8.32 Unconsolidated Bottom
			8.33 Aquatic Bed
			8.34 Open Water
		8.4 Riverine (lower perennial)	8.41 Unconsolidated Bottom
			8.42 Aquatic Bed
			8.43 Open Water
		8.5 Riverine (upper perennial)	8.51 Rock Bottom
			8.52 Unconsolidated Bottom
			8.53 Aquatic Bed
			8.54 Open Water
		8.6 Riverine (unknown perennial)	8.61 Rock Bottom
			8.62 Unconsolidated Bottom
			8.63 Aquatic Bed
			8.64 Open Water
		8.7 Lacustrine (limnetic)	8.71 Rock Bottom
			8.72 Unconsolidated Bottom
			8.73 Aquatic Bed
			8.74 Open Water

appendix E
SELECTED SOURCES

Aeronautics and Space Report of the President, Fiscal Year 1994 Activities. NASA, Washington, D.C., 1995.

Avery, Thomas E., and Graydon Lennis Berlin. *Fundamentals of Remote Sensing and Airphoto Interpretation.* 5th ed., Macmillan, New York, 1992.

Blair, C.L., and B.V. Gutsell. *The American Landscape, Map and Air Photo Interpretation.* McGraw-Hill, New York, 1974.

Cowardin, L.M., V. Carter, F.C. Golet, and E.T. LaRoe. *Classification of Wetlands and Deepwater Habitats of the United States.* (Washington, D.C.: Office of Biological Services, USF & WS, US Dept. of the Interior, 1979).

Curran, H. Allen, E.L. Perdew, P.S. Justus, and M.B. Prothero. *Atlas of Landforms.* 2nd ed., Wiley, New York, 1974.

Earth Observation Satellite Company. 4300 Forbes Blvd., Lanham, MD 20706.

Gurney, R.J., J.L. Foster, and C.L. Parkinson, eds. *Atlas of Satellite Observations Related to Global Change.* Cambridge Univ. Press, New York, 1993.

Hamblin, W. Kenneth. *Atlas of Stereoscopic Aerial Photographs and Landsat Imagery of North America.* TASA Pub. Co., Minneapolis, MN, 1980.

_____ , and J. D. Howard. *Exercises in Physical Geology.* 8th ed., Macmillan, New York, 1992.

Image, Object, and Illusion, Readings from Scientific American. W.H. Freeman, San Francisco, CA, 1971.

Jensen, John R. *Introductory Digital Image Processing, A Remote Sensing Perspective.* 2nd ed., Prentice Hall, Upper Saddle River, NJ, 1996.

Knowles, R., and P.W.E. Stowe. *Europe in Maps, Topographical Map Studies of Western Europe.* Books One & Two. Longman, London, 1969 & 1971.

_____ . *North America in Maps, Topographical Map Studies of Canada and the USA.* Longman, London, 1976.

Lillesand, Thomas M., and R.W. Kiefer. *Remote Sensing and Image Interpretation,* 3rd ed., Wiley, New York, 1994.

Manual of Photogrammetry. 4th ed., Chester C. Slama, ed-in-chief, ASP&RS, Falls Church, VA, 1980.

Manual of Remote Sensing. Robert G. Reeves, ed-in-chief, ASP&RS, Falls Church, VA, 1975.

Mausel, Paul W., ed. *15th Biennial Workshop on Videography & Color Photography in Resource Assessment.* Terre Haute, IN 1-3 May, 1995. ASP&RS, Bethesda, MD, 1995.

Mission to Planet Earth, Task Force Report. Dept. of the Army, U.S. Army Corps of Engineers, Washington, D.C., 1993.

Muehrcke, Phillip C., and J.O. Muehrcke. *Map Use, Reading, Analysis, and Interpretation.* 3rd ed., JP Publications, Madison, WI, 1992.

National Archives (airphotos) Trust Fund Board, P.O. Box 100793 Atlanta, GA 30384 (202) 501-5170

NESDIS Programs, NOAA Satellite Operations. NESDIS, Washington, D.C., 1985.

NOAA Coastwatch Change Analysis Project, Guidance for Regional Implementation. J.E. Dobson et. al., eds. (Washington, D.C.: U.S. Dept. of Commerce, NOAA, 1992).

Photo Interpretation Keys. Defense Mapping Agency Aerospace Center, St. Louis, MO, 1977.

Product Development Plans for Operational Satellite Products for the NOAA Climate and Global Change Program. NOAA Special Report No. 5. University Corporation for Atmospheric Research, Washington, D.C., 1991.

Rabenhorst, Thomas D., and P.D. McDermott. *Applied Cartography, Introduction to Remote Sensing.* Merrill, Columbus, OH, 1989.

Rasher, Michael E., and Wayne Weaver. *Basic Photo Interpretation, A Comprehensive Approach to Interpretation of Vertical Aerial Photography for Natural Resource Applications.* USDA, SCS, Fort Worth, TX, 1990.

Ray, Richard G. *Aerial Photographs in Geologic Interpretation and Mapping.* Geological Survey Professional Paper 373. Dept. of the Interior, Washington, D.C., 1960.

Richason, Benjamin F. *Atlas of Cultural Features, A Study of Man's Imprint on the Land.* Hubbard Press, Northbrook, IL, 1972.

Russ, John C. *The Image Processing Handbook.* 2nd ed., CRC Press, Ann Arbor, MI, 1995.

Sabins, Floyd F., Jr. *Remote Sensing, Principles and Interpretation.* 2nd ed., W.H. Freeman, New York, 1987.

Short, Nicholas M. *The Landsat Tutorial Workbook, Basics of Satellite Remote Sensing.* NASA Ref.Pub. 1078. NASA, Washington, D.C., 1982.

Soil Survey of Essex County, Massachusetts, Northern Part and Southern Part. U.S. Department of Agriculture, Soil Conservation Service with Massachusetts Agricultural Experiment Station, Washington, D.C., 1981.

Sources of Earth and Planetary Photography. Compiled by Rose Steinat, Regional Planetary Image Facility, Center for Earth and Planetary Studies, National Air and Space Museum, Smithsonian Institution, Washington, D.C., July, 1993.

SPOT Image Corp., 1897 Preston White Dr., Reston, VA 22091.

Stereo Atlas. Earth Science Curriculum Project, American Geological Institute, Hubbard, Northbrook, IL, 1968.

Stroebel, Leslie, J. Compton, I. Current, and R. Zakia. *Photographic Materials and Processes.* Focal Press, Boston, 1986.

Thematic Map Design. Harvard Papers in Theoretical Cartography. Harvard University, Laboratory for Computer Graphics and Spatial Analysis, Cambridge, MA, 1979.

Upton, William B., Jr. *Landforms and Topographic Maps.* Wiley, New York, 1970.

U.S. Dept. of Agric., Agric. Stabilization and Conserv. Serv., Aerial Photography Field Office, 2222 W. 2300 S.-84119, P.O. Box 30010 84130-0010, Salt Lake City, UT (801) 975-3503.

Wanless, Harold R. *Aerial Stereo Photographs.* Hubbard, Northbrook, IL, 1973.

Weldon, Roger B., and S.J. Holmes. *Water Vapor Imagery, Interpretation and Applications to Weather Analysis and Forecasting.* NOAA Technical Report NESDIS 57. NOAA, Washington, D.C., 1991.

GLOSSARY OF SELECTED TERMINOLOGY

A,B,C location method: A method for indicating locations within a map or image that was utilized by the USGS in the *Series of 100 Topographic Maps.*

Absorption: The process in which incident radiant energy is absorbed and converted to other forms of energy.

Active system: This is a type of remote sensor, such as radar or sonar, that sends out some sort of signal to "paint" the target.

Aerosols: Tiny liquid or solid particles that exist suspended in the atmosphere.

Airphoto interpretation: A sub-field of remote sensing that deals with the extraction of information from photographs that acquired reflected energy from above the Earth's surface.

Albedo: The ratio of the amount of EMR reflected by a body to the amount incident upon it, expressed as a percentage.

Alluvial fans: Depositional landform features produced by surface streams laden with sediment, usually when exiting a narrow valley onto a less steeply sloping surface. Semi-circular fan shape occurs as stream travels to left and right of deposited debris.

Anticlinal fold: Horizontal rock strata when folded into an upfold produce an anticline which may appear at the surface as a ridge.

Apollo, Mercury, Gemini: NASA's Mercury Project (May, 1961 to May, 1963) demonstrated man's ability to withstand space travel. Gemini Project (1964 to 1966) was the beginning of sophisticated manned flight. The Apollo Project (1961 to 1973) achieved its major goal of placing a man on the Moon on July 20, 1969.

Area-specific stretch: The enhancement of a particular part of an image by means of modifying the histogram at specific gray levels.

Areal units of measure: Units used to designate areas, e.g., square foot, acre, square mile, square centimeter, square meter, square kilometer.

ARGOS DCS: A data collection and platform location system on the TIROS-N polar orbiter environmental satellite.

Arithmetic proportion: A mathematical relationship wherein one ratio equals another ratio, e.g., 1/2 = 2/4.

Atmospheric transmission: The movement of radiation through the atmosphere.

Atmospheric windows: Specific wavelengths or band widths that allow radiant energy to be transmitted.

Attentuation: The loss of energy moving through space by absorption, scattering, and reflectance.

Average photo base length: The average of the distance between the principal point and the corresponding principal point on successive photos within a flight line.

AVHRR: Advanced Very High Resolution Radiometer sensor on the TIROS-N polar orbiter satellite is a 5 channel scanning radiometer which records imagery and quantitative radiance data at 1 and 4 km resolution.

Azimuth direction: Parallel to the line of flight in radar imaging from aircraft.

Azimuth resolution: The quality of resolving or seeing objects in radar images parallel to the line of flight.

Bajada: A bench-like surface formed at the base of mountains by the coalescing of alluvial fans.

Banding: Cross-polarized radar images, HV or VH, are often affected by this phenomenon which produces parallel bands on the image parallel to the line of flight.

Banding (striping): When a sensor in a MSS system malfunctions, it will not provide the proper values, leading to stripes of constant gray values (black or other level) each time that sensors values are recorded.

Bands: A selection or range of wavelengths.

Barrier islands: Islands that are formed parallel to a mainland coast where breaking ocean waves create sandbars that increase in size due to the impact of wind, waves, and ocean currents.

Beam width: In radar imaging this is the width of the area "painted" by the radar energy. It is constant in synthetic aperture radar and becomes larger with distance from the plane in "brute force" radar.

Binary digits (bits): The most basic language of the computer is to recognize "0" electric circuit off, or "1" electric circuit on; the "0" and "1" are binary digits or bits.

Black and white infrared: A single emulsion layer film that is sensitive to light from 2.5 to 9.0 micrometers and records images as black & white. A red filter is required for proper exposure.

Black and white photography: The production on film and paper print material of reflected energy as shades of gray (from black to white) within the visible and near infrared portions of the EMS.

Broad clusters: Refers to an unsupervised classification scheme for a digital false color composite image where pixels are grouped with the peak of a three dimensional histogram that is closest in value; nearest neighbor pixels are considered in this Idrisi module.

Camera obscura: A precursor to the modern camera; initially operated similar to a pinhole camera, projecting images through a pinhole without a lens.

CD-ROM: Compact disc, read only memory; one disc holds up to approximately 540 megabytes of data, greatly increasing the accessibility of a PC or Mac computer for large images.

Central pivot agriculture: A type of agriculture that uses an irrigation system that distributes water through a pipe that travels in a circle like the hands on a clock and produces distinctive, circular field patterns.

Change through time: A powerful tool of analysis that can be documented by photographic or other image records of the Earth's surface.

CIR film dye lot: This refers to the number of the batch of dye used in producing the film. Each batch of dye is a factor in the particular range of colors recorded by the film. The number is printed on each box containing CIR film.

CIR signatures: Objects recorded on CIR film will display particular types of false color patterns that are replicable, especially with regard to types of vegetation, and can be used to aid identification.

CIR videography: The process of recording a scene or area on videotape in the color signatures common to color infrared film.

CIR window: The period of time in the spring and fall seasons of the middle latitudes when some but not all vegetative growth is ongoing. It is the time when the widest range of colors and hues may be recorded on CIR film and make differentiation of phenomena easier.

Civil Engineer's scale: A measuring rule of six scales showing inches divided into 10, 20, 30, 40, 50, and 60 parts.

Classification factors: Some of the parameters that impact the classification of similar areas of land use or land cover are that classes be mutually exclusive, homogeneous, and consistently interpretable within the limitations of scale and seasonal variation in cover.

Cluster analysis: A technique of image classification in which histogram peaks of pixel values are identified by the computer and which are then used to group pixels with similar values.

Coastal Zone Color Scanner (CZCS): An instrument on NIMBUS 7 designed to measure chlorophyll concentrations and color variations in ocean water around the world.

Color infrared (CIR): A false color rendition of light recorded within the visible and reflected infrared portions of the EMS.

Color infrared photography (CIR): The production on film and paper print material of reflected energy in the visible and near IR portions of the EMS as false color images.

Color photography: The production on film and paper print material of reflected energy in the visible portion of the EMS of colors as seen by the human eye through the use of blue, green, and red-sensitive dye layers.

Compatible land uses: In planning theory, the concept that certain types of use of the land can exist in economic and social harmony, e.g., residential areas and cemeteries, whereas other combinations are incompatible, e.g., country club and an active landfill.

Compensating polar planimeter: A device used to trace the circumference of a feature (lake, forest, etc.) on a map or photo and yields the area in square inches or square centimeters. May be mechanical or digital.

Complex dielectric constant: A value controlled by the electrical properties of a surface material and is an indication of likely reflectance or penetration capabilities.

Contact prints: A photographic print made at the same size as a film negative or film positive through contact during reproduction.

Contrast stretching: Refers to a group of image enhancement techniques that are used to increase contrast in various parts of a histogram of pixel values, thus rendering greater clarity of detail.

Controlled mosaic: A series of overlapping airphotos that have been rectified and aligned with ground control points so that accurate measurements of distance may be made directly on the mosaic.

Corresponding principal point (CPP): Also referred to as conjugate principal point. The location of a principal point from one photo on the succeeding photo in a flight line.

Cross polarization: In radar imaging, energy that is transmitted in vertical or horizontal wave motion returns to the aircraft in the opposite configuration, H to V or V to H.

Culture: Any artifacts or works of people as seen on maps or photos.

Daily Weather Maps: Surface weather maps of the United States that are a compilation of 7 AM EST conditions of the atmosphere across the USA.

Density slicing: A computer process that converts an image into a series of levels or steps of gray tones so that patterns or shapes are more easily recognized.

Depression angle: The angle between a horizontal at the aircraft and the target of a radar beam.

Desktop publishing: The use of a computer to prepare text, graphics, and pictures in digital form for the printing process.

Detection: Quality of a remote sensor to distinguish an object on an image; the smallest feature that can be distinguished.

Dichotomous key: A form of elimination key in airphoto identification wherein the user is faced with a series of choices between two alternatives until the correct identification is made.

Differential parallax: The difference in displacement between ground points of different elevation on overlapping photos, e.g., the measured distance between the top of a chimney on one photo to another subtracted from the distance between the bottom on one to the bottom on the other.

Diffuse reflection: A radar beam striking a rough surface will produce reflection in many directions and yield a strong return.

Digital numbers (DNs): The values of each pixel in an image based upon the intensity of the field recorded.

Digital terrain maps: Topography on tapes in digital form that was derived from 1:250,000 scale USGS maps; useful in applications that employ slope maps and terrain profiles; available from USGS.

Directional filters: Computer operations designed to enhance specific linear features in an image.

DMSP Visible-IR: Defense Meteorological Satellite Program of the U.S. Air Force acquires images 24 hours a day in the visible-near IR as well as a thermal band.

Doppler principle: A statement of the changes in wavelength or frequency of energy generated by a moving object with reference to a target.

Dot grids: An overlay device for arriving at a quick estimate of area of some feature on a photo by counting dots. Dots may be regularly spaced or random.

Drafting tape: A tape used in cartographic and architectural drafting which appears similar to masking tape, but has less adhesive material and is less damaging to surfaces.

Dry plate process: The production of a photographic film emulsion by George Eastman that used gelatin in place of wet emulsions.

Earth as closed system: The concept that the Earth and its environment are in reality one large ecosystem; actions that impact one part of the system will eventually impact the remaining parts.

Earth Observing System (EOS): Part of the Mission to Planet Earth Program; includes 25 NASA sponsored instruments and others scheduled for deployment from 1998–2015.

Edge enhancement: A group of image enhancement techniques that are designed to increase the contrast along the edges of linear features so that they are easier to see.

8-bit, 256 levels: Currently the most common form of digital images is an 8 bit system which provides 256 levels of gray, from 0 = black to 255 = white.

Electromagnetic radiation (EMR): Energy that travels through space or materials in harmonic wave patterns (of consistent length) including the interaction at right angles of electric and magnetic fields.

Electromagnetic spectrum (EMS): A continuum of energy or an ordered array that extends from shortest cosmic rays, gamma rays, and X-rays to visible radiation, infrared radiation, and the longest microwaves and radio waves.

Elimination key: A form of airphoto identification in which the user is confronted with general choices for identification and then another set of choices until only one type of identification remains.

EOSAT: Earth Observation Satellite Company is the quasi-public firm that distributes image products for the Federal government at EROS Data Center, Sioux Falls, SD and other centers.

EROS Data Center (EDC): The main distribution center and archive for airphotos and satellite imagery maintained by the U.S. Department of the Interior, U.S. Geological Survey, at Sioux Falls, SD 57198.

Erosional remnants: Resistant landform features that remain after an area has been subjected to weathering and erosion, e.g., monadnocks, nunataks, and inselbergs.

ERTS: Earth Resources Technological Satellite, the early name for what came to be known as Landsat.

False color: The range of colors used to represent a scene in a form other than what would normally be perceived by the human eye. The most common example is CIR, color infrared photography.

False color composite: Refers to the combining of several bands of photography or imagery with colors that do not replicate normal human vision. The most common form produces color responses similar to color infrared.

Fiducial marks: Marks that are built in to aerial cameras and appear on each photo acquired in the corners or sides or both, and are used to determine the precise location of the principal point.

Fine clusters: Refers to the grouping of pixels in an unsupervised classification of a digital false color composite image by comparison of pixel DNs with histogram peaks in a relaxed nearest neighbor approach; an Idrisi module as applied here.

Forward overlap: The amount of overlap between successive photos in a flight line to allow for stereoviewing; usually 60–70%.

French long lot: Also called arpent system; a form of land partition in which long, narrow pieces of property had their frontage on a river rather than on a street; distinctive pattern found where French colonized during the past 200 years.

Frequency: The number of wave crests passing a given point in a specific period of time, e.g., cycles per second.

Full disc infrared image: A thermal image of all of the Earth that can be viewed from the GOES satellite.

Full disc visible image: A visible image of the entire Earth as seen from the GOES satellite.

Gaussian stretch: A form of image enhancement that forces pixel DNs into a normal curve thus increasing overall scene contrast.

Geographic registration: Process of superimposing two or more images so that points on the Earth's surface coincide on the final image product.

Geostationary: A satellite whose orbit and speed allows it to maintain a position above the same location on the Earth's surface.

Global Change Research Program: An international effort to study changes in the Earth's atmosphere and surface conditions that includes the efforts of several U.S. agencies such as NASA and NOAA.

GEOS satellite: Normal operation calls for 2 Geostationary Operational Environmental Satellites positioned 22,300 miles above the Equator at 75 and 135 degrees west longitude. They provide continuous weather imagery of the entire conterminous United States.

Graphic or bar scale: One manner of presenting linear scale on a map; a line is divided into scaled units for direct measure on the map, e.g., feet, miles, kilometers.

Great valley: The open portion of the Ridge and Valley Province that extends from the Hudson River south of Albany, NY to Birmingham, AL.

Grazing angle: In radar imaging, the angle at which the beam strikes the ground.

Ground control: The location of specific points on the ground in terms of their horizontal and vertical location by means of ground survey.

Ground resolution cell: The area on the ground that is covered by a pixel in a remotely sensed image.

Hand-held photography: Photographs taken from space on any of the manned missions, most often using a modified Haselblad 70 mm camera.

Haze correction: The visible bands of the EMS are subject to attenuation due to haze whereas the reflected IR are less affected. The histogram of the latter is used to correct that of the former.

Haze penetration: The shorter wavelengths of visible light are partially attenuated by atmospheric haze whereas wavelengths of the reflected IR are able to pass through haze and be recorded.

High frequency areas: Areas of an image with great detail or many changes in brightness values.

High oblique: An airphoto which shows the horizon line due to angle of view.

High-pass filters: An electronic enhancement that enhances the detail in an image.

High resolution visible (HRV): Refers to the three multispectral bands at 20 meter resolution and the panchromatic band at 10 meter resolution of the SPOT program.

Histogram: A graphic arrangement of the pixel DNs in a band of an image, usually with the DN range on the X axis and the frequency on the Y.

Histogram equalization: Is the uniform distribution stretch of the DNs in a histogram to achieve uniform density of pixels in all gray levels.

Histogram peaks: Pixels tend to be grouped in clusters in a histogram leading to the formation of peaks representing areas in a digital image where values are similar, e.g., water area, forest area, etc.

Homogeneous sites: refers to the selection of training sites in creating a supervised classification of a composite digital image; groups of pixels are selected by the interpreter based upon knowledge of the study area; the pixels in each group should be homogeneous.

Horizontal polarization: a radar pulse of energy that is transmitted and received from the target with horizontal orientation of microwaves.

Hudson Highlands: The name given for a section of the Taconic Mountains in lower New York state bordering the Hudson River.

Human/environment interface: A focus for study in many disciplines, including the interpretation of airphoto and images, is the impact of the natural environment on human activity and of human actions upon the environment.

IBM-compatible, PC environment: Refers to one of the two major versions of individual computers, the system that was created by International Business Machines, Inc. and the clones that are compatible with that system.

IBM precision processing: An image processing system developed by IBM in the early 1970s that produced sharper images with a much wider array of colors than was available as a standard product from EDC at the time.

Image: The representation of an object by means of a sensing system, a focusing system, and a recording medium and produced by optical, electro-optical, optical mechanical, or electronic means.

Image enhancement: The manipulation of an image in order to extract added data with a resulting loss of data elsewhere in the image.

Image "noise" or static: Non-valid data that is introduced to recorded information during recording or transmission, e.g., banding in radar or line dropouts in MSS imagery.

Image restoration: Image enhancement procedures that are designed to remove the effects of "noise," the imperfect function of the sensing system, and the effects or vehicle instability.

Industrial Revolution: The period from the middle of the 18th century through most of the 19th century when the methods of production, transportation, and communication were totally revamped.

Inselberg (Bornhardt): The name of an erosional remnant in a mountainous, arid landscape.

Intermittent stream: A river or stream that flows for only part of the year.

Intermontane region: In the United States that area that lies between the Cascade and Sierra Nevada Mountains on the west and the Rocky Mountain system to the east.

Internet: An international computer network maintained by national governments and easily accessible by an individual with a computer and a modem.

Interpretation log: A record of image signatures and other information that has been derived from an airphoto or stereopair, stereotriplet.

Ka, X, L band radars: Refers to wavelengths of side-looking airborne radar systems of 0.8–1.1, 2.4–3.8, and 15–30 cm., respectively.

Karst topography: A type of landscape that develops in areas of the world where there are layers of limestone rock underlying the surface and sufficient rainfall or groundwater flow to dissolve the rock.

Kodacolor Aero Reversal Film: An aerial film that is used in a positive to positive photographic process, similar to the production of 35 mm slides.

Kodachrome CIR: A 35 mm film that may be used to produce positive slides that portray false color infrared display. Requires the use of a wratten #12 minus blue filter.

Kodak No. 1: The camera produced by George Eastman in 1988 that opened the field of photography to the layman.

Land cover: Usually used to denote the nature of the Earth's surface in areas where the natural environment is dominant.

Landsat: A series of unmanned satellites that began recording the Earth's surface in 1972 in multiple wavelength bands. It is ongoing.

Land use: The nature of the Earth's surface in areas where the human imprint upon land is dominant.

Land use/land cover maps: Maps produced by the USGS and other agencies that delineate areas of common use or cover.

LU/LC signatures: The collection of characteristics related to the land use and/or land cover of a feature that together allow its identification.

Large format camera: A special camera created for the Space Shuttle that acquires 9″× 18″ contact photographs of the Earth.

Large scale vs. small scale: Large scale maps or images show great detail for a small ground area, whereas small scale products portray less detail for a larger area, e.g., RF 1:10,000 = large scale; small scale = RF 1:500,000.

Layover: Radar images portray the displacement of the tops of mountains and features projecting above the landscape towards the sensor.

Leaf on/leaf off: Refers to aerial photography taken to portray an area when vegetated or when trees are bare of leaves, one or the other is required depending upon the ultimate use of the photography.

Linear distance vs. area: Two methods of relating the scale of a map to another map or to the surface represented, e.g., RF 1:5,000 represents linear distance between objects twice as long as on a map of RF 1:10,000, but covers one-fourth the area.

Linears (lineaments): Line features on an image that may be faults, roads, or railroads, etc., on the surface recorded.

Linear stretch: A group of enhancement procedures that allows for the increase in contrast of a digital image while maintaining the original brightness relationships.

Linear units of measure: Units of measurement along a line, e.g., inch, foot, mile, centimeter, meter, kilometer.

Line dropouts: When a sensor fails or sticks, no DN values are recorded and this produces recurring black lines in an MSS image.

Loessial soil: A fine, silt-loam deposited by wind that has the characteristic of forming steep slopes when eroded.

Look direction: In radar imaging, the direction in which the beam is transmitted and at a right angle to the flight path.

Low frequency areas: Parts of an image characterized by few changes in brightness values or little change in detail.

Low oblique: An airphoto that is tilted from the vertical view, but not enough to include the horizon in the photo.

Low-pass filter: An electronic enhancement of an image that increases the general patterns or forms by suppressing detail.

MacConnell's classification: William P. MacConnell of the Department of Forestry and Wildlife Management at the University of Massachusetts created a classification of land use/land cover at scale of RF 1:24,000 based upon airphoto interpretation in order to map the state's wildlife habitat.

Macintosh environment: One of the two major versions of individual computers; this one created by Apple and dominant in education below college level.

Map measurer: An instrument used to trace an irregular linear distance. Reads in inches or centimeters. Also known as an opisometer.

Maximum likelihood: A method of supervised classification of a digital image false color composite related to the probability that a pixel DN would be assigned to a particular classification, in this instance as performed by Idrisi software.

Medical imaging: An extremely sophisticated field that borrowed early techniques and algorithms from those developed in remote sensing. It is now possible to record images of the brain, other organs, and even the valves of an operating heart without invasive procedures.

Mental library: Refers to the remembered image signatures which assist someone in identifying features on airphotos or other images.

Minimum distance to means: A method of supervised classification of a digital image false color composite where each pixel DN is placed in a class based upon its closeness to the mean of the class, in this instance as performed by Idrisi software.

Mirror stereoscope: An optical instrument that is equipped with first surface mirrors and prisms to allow the viewing of successive photos in a flight line without overlap, to produce a mental 3D image.

Mission to Planet Earth (MTPE): A NASA program that is part of an international effort to study processes that determine Earth climatic change.

Modem: A device used to connect a computer with another computer or a network via a telephone connection; may be external or internal to the computer.

Modification of physical and cultural landscapes: Changes in physiography and cultural landscapes such as land use, urban systems, and economic distributions are often revealed by the study of airphotos and satellite images covering a period of time.

Mosaic: The arrangement of airphotos in a series of overlapping exposures within a number of overlapping flight lines.

Multi-band sensing: Acquiring several bands of reflected/emitted data from a target greatly increases the amount of information available and the analytic capability.

Multi-sensor approach: Acquiring several bands of reflected and/or emitted energy from a target on a single overflight.

Multispectral classification: A procedure that deals with information extraction from a digital image; usually three or more bands that hold most of the variation in an image are used to create a false-color composite which is subjected to a variety of statistical applications.

Multispectral scanner (MSS): The scanning system of Landsat that acquires four registered bands of reflected energy in the visible and near IR.

Nadir: The point on the ground that is directly beneath the aerial camera and lies on a vertical between the camera and the center of the Earth.

NAPP photography: National Aerial Photography Program began in 1987 and is ongoing. All of the contiguous United States will be photographed at RF 1:40,000 scale in CIR. Each photo covers one-half of a USGS topographic quadrangle.

NHAP photography: National High Altitude Program extended from 1980–1987 and was intended to cover the USA with CIR at RF 1:58,000, each photo records the area of one USGS topographic quadrangle. Black & white photography at 1:80,000 scale was also acquired.

NASA: National Aeronautics and Space Administration.

Natural color: The range of colors that would be perceived by the human eye as replicated in a photo or image.

Nearest neighbor analysis: A type of matrix or digital image analysis that examines the DN of a pixel and those of nearby pixels to determine if there is a statistical relationship that makes clustering of pixels logical, or leads to some other modification of pixels.

Negative-positive process: The process of recording a negative image of some object or scene and then producing a positive image through chemical steps; originated in 1830s by William Henry Fox Talbot and is the forerunner of modern photography.

Nephanalysis: The study of cloud patterns and forms as a means of understanding and predicting weather.

NIMBUS: This Latin word meaning cloud identifies an experimental research and developmental spacecraft.

NOAA: National Oceanic and Atmospheric Administration.

NOAA, NESDIS: NOAA, National Environmental Satellite, Data, and Information Service with headquarters in Suitland, MD.

Non-linear stretch: An image enhancement technique that allows for the variable modification of pixel DNs as opposed to linear stretch.

Old age desert mountains: Also referred to as inselbergs or bornhardts; erosional remnants buried in their own eroded debris.

Old age river: A river system that has eroded downwards to near base level and exhibits features such as a floodplain several times wider than the meander belt, meander scars, oxbow lakes, levees, and Yazoo streams.

Oblique airphoto: An airphoto that was acquired with a view of the ground that was not vertical.

Operational Line-scan System (OLS): The scanning system on board the U.S. Air Force DMSP satellite operates in a visible-near IR band and a TIR band.

Orthophoto: A vertical aerial photograph which has been rectified to remove parallax.

Orthophoto mosaic: A series of overlapping orthophotos that photographically cover a specific ground area.

Orthophotoscope: The optical instrument that is used to rectify vertical airphotos in order to remove parallax.

Palisades: An area of nearly vertical basalt columns exposed along the west bank of the Hudson River just north of New York City.

Panchromatic: Black & white film that is sensitive to all visible wavelengths of light.

Paper straight edge: Refers to using the straight edge of a piece of paper to mark the straight line distance between two points on a map; can then be converted to miles or kilometers by direct comparison to the map's graphic scale.

Parallax: The displacement of objects on airphotos as a result of their position with reference to the ground datum and the center of the photo.

Passive system: A remote sensing system that records the existing reflected or emitted energy leaving a target.

Patterns and processes: A frame of reference that is useful when interpreting landforms and natural features is to relate the features to noteworthy patterns or distinctive steps in a process of change.

PE&RS journal: The official journal of the American Society for Photogrammetry and Remote Sensing and formerly titled, *Photogrammetric Engineering and Remote Sensing.*

Photogrammetry: The science of obtaining reliable measurements through the medium of photography.

Photograph: An image formed on a base material through the action of light on a sensitized emulsion and produced by means of chemical steps.

Photographic process: The procedure of recording reflected light on a lightsensitive emulsion and then producing an image by means of chemical steps.

Photographic spectrum: That portion of the electromagnetic spectrum that may be recorded on film.

Photo index mosaic: An uncontrolled mosaic of airphotos arranged to show the coverage of the ground while illustrating exposure numbers of the various parts of the area.

Physiographic regions: Areas of the Earth's surface that are homogeneous in rock materials, formations, or processes that produced the landforms therein.

Picture elements: The smallest units of data in a digital image, pixels.

Pixel: Short for "picture element," the smallest unit of data in a scanned satellite image.

Plan position indicator (PPI): Radar images produced by a rotating antenna so that a cathode ray tube displays return from an area surrounding the radar system, e.g., local weather radar.

Plunging anticline: An anticline that when folded was also tilted so that it disappears beneath the present surface at one end of the ridge.

Plunging syncline: The remnants of a downfold that now extends above the surface; when folded it was also tilted so that the feature disappears beneath the existing ground surface at one end.

Pocket stereoscope: An optical device that allows one to view the same ground area on adjacent, overlapping airphotos to produce a mental 3D image.

Polarity of gray tones: It is possible to create thermal images with either white or black representing coldest or warmest temperatures and so caution in interpretation is required.

Polar orbiters: Satellites that follow a near polar orbit and as the Earth rotates beneath the satellite it scans new areas on each pass.

Positive identification: Sufficient evidence has been gathered to allow for certain identification of objects or features on an image.

Pre-press image processing: Techniques of image enhancement used in the publishing industry; changes in contrast, color, or any element of an image designed to attain a particular effect or visual impact.

Principal components analysis: A statistical technique that compares the bands in a digital image to determine which ones portray most of the variation in data in the image; fewer bands may then be used in image enhancement/classification, saving time and reducing costs.

Principal point (P): The center of a vertical airphoto; it is assumed to be the photo location of the nadir.

Radar: Radio Detection And Ranging, an active remote sensing system that sends out an electronic pulse and records returned energy.

Radar return: The energy from a radar beam that strikes an object and is reflected back to the sensing device.

Range resolution: The quality of distinguishing between objects on a radar system in the look direction.

RASCAL, Flight Landata, Inc.: Remote Airborne Sensor Computer Analysis Link; a multispectral video imaging system that produces real-time CIR video as well as the individual bands; can be fine tuned for very specific band selection.

Raster image: A digital image consisting of pixels arranged in rows and columns.

Ratio images: An image produced by dividing the pixel DNs of one band by the DNs of the same pixels in another band; enables the effects of topography (shadow vs. bright) to be minimized.

Rayleigh criterion: A statement of the relationship between radar wavelength, surface roughness, and depression angle that affects the nature of the radar return.

Real aperture radar: Also known as "brute force" radar in which the azimuth resolution is related to the length of the antenna and increasing beam width.

Real time CIR video: This refers to videography that produced CIR color rendition as it was being acquired.

Rectified airphotos: Photos that have had the effects of tilt and terrain relief removed.

Reflected infrared: The near and middle infrared portion of the EMS, 0.7 to 3.0 micrometers.

Remote sensing: The measurement and analysis of objects without physical contact between the sensing device and the target; includes all the areas of investigation in this publication.

Representative Fraction: The relation between linear distance on a map or photo and ground distance expressed as a fraction, e.g., 1/25,000 or 1:25,000 (where 1 inch on the map represents 25,000 inches on the ground).

Resolution: The ability of a remote sensor system to separate or distinguish objects on an image; sometimes expressed as line pairs per inch or per millimeter.

Return Beam Vidicon (RBV): A sensing system on the first 3 Landsat vehicles that recorded black & white visible information in a form somewhat similar to a television picture.

Ridge and furrow agriculture: An intensive form of agriculture of southeast Asia in which trenches are dug for irrigating rice and the soil removed from the furrow is piled on an adjacent ridge where crops such as plantain are grown.

Ridge and valley: A physiographic region of the Appalachian area of the United States lying between the Appalachian Plateau to the west and the Great Valley to the east.

Rotures: The divisions of land in the French long lot system in which land had narrow frontage on a river.

Sampling procedures: Refers to a large body of scientific methodology in which a sample of a larger matrix of data or of a larger area is used to arrive at conclusions about the whole.

SARSAT: Search and Rescue System is one of the sensor systems on board TIROS-N.

Scale: The relationship between distance on a map or photo to the corresponding distance between the same points on the Earth's surface.

Scale by comparison: A method of determining the unknown scale of a map or photo by comparison with a map of known scale of the same area.

Scanner: An imaging system that sweeps across the terrain in across-track or along-track direction, sensing a stream of pixel DNs.

Seamless mosaicing: A form of splicing contiguous digital images together using geometric orientation and contrast stretching so that the boundary between scenes is not apparent.

Seasat: A NASA satellite that was designed to provide data about ocean areas and operated for part of 1978 recording and sending back to Earth radar images and other information.

Section: A square mile parcel that is the basic unit of a township in the US Public Land Survey system.

Selective key: A means of identification of objects on airphotos by comparison with representative images, drawings, stereograms, etc. of typical features.

SEM: Space Environment Monitor is a sensor on board TIROS-N.

Sequent occupance: A geographic concept from the early decades of the 1900s in which the successive foci of economic activity in a region was studied in analyzing change, e.g., farming, shipbuilding, textiles and leather manufacturing, to electronic industry in coastal Mass.

Shuttle imaging radar (SIR-A): An L-band imaging radar that was operated from the NASA Space Shuttle in 1984.

Sidelap: The amount of overlap between airphotos in adjacent flight lines, usually presented as a percentage.

Side Looking Airborne Radar (SLAR): Airborne radar that transmits a microwave burst of energy sideways from the plane and records the energy that is reflected back by surface targets.

Signature characteristics: Defining parameters of a feature on an image that collectively allow for its identification, e.g., size, shape, pattern, contrast, texture, shadow, color, reflectance, et. al.

Signature identification: Using characteristics of a feature on an image to identify it on subsequently examined images.

Sinkholes: Holes in areas of karst topography where solution of underlying limestone has led to cave in of surface materials.

Sinusoidal stretch: A form of image enhancement that is designed to provide more detail in areas of an image that appear to have similar DNs.

Site factors: Those characteristics of a specific location that have an impact on the nature of an image of the area, e.g., roughness, shapes, texture, slope, etc.

Situation factors: Those characteristics of the area surrounding a target that might give insight into the interpretation of an image of the target, e.g., vegetative cover, landforms, cultural artifacts, etc.

Skylab: An orbiting workstation that was utilized by 3 crews in 1973–74.

Slant range distance: The straight line distance from a radar to its target.

Slope maps: Selected locations in the United States have had slope maps produced by the USGS; they are useful in evaluating danger of flooding, soil erosion, landform stability, etc.

Soil marks: In archaeology some vestiges of earlier use of land can be seen from airphotos where buried landscapes impact and produce patterns in the present surface of the ground.

Space Shuttle: The commonly used name for NASA's Space Transportation System.

Space Transportation System (STS): The official name for the U.S. Space Shuttle system.

Spatial filtering: A type of image enhancement technique in which groups of pixels in close proximity have their DNs statistically examined to determine what image modification may be made (see low-pass & high-pass filters).

Spectral bands: Refers to specific wavelength values that make up one band to be recorded or sensed by a satellite system.

Specular reflection: A smooth surface produces a mirror-like reflection and in radar imaging results in no return and a dark area on images.

SPOT progam: The unmanned French satellite system, Systeme Probatoire d'Observation de la Terre, that began operations in 1986.

Standard deviation units: A statistical value that is used to describe the distance above or below the mean of a distribution that a particular observation lies.

Standard IR: Refers to the TIR images most commonly acquired by the GOES satellite system.

Standard RGB array: Is the typical means of producing color as it would be seen by the human eye, as opposed to false color.

Stereogram: A stereopair or stereotriplet of airphotos mounted for proper stereovision.

Stereogram cards: Card stock designed to be used for creating a stereogram file. See Appendix C.

Stereopair: Two overlapping, successive photos in a flight line aligned for stereoviewing.

Stereovision: Viewing two overlapping, successive airphotos in a flight line so that a three dimensional effect is produced.

Sunsynchronous: A condition in which a satellite is placed in a near polar orbit such that it passes over all places on the Earth's surface at the same latitude twice a day at the same local time.

Sun angle correction: When joining images in a mosaic the sun angle in all the images can be made the same through mathematical manipulation of pixel DN values.

Superposed stream: In geomorphology when a river has been flowing on a surface and is able to downcut fast enough to erode an underlying subsurface feature onto which it cuts down, it is termed superposed.

Supervised classification: An image interpreter identifies pixels or groups of pixels that are known to represent a specific type of land use, vegetation, etc., and then gives directions to the computer to locate all such pixels in the image.

Synclinal fold: In folded mountain topography the downfold or syncline can be isolated on a later erosional surface as a synclinal ridge or a synclinal valley and the forms may be interpreted on topographic maps and airphotos.

Synthetic Aperture Radar (SAR): The Doppler principle is employed to produce a synthetic antenna that operates as though it were much longer than a typical SLAR antenna, resulting in consistent resolution of objects in the azimuth direction.

Taconic Mountains: A section of the New England physiographic province that extends north–south on the east side of the Hudson Valley and reaches elevations of 2000 feet.

Thematic Mapper (TM): The sensing system on Landsat 4 and after that acquired reflected and thermal data from the Earth's surface in 7 bands.

Thermal infrared: The far infrared portion of the EMS, 3.0 to 15 micrometers, TIR.

Thermal infrared images: Images that display variations in tone or color that represent temperature differences.

TIROS, TIROS-N: Television Infrared Operational Satellite operated by NOAA. The latest version is TIROS-N or ATN, an advanced version. They are polar orbiters designated as NOAA-1, 2, etc. after they become operational. Many sensing systems on board.

TOMS: Total Ozone Mapping Spectrometer sensor on board Nimbus-7 that studied the total column of ozone in the atmosphere.

Topographic inversion: An optical illusion produced on images with shadows where ridges appear to be valleys and vice versa; usually corrected by orienting an image so that shadows face the viewer.

TOVS: TIROS Operational Vertical Sounder includes sensors that monitor atmospheric radiance in a number of bands to allow mapping of temperature and humidity levels.

Transects: A method of sampling in which a set of parallel, equidistant lines is drawn across a map or photo. Summation of image information beneath the lines is used to predict values for the whole photo.

Triassic lowland: A small physiographic province of easily eroded sedimentary rocks that separates the New England Province from the older Appalachians in the area between the Hudson River across New Jersey to Virginia.

Training sites: Areas of a digital image that are identified by an interpreter as representing a specific type of use, soil, rock, or vegetation type; used to give instructions to computer in supervised classification.

Uncontrolled mosaic: The overlapping of airphotos by means of aligning objects on the photos without the use of ground control points. Whole photos are used in index mosaics, while only the central portions of photos are used when more accurate representation of surface features is desired.

Uniform distribution stretch: An image enhancement technique that is non-linear in form; results in the redistribution of DN values based upon frequency of occurrence.

Unsupervised classification: A group of digital image enhancement techniques based upon statistical manipulation without guidance from the interpreter using training sites or similar information.

USGS Topographic Maps: The conterminous United States has been mapped in contiguous quadrangles whose bounds are specific latitude and longitude values allowing for creation of a mosaic of maps for an area; several map scales are available for maps that display topographic surfaces and almost all other surface features except movable or non-permanent phenomena; contact the nearest Federal office bldg.

USGS Land Use/Land Cover Classification: A classification published by the US Geological Survey designed to facilitate LU/LC mapping of products produced by aircraft and satellites; 3 levels of generalization.

US Public Land Survey: The Land Ordinance of 1785 that divided the lands of the United States, other than the original 13 colonies, into Townships and Ranges based upon a series of north–south principal meridians and east–west base lines.

Vector image: A digital image formed by lines that connect points of specific X and Y values, e.g., road or railroad lines, property boundaries.

Verbal scale: A statement of the scale relationship of a map or photo to the area it portrays, e.g., "1 inch equals 100 feet," or "1 centimeter represents 5 kilometers."

Vertical airphoto: An airphoto acquired with the camera pointed directly at the nadir.

Vertical polarization: The microwave energy that is transmitted from a radar system and received from the target is oriented in waves in a vertical plane.

Video disc reader: A device that captures image data from a video disc (similar in size to an old 78 rpm phonograph record) and feeds it to VCR/television setup.

VIFIS system: Variable Interference Filter Imaging Spectrometer; a video recording system that allows for one flyover of a target and yields multiple bands of data from one videotape.

Visible spectrum: That portion of the electromagnetic spectrum to which the human eye is sensitive.

VISSR: Visible Infrared Spin-Scan Radiometer is the sensing system on the GOES geostationary satellite and monitors in visible and TIR wavelengths.

Water vapor imagery: The 6.7 and 7.3 micrometer bands are greatly affected by the presence of water vapor in the atmosphere and so are used to monitor it from the GOES satellites. The imagery data is complex and requires sophisticated interpretation.

Wavelength: The distance from the crest of one wave to the crest of the next wave (or trough to trough); used as a means of ordering radiation in the electromagnetic spectrum.

Wet plate photography: During the middle and late 1800s this was the most common photographic procedure; glass plates were coated with a moist emulsion and the image had to be recorded before drying was complete.

MILE SCALE 1:62 500

UNITED STATES
DEPARTMENT OF THE INTERIOR
GEOLOGICAL SURVEY

TOPOGRAPHIC
MAP INFORMATION AND SYMBOLS
MARCH 1978

QUADRANGLE MAPS AND SERIES

Quadrangle maps cover four-sided areas bounded by parallels of latitude and meridians of longitude. Quadrangle size is given in minutes or degrees.

Map series are groups of maps that conform to established specifications for size, scale, content, and other elements.

Map scale is the relationship between distance on a map and the corresponding distance on the ground.

Map scale is expressed as a numerical ratio and shown graphically by bar scales marked in feet, miles, and kilometers.

NATIONAL TOPOGRAPHIC MAPS

Series	Scale	1 inch represents	1 centimeter represents	Standard quadrangle size (latitude-longitude)	Quadrangle area (square miles)
7½-minute	1:24,000	2,000 feet	240 meters	7½× 7½ min.	49 to 70
7½× 15-minute	1:25,000	about 2,083 feet	250 meters	7½× 15 min.	98 to 140
Puerto Rico 7½-minute	1:20,000	about 1,667 feet	200 meters	7½× 7½ min.	71
15-minute	1:62,500	nearly 1 mile	625 meters	15× 15 min.	197 to 282
Alaska 1:63,360	1:63,360	1 mile	nearly 634 meters	15× 20 to 36 min.	207 to 281
Intermediate	1:100,000	nearly 1.6 miles	1 kilometer	30× 60 min.	1568 to 2240
U. S. 1:250,000	1:250,000	nearly 4 miles	2.5 kilometers	1°× 2° or 3°	4,580 to 8,669
U. S. 1:1,000,000	1:1,000,000	nearly 16 miles	10 kilometers	4°× 6°	73,734 to 102,759
Antarctica 1:250,000	1:250,000	nearly 4 miles	2.5 kilometers	1°× 3° to 15°	4,089 to 8,336
Antarctica 1:500,000	1:500,000	nearly 8 miles	5 kilometers	2°× 7½°	28,174 to 30,462

CONTOUR LINES SHOW LAND SHAPES AND ELEVATION

The shape of the land, portrayed by contours, is the distinctive characteristic of topographic maps.

Contours are imaginary lines following the ground surface at a constant elevation above or below sea level.

Contour interval is the elevation difference represented by adjacent contour lines on maps.

Contour intervals depend on ground slope and map scale. Small contour intervals are used for flat areas; larger intervals are used for mountainous terrain.

Supplementary dotted contours, at less than the regular interval, are used in selected flat areas.

Index contours are heavier than others and most have elevation figures.

Relief shading, an overprint giving a three-dimensional impression, is used on selected maps.

Orthophotomaps, which depict terrain and other map features by color-enhanced photographic images, are available for selected areas.

COLORS DISTINGUISH KINDS OF MAP FEATURES

Black is used for manmade or cultural features, such as roads, buildings, names, and boundaries.

Blue is used for water or hydrographic features, such as lakes, rivers, canals, glaciers, and swamps.

Brown is used for relief or hypsographic features—land shapes portrayed by contour lines.

Green is used for woodland cover, with patterns to show scrub, vineyards, or orchards.

Red emphasizes important roads and is used to show public land subdivision lines, land grants, and fence and field lines.

Red tint indicates urban areas, in which only landmark buildings are shown.

Purple is used to show office revision from aerial photographs. The changes are not field checked.

INDEXES SHOW PUBLISHED TOPOGRAPHIC MAPS

Indexes for each State, Puerto Rico and the Virgin Islands of the United States, Guam, American Samoa, and Antarctica show available published maps. Index maps show quadrangle location, name, and survey date. Listed also are special maps and sheets, with prices, map dealers, Federal distribution centers, and map reference libraries, and instructions for ordering maps. Indexes and a booklet describing topographic maps are available free on request.

HOW MAPS CAN BE OBTAINED

Mail orders for maps of areas east of the Mississippi River, including Minnesota, Puerto Rico, the Virgin Islands of the United States, and Antarctica should be addressed to the Branch of Distribution, U. S. Geological Survey, 1200 South Eads Street, Arlington, Virginia 22202. Maps of areas west of the Mississippi River, including Alaska, Hawaii, Louisiana, American Samoa, and Guam should be ordered from the Branch of Distribution, U. S. Geological Survey, Box 25286, Federal Center, Denver, Colorado 80225. A single order combining both eastern and western maps may be placed with either office. Residents of Alaska may order Alaska maps or an index for Alaska from the Distribution Section, U. S. Geological Survey, Federal Building-Box 12, 101 Twelfth Avenue, Fairbanks, Alaska 99701. Order by map name, State, and series. On an order amounting to $300 or more at the list price, a 30-percent discount is allowed. No other discount is applicable. Prepayment is required and must accompany each order. Payment may be made by money order or check payable to the U. S. Geological Survey. Your ZIP code is required.

Sales counters are maintained in the following U. S. Geological Survey offices, where maps of the area may be purchased in person: 1200 South Eads Street, Arlington, Va.; Room 1028, General Services Administration Building, 19th & F Streets NW, Washington, D. C.; 1400 Independence Road, Rolla, Mo.; 345 Middlefield Road, Menlo Park, Calif.; Room 7638, Federal Building, 300 North Los Angeles Street, Los Angeles, Calif.; Room 504, Custom House, 555 Battery Street, San Francisco, Calif.; Building 41, Federal Center, Denver, Colo.; Room 1012, Federal Building, 1961 Stout Street, Denver Colo.; Room 1C45, Federal Building, 1100 Commerce Street, Dallas, Texas; Room 8105, Federal Building, 125 South State Street, Salt Lake City, Utah; Room 1C402, National Center, 12201 Sunrise Valley Drive, Reston, Va.; Room 678, U. S. Court House, West 920 Riverside Avenue, Spokane, Wash.; Room 108, Skyline Building, 508 Second Avenue, Anchorage, Alaska; and Federal Building, 101 Twelfth Avenue, Fairbanks, Alaska.

Commercial dealers sell U. S. Geological Survey maps at their own prices. Names and addresses of dealers are listed in each State index.

INTERIOR—GEOLOGICAL SURVEY, RESTON, VIRGINIA—1978

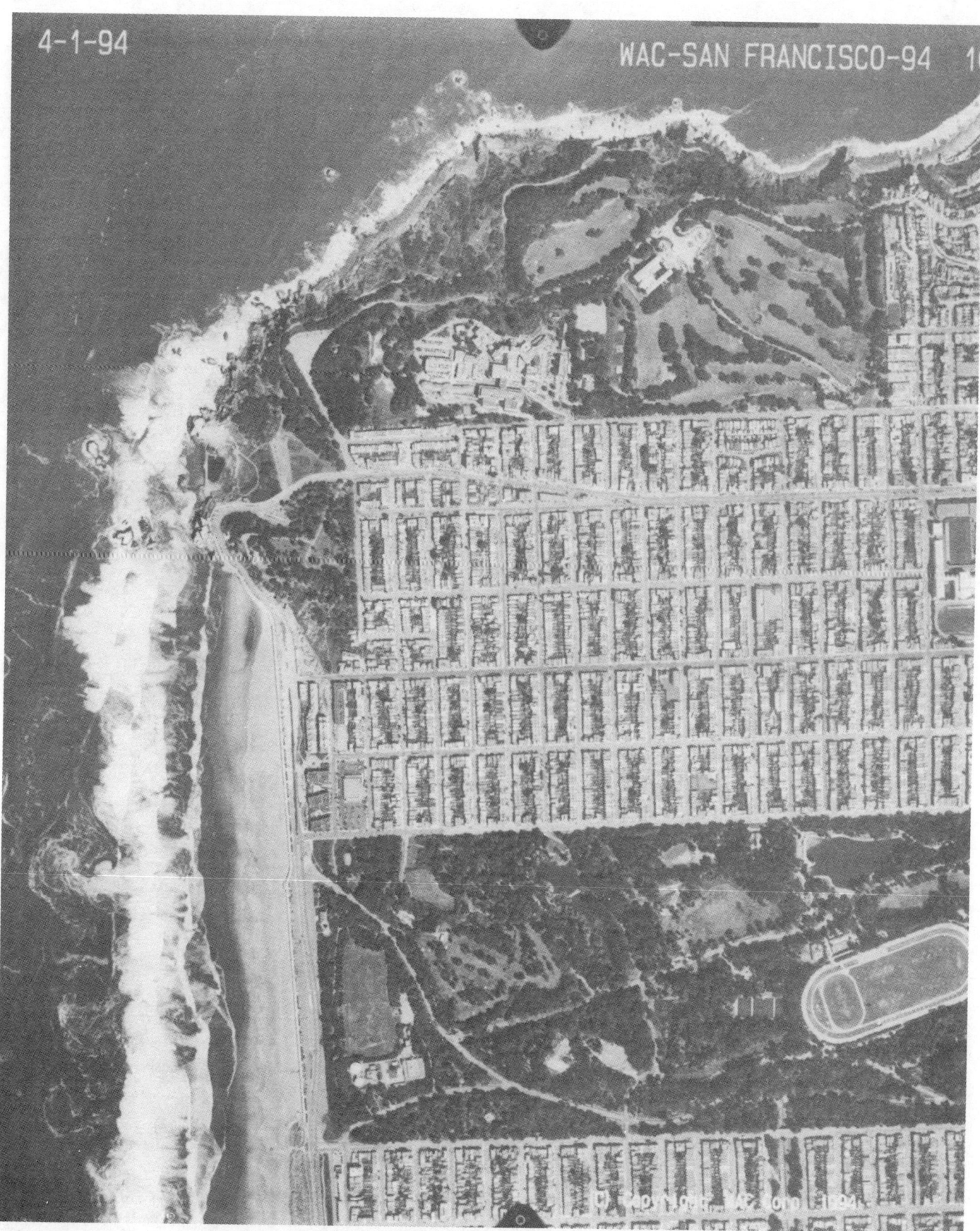

Figure 3–4. Aerial Photograph–Northwest San Francisco, CA.
Courtesy WAC Corporation, Eugene, OR.

Plate I

Plate II

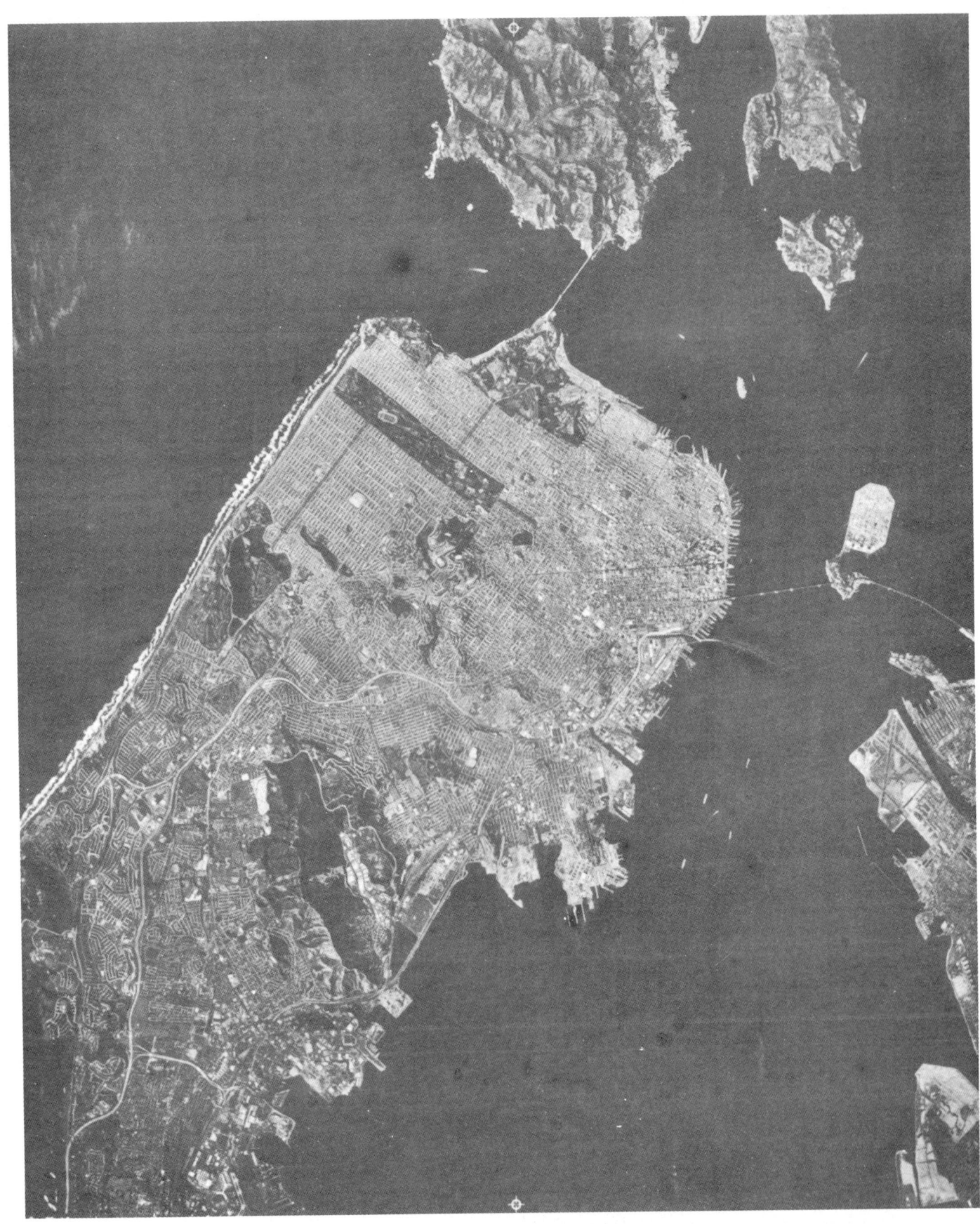

Figure 3–5. NASA high altitude photograph, San Francisco, CA. Circa 1980.

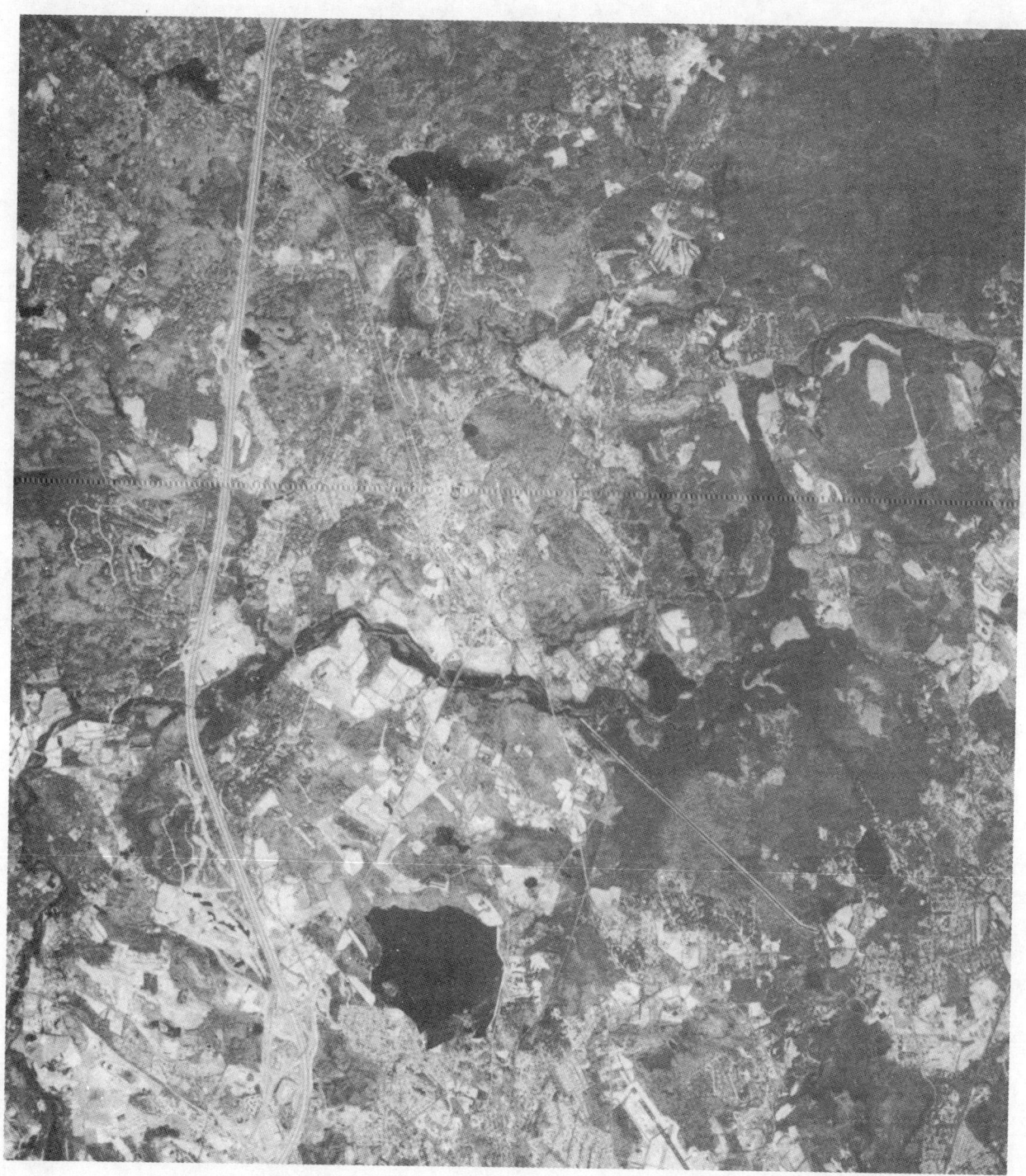

Figure 7–3. NHAP CIR airphoto #69-210. Topsfield, MA acquired 4/1/86.

Plate III

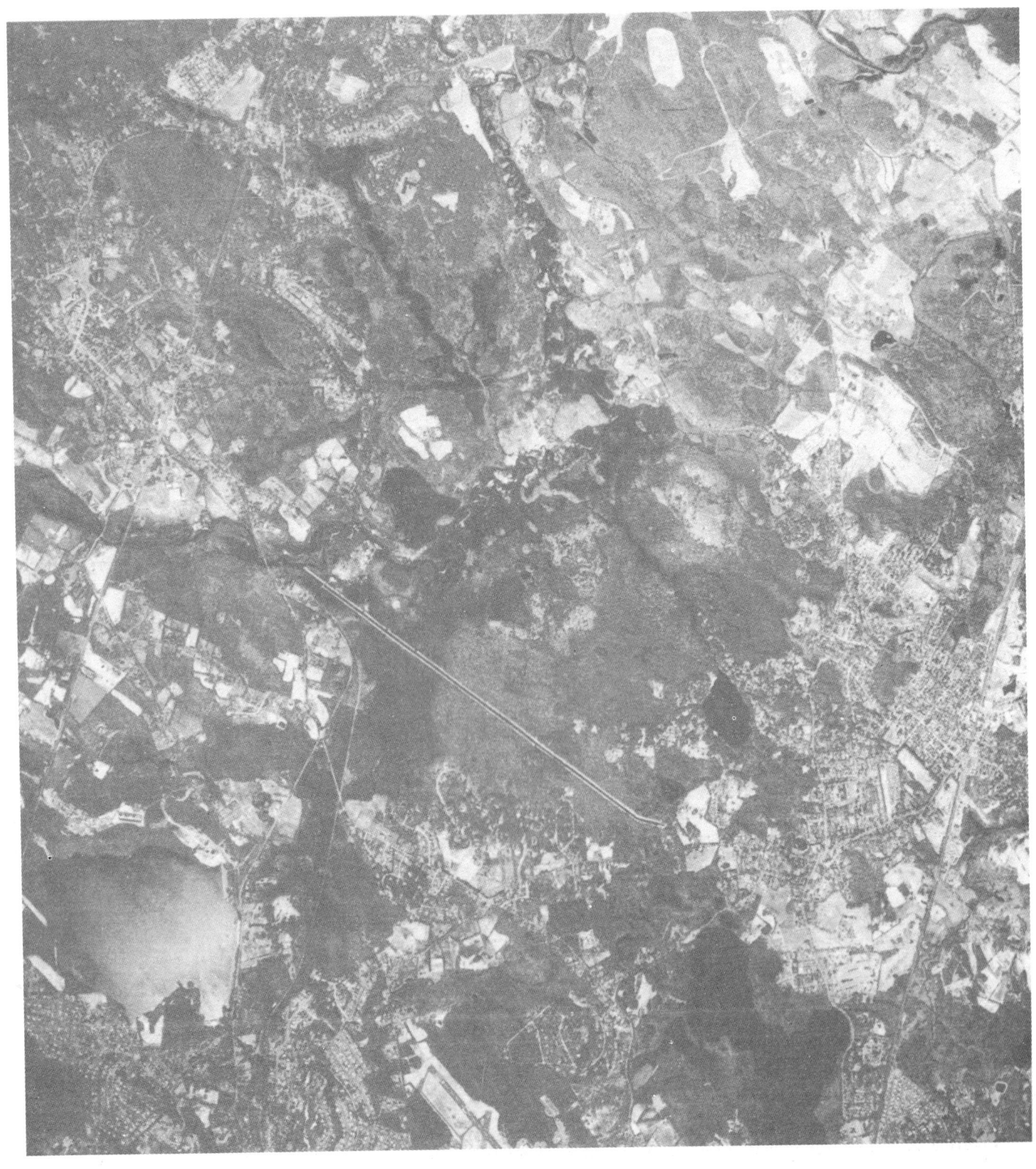

Figure 7–4. NAPP CIR airphoto #2023-108. Topsfield, MA acquired 4/13/92.

Plate **IV**

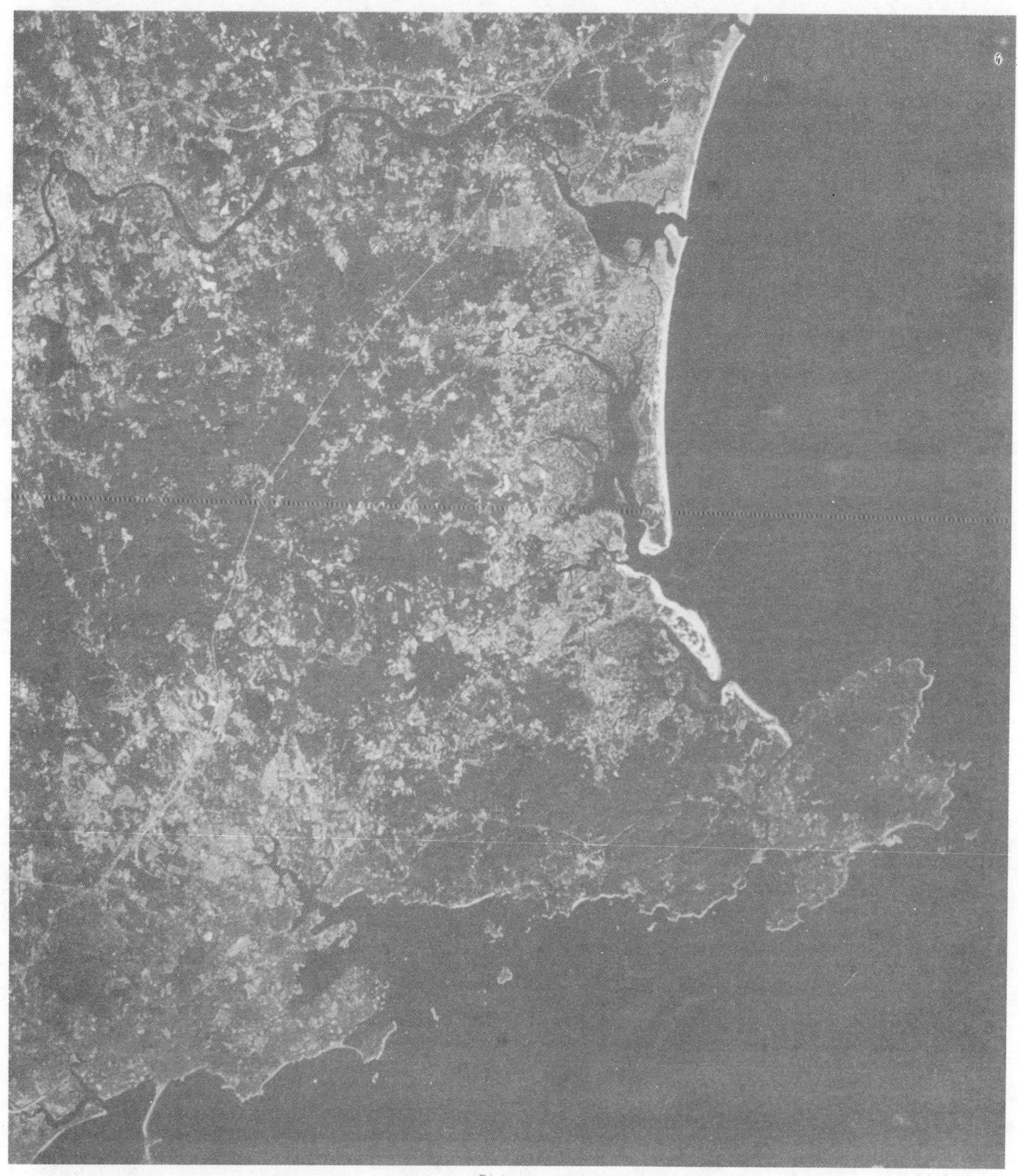

Figure 7–6. NASA Skylab CIR airphoto circa spring 1974. RF 1:230,000.

Plate **V**

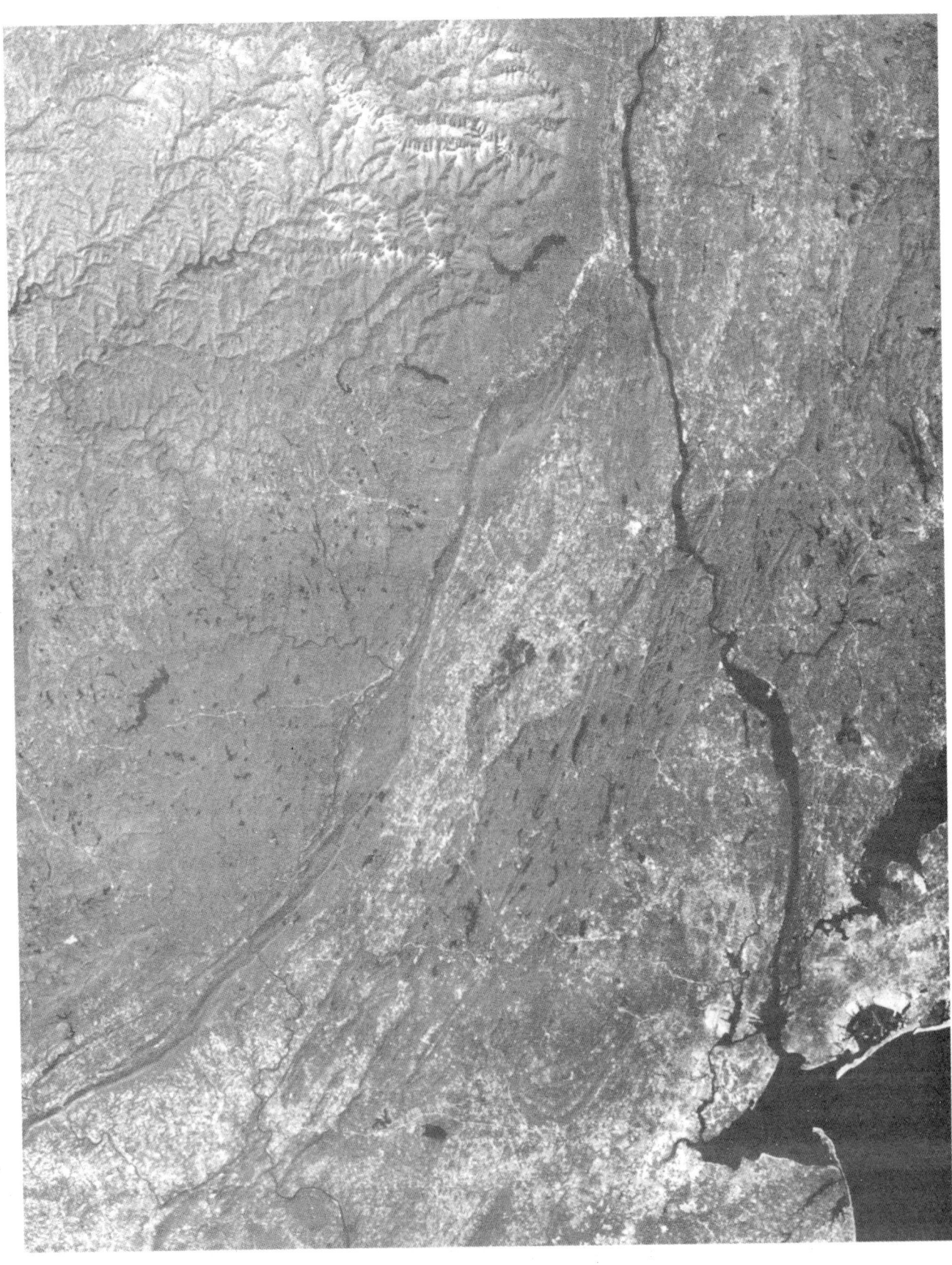

Figure 9–4. IBM Precision Processed Mosaic Landsat MSS Bands 4, 5, 7 NYC, Hudson River, and downstate New York. NASA ERTS1 Scenes 1079-15124, 15131, and 15133. 10/10/72.
Courtesy IBM Corp.

Plate **VI**

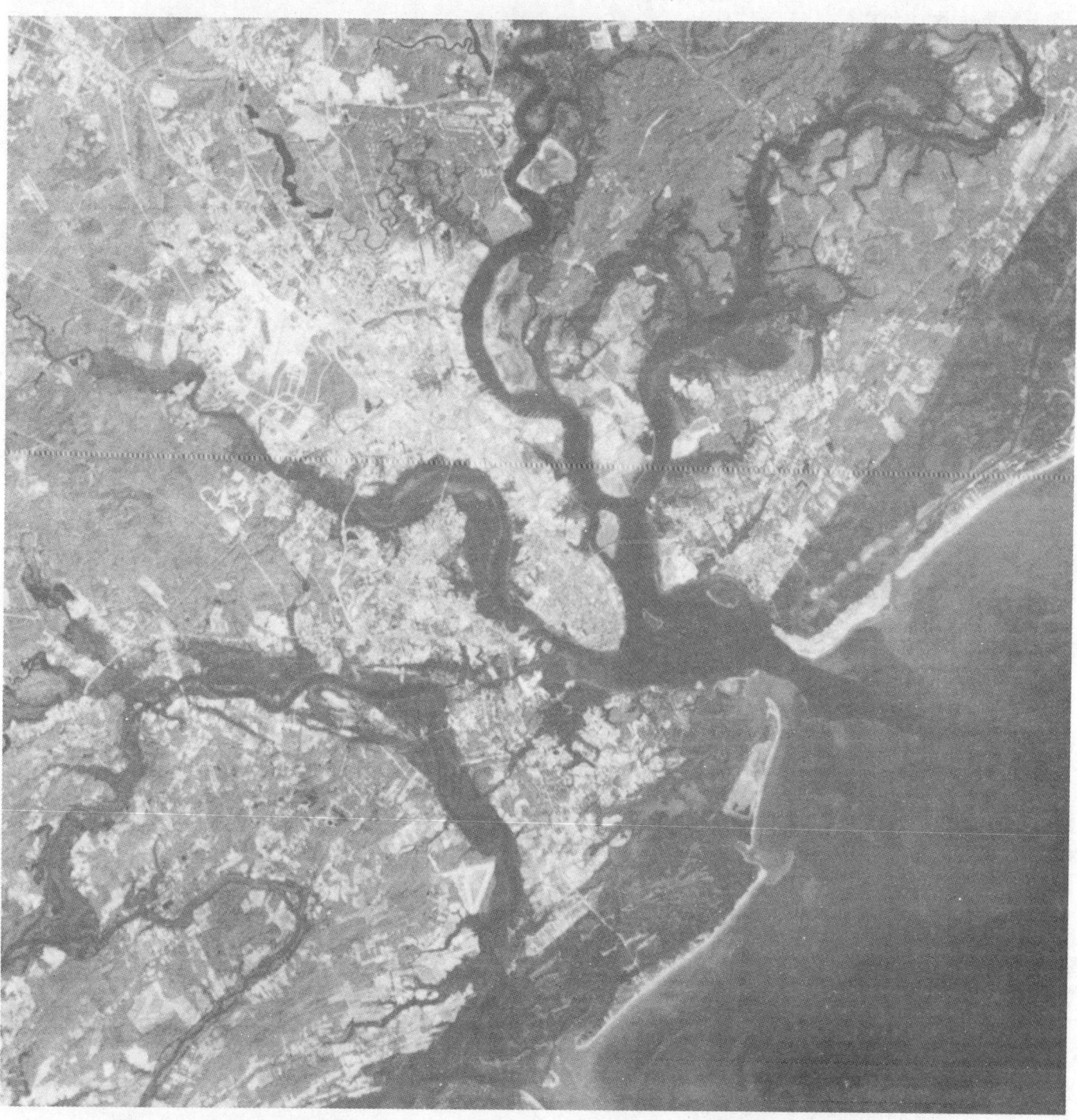

Figure 9–11. Landsat 4 TM Bands 2, 3, 4. Charleston, SC Subscene. Acquired 11/9/82. Courtesy NASA/GSFC.

Plate **VII**

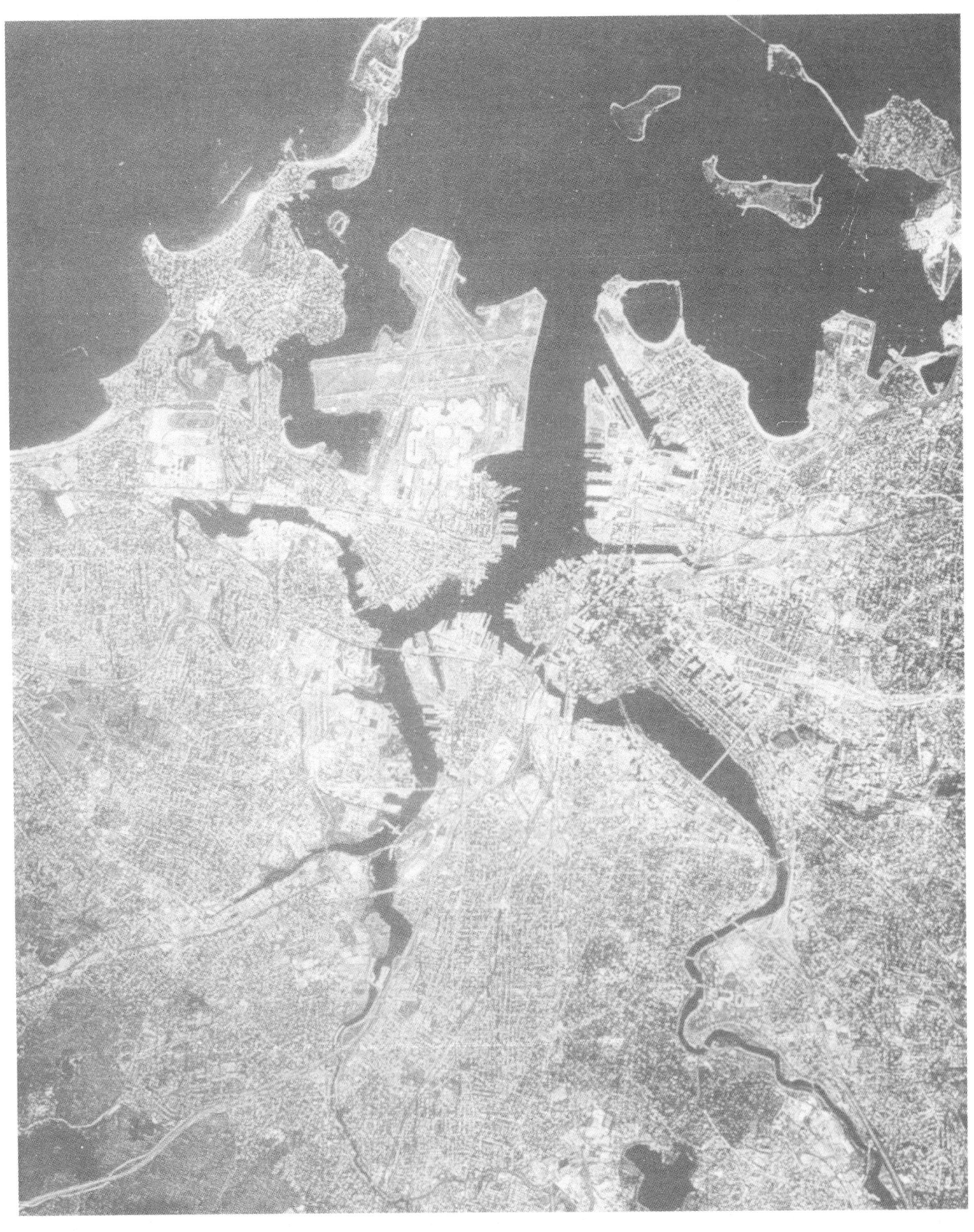

Figure 9–20. SPOT Multispectral Image (part). Boston, MA. 11/7/88 Data courtesy SPOT Image Corporation.
Image processed by TASC of Reading, MA. Laser writing by Cirrus Technology, Inc. Nashua, NH.

Plate **VIII**

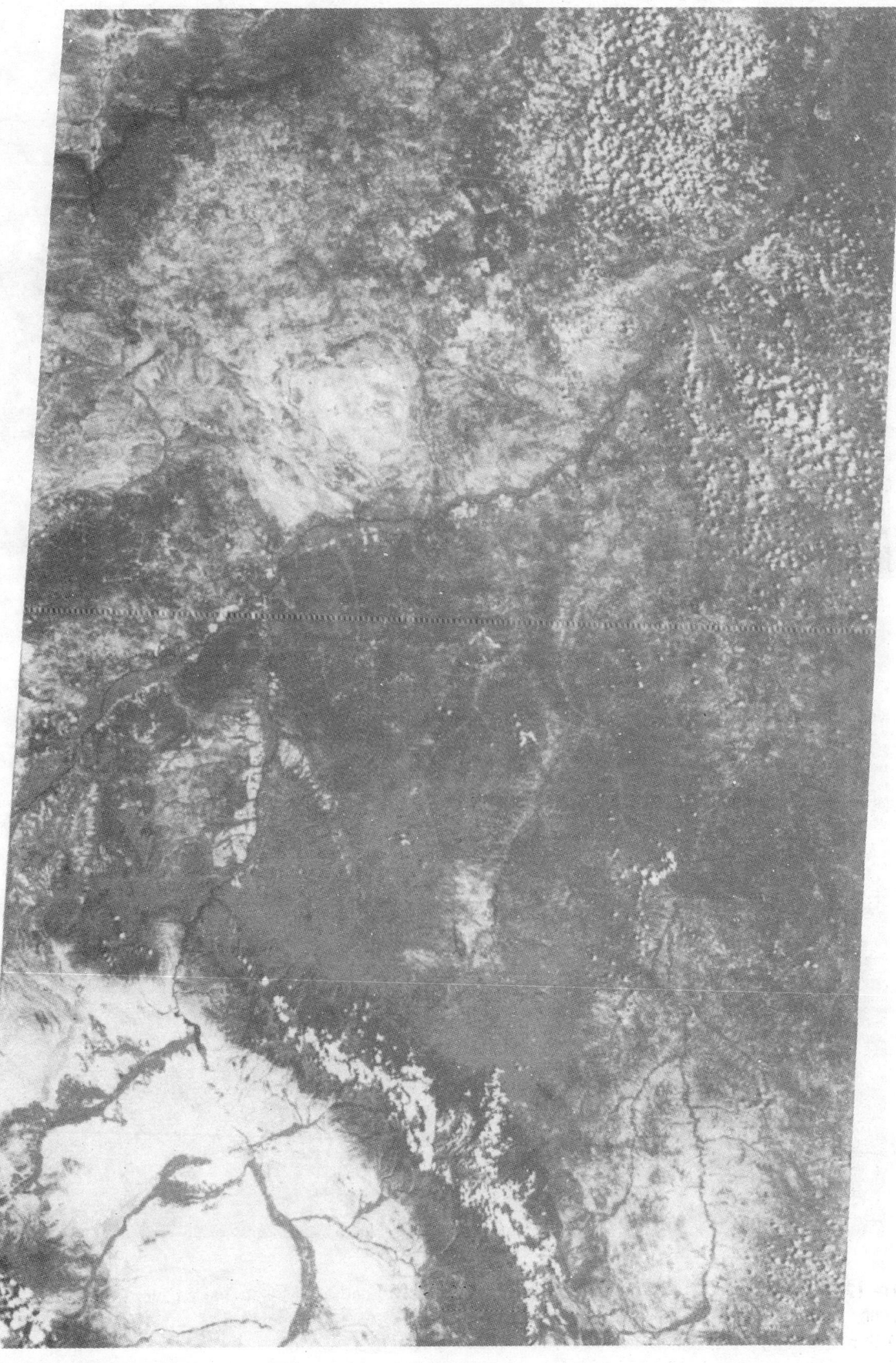

Figure 13–9. IBM Precision Processed Mosaic Landsat MSS Bands 4, 5, 7. Portion of Eight-Image Mosaic over SE Montana and NE Wyoming. Acquired 7/30-31/73.
Courtesy R. Bernstein, IBM.

Plate **IX**

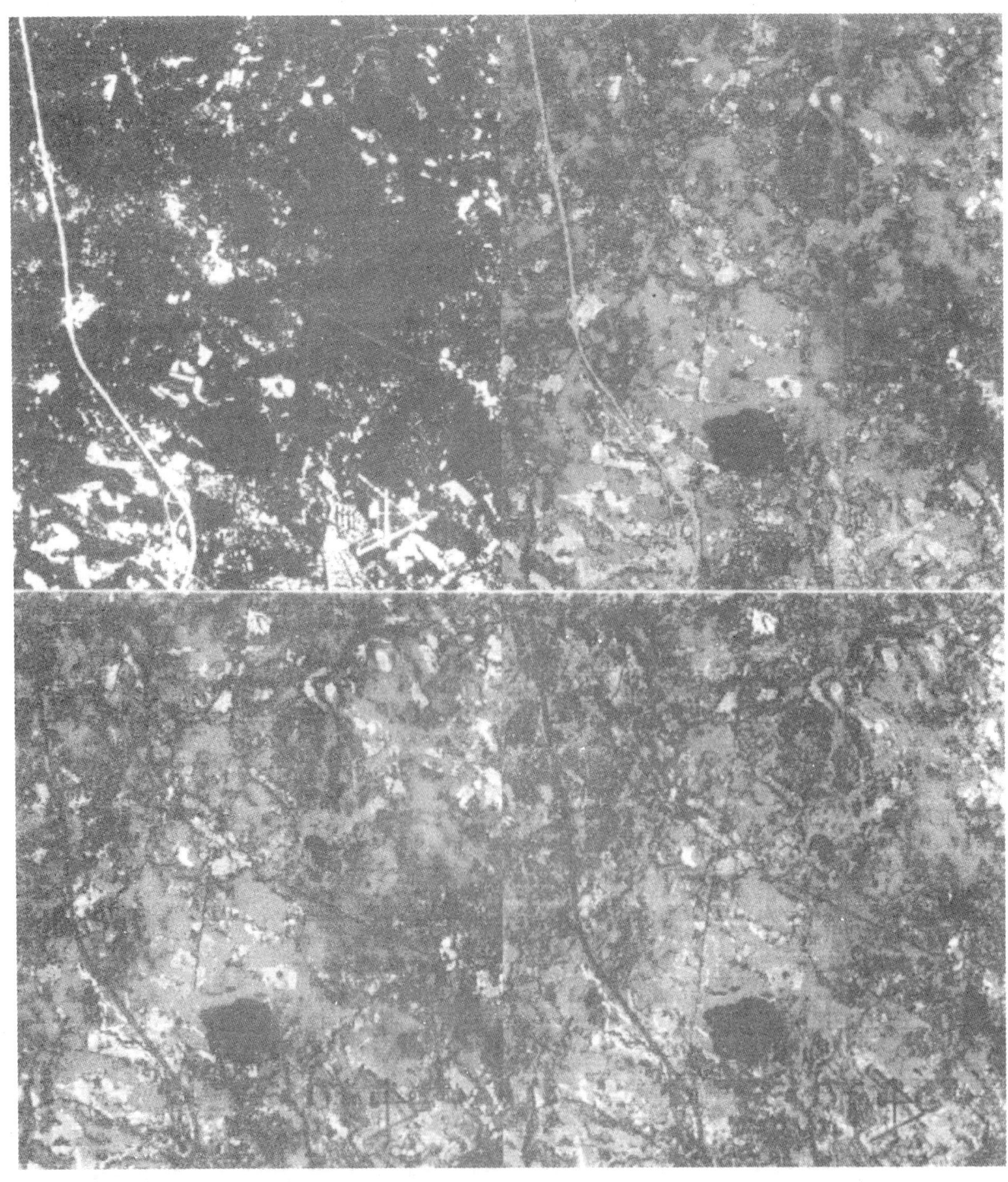

Figures 13–10, 13–11, 13–12, 13–13. (clockwise from upper left) Landsat TM Data False Color Composites of Subscene. Acquired 6/14/88. Topsfield, MA. 13–10 Bands 1, 2, 3. 13–11 Bands 2, 3, 4. 13–12 Bands 1, 4, 5. 13–13 Bands 3, 5, 4. Generated by Idrisi software. Data Courtesy EOSAT Co.

Plate **X**

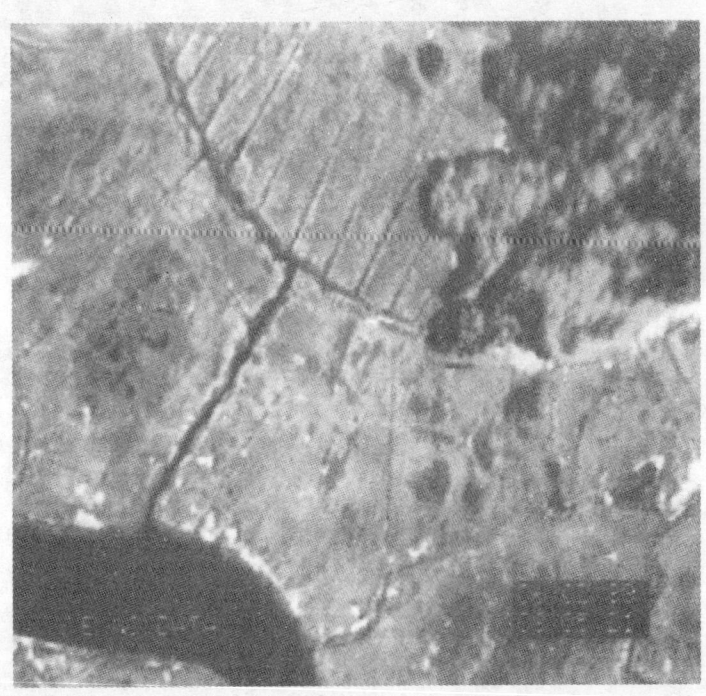

Figure 14–3. RASCAL CIR False Color Composite Area in
Newbury, MA. 6/2/88 by Flight Landata, Inc.
Flight Landata, Inc. P.O. Box 528, Newburyport, MA 01950

Plate **XI**

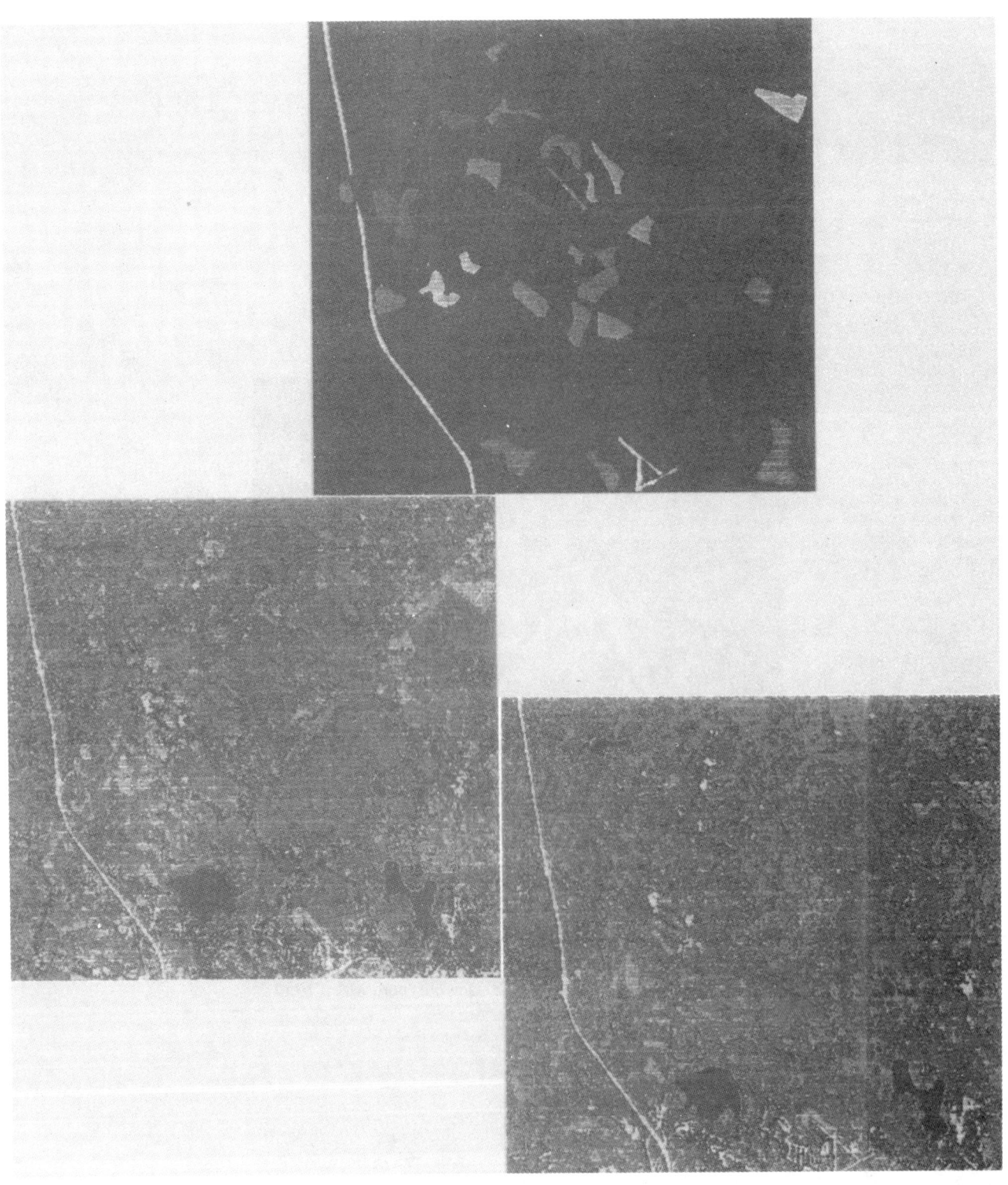

Figure 14–10, (top) Training Sites. **Figure 14–11,** Maximum Likelihood Classification. **Figure 14–12,** (bottom) Minimum Distance to Means using Standard Deviation. Generated by Idrisi Software. Clark University, Worchester, MA.

Plate **XII**

TOPOGRAPHIC MAP SYMBOLS

VARIATIONS WILL BE FOUND ON OLDER MAPS

Primary highway, hard surface . . . .

Secondary highway, hard surface . . . .

Light-duty road, hard or improved surface . . . . . .

Unimproved road . . . .

Road under construction, alinement known . . . .

Proposed road . . . .

Dual highway, dividing strip 25 feet or less . . . .

Dual highway, dividing strip exceeding 25 feet . . . .

Trail . . . .

Railroad: single track and multiple track . . . .

Railroads in juxtaposition . . . .

Narrow gage: single track and multiple track . . . .

Railroad in street and carline . . . .

Bridge: road and railroad . . . .

Drawbridge: road and railroad . . . .

Footbridge . . . .

Tunnel: road and railroad . . . .

Overpass and underpass . . . .

Small masonry or concrete dam . . . .

Dam with lock . . . .

Dam with road . . . .

Canal with lock . . . .

Buildings (dwelling, place of employment, etc.) . . . .

School, church, and cemetery . . . .

Buildings (barn, warehouse, etc.) . . . .

Power transmission line with located metal tower . . . .

Telephone line, pipeline, etc. (labeled as to type) . . . .

Wells other than water (labeled as to type) . . . . oOil . . . . . oGas

Tanks: oil, water, etc. (labeled only if water) . . . . ● ● ● Water

Located or landmark object; windmill . . . . o

Open pit, mine, or quarry; prospect . . . . x x

Shaft and tunnel entrance . . . . ■ . . . . . Y

Horizontal and vertical control station:

Tablet, spirit level elevation . . . . BM △ 5653

Other recoverable mark, spirit level elevation . . . . △ 5455

Horizontal control station: tablet, vertical angle elevation VABM △ 95I9

Any recoverable mark, vertical angle or checked elevation △3775

Vertical control station: tablet, spirit level elevation . . . . . BM X 957

Other recoverable mark, spirit level elevation . . . . X 954

Spot elevation . . . . x 7369   x 7369

Water elevation 670   670

Boundaries: National . . . .

State . . . .

County, parish, municipio . . . .

Civil township, precinct, town, barrio . . . .

Incorporated city, village, town, hamlet . . . .

Reservation, National or State . . . .

Small park, cemetery, airport, etc. . . . .

Land grant . . . .

Township or range line, United States land survey . . . . . .

Township or range line, approximate location . . . .

Section line, United States land survey . . . .

Section line, approximate location . . . .

Township line, not United States land survey . . . .

Section line, not United States land survey . . . .

Found corner: section and closing . . . .

Boundary monument: land grant and other . . . .

Fence or field line . . . .

Index contour . . . . Intermediate contour . . .

Supplementary contour . . . . Depression contours . . .

Fill . . . . Cut . . . .

Levee . . . . Levee with road . . . . . .

Mine dump . . . . Wash . . . .

Tailings . . . . Tailings pond . . . .

Shifting sand or dunes . . . . Intricate surface . . . .

Sand area . . . . Gravel beach . . . .

Perennial streams . . . . Intermittent streams . . . .

Elevated aqueduct . . . . Aqueduct tunnel . . . .

Water well and spring . o . . . . . . o~ Glacier . . . .

Small rapids . . . . Small falls . . . .

Large rapids . . . . Large falls . . . .

Intermittent lake . . . . Dry lake bed . . . .

Foreshore flat . . . . Rock or coral reef . . . .

Sounding, depth curve . . . 10 . . . . Piling or dolphin . . . .

Exposed wreck . . . . Sunken wreck . . . .

Rock, bare or awash; dangerous to navigation . . . .

Marsh (swamp) . . . . Submerged marsh . . .

Wooded marsh . . . . Mangrove . . . .

Woods or brushwood . . . . . . Orchard . . . .

Vineyard . . . . Scrub . . . .

Land subject to controlled inundation . . . . Urban area . . . .